Library of
Davidson College

CULTURE OF
MARINE INVERTEBRATE ANIMALS

CULTURE OF MARINE INVERTEBRATE ANIMALS

Edited by
Walter L. Smith
Head, Department of Marine Science
and Technology
Suffolk County Community College
Selden, New York

and
Matoira H. Chanley
Shelter Island Oyster Company
Greenport, New York
and
Marine Sciences Research Center
State University of New York
Stony Brook, New York
and
Suffolk County Community College
Selden, New York

PLENUM PRESS • NEW YORK AND LONDON

Library of Congress Cataloging in Publication Data

Conference on Culture of Marine Invertebrate Animals, Greenport, N.Y., 1972.
  Culture of marine invertebrate animals.

  Includes bibliographies.
  1. Marine invertebrates—Cultures and culture media—Congresses. 2. Shellfish culture—Congresses. I. Smith, Walter Leonard, 1918-   ed. II. Chanley, Matoira H., ed. III. Title. [DNLM: 1. Invertebrates—Congresses. 2. Marine biology—Congresses. QL362 C968 1972]
QL362.8.C66   1972   639'.4   74-11367
ISBN 0-306-30804-5

639.4
C748c

Proceedings of the conference on Culture of Marine Invertebrate Animals
held in Greenport, New York, in October 1972

© 1975 Plenum Press, New York
A Division of Plenum Publishing Corporation
227 West 17th Street, New York, N.Y. 10011

United Kingdom edition published by Plenum Press, London
A Division of Plenum Publishing Company, Ltd.
4a Lower John Street, London W1R 3PD, England

All rights reserved   76-5912

No part of this book may be reproduced, stored in a retrieval system, or transmitted, in any form or by any means, electronic, mechanical, photocopying, microfilming, recording, or otherwise, without written permission from the Publisher

Printed in the United States of America

Preface

This volume is based on presentations at the conference on Culture of Marine Invertebrate Animals which was held in Greenport, New York in October, 1972. The conference was sponsored by the Middle Atlantic Natural Sciences Council, Inc., a nonprofit educational corporation, together with the Marine Science Centers of Adelphi University, the State University of New York at Stony Brook, Long Island University, Suffolk County Community College, and the Shelter Island Oyster Company.

The purpose of the conference was to provide a needed exchange of knowledge among scientists of various specialties whose information would be invaluable to others confronted with similar problems, even with different marine animals.

Part I considers supportive techniques -- general isolation and culture methods, problems of disease and feeding. Specific techniques employed in the culture of a wide range of invertebrate organisms is covered in Part II.

We want to thank the contributors for their cooperation in preparing the manuscripts based on their conference presentations.

                                  Walter L. Smith

                                  Matoira H. Chanley

Contents

PART I

Recirculating System Culture Methods for Marine
    Organisms 3
    John M. King

Maintenance of Some Marine Filter Feeders on Beef
    Heart Extract 15
    Helen M. McCammon

Culture of Phytoplankton for Feeding Marine
    Invertebrates 29
    Robert R.L. Guillard

Bacterial Pathogens Associated with Cultured Bivalve
    Mollusk Larvae 61
    Haskell S. Tubiash

Marine Microbiology: Some Practical Aspects for
    Aquaculture 73
    Joseph M. Cassin, Patricia E. Cassin, Elsa Brunn,
    Kenneth Frenke, Michael Priano, Heidi Wetherall,
    Neil Wetherall

Culture of Salt Marsh Microorganisms and Micro-
    metazoa 87
    John J. Lee and William A. Muller

Antibiotics in Cultures of Invertebrates 109
    Anthony D'Agostino

PART II

A Review of Coelenterate Laboratory Culture 137
    A. Harry Brenowitz

Comments on the Laboratory Culture of Scyphozoa David G. Cargo	145
Bryozoa Marie B. Abbott	155
Methods of Culturing Polychaetes David Dean and Michael Mazurkiewicz	177
Problems Associated With Culture of Marine Copepods J.G. Gonzalez, P.P. Yevich, J.H. Gentile, and N.F. Lackie	199
Culture Techniques for Decapod Crustacean Larvae Morris H. Roberts	209
Lobster Culture Based on the conference presentation of John T. Hughes	221
Echinodermata George D. Ruggieri, S.J.	229
Opisthobranch Culture David R. Franz	245
New Approaches and Techniques for Studying Bivalve Larvae J.L. Culliney, P.J. Boyle, and R.D. Turner	257
The Development of Methods for Rearing the Coot Clam, Mulinia lateralis, and Three Species of Coastal Bivalves in the Laboratory Edwin W. Rhodes, Anthony Calabrese, Wayne D. Cable and Warren S. Landers	273
Culture of American and European Oysters Herbert Hidu	283
Laboratory Cultivation of Assorted Bivalve Larvae Paul Chanley	297
Contributors	319
Animal and Plant Index	323
General Index	329

PART I

RECIRCULATING SYSTEM CULTURE METHODS FOR MARINE ORGANISMS

John M. King

Director of Research, Aquarium Systems, Inc.

Eastlake, Ohio

The cultivation of marine organisms in recirculating systems (RS) offers several advantages to the investigator (Sandt, 1968):

1. RS are independent of the effects of nature (storms, floods, etc.) and man (pollution).
2. RS can be established away from the sea.
3. RS offer control of various environmental parameters, such as salinity, temperature, photoperiod.
4. RS are easily monitored.
5. RS are convenient to sample.
6. RS offer the availability of continuous visual observation.
7. Different systems can be simultaneously monitored.
8. The biota can be preferentially selected without danger of contamination of planktonic forms.

It must be kept in mind, however, that a recirculating system is an artificial environment. It is a captive body of water and not a duplication of part of the ocean. In fact, chemical changes can be detected in a captive volume of sea water within minutes after its removal from the sea (Collier and Marvin, 1953). Moreover, there are chemical reactions and accumulations in recirculating systems that do not occur naturally in the oceans (Honig, 1934; Atz, 1964b).

The basic prerequisite for the successful cultivation of marine organisms in recirculating systems is an awareness of the changes occurring in a captive body of water.

CHANGES IN SEAWATER IN RECIRCULATING SYSTEMS

Nitrogen

Nitrogen usually makes its first appearance in the recirculating systems as ammonia, the major metabolic waste of most aquatic animals (Emerson, 1969; Foster and Goldstein, 1969). Ammonia is extremely toxic to all life forms, even in small concentrations, e.g., salmonid fishes are sensitive to concentrations as low as 0.006 parts per million (Burrows, 1964). Fortunately, several species of chemoautotrophic bacteria, called nitrifiers are almost universally present in recirculating systems. They oxidize this toxic ammonia to nitrite and finally to relatively non-toxic nitrate (figure 1).

In the open oceans, the concentration of nitrate ions is very low, usually 0.1 - 0.6 mg/L (Waksman et al. 1933; Honig, 1934; Harvey, 1955; Atz, 1964b). This is due to its assimilation by planktonic organisms and to bacterial denitrification. The latter process does not occur to any great extent in recirculating systems -- at least under aerobic conditions (Honig, 1934). In the recirculating systems the concentration of nitrates therefore continually increases (Oliver, 1957).

After twenty years of use an analysis of the water in the seawater systems of the New York Aquarium showed a 250-fold increase (Atz, 1964b) and the Aquarium of the Koninklyk Zoologisch Genootschap Natura Artis Magistra in Amsterdam once reported a nitrate concentration of 730 parts per million! (Honig, 1934). Comparing such concentration with the trace amounts of nitrate present in natural seawater, or with the medium made from synthetic sea salts, shows that the nitrate ion becomes a major factor of difference (figure 2).

Although Kelley (1965) points out that nitrate in high concentrations is less toxic than ammonia by approximately three orders of magnitude, it does have a marked effect on the organisms in recirculating systems. Physiologically, increased levels of nitrate interfere with respiration, particularly in invertebrates (Oliver, 1957; Hirayama, 1966a; Kuwatani, Nishii, and Isogai, 1969a). De Graff (1964) stated that *Pagurus* showed "discomfort at nitrate levels of 75-100 mg/L and that at 250-350 mg/L many fishes die." Chemically, an abundance of nitrate ions causes an undesirable decrease in pH by replacing carbonate and bicarbonate ions and forming nitric acid (Honig, 1934).

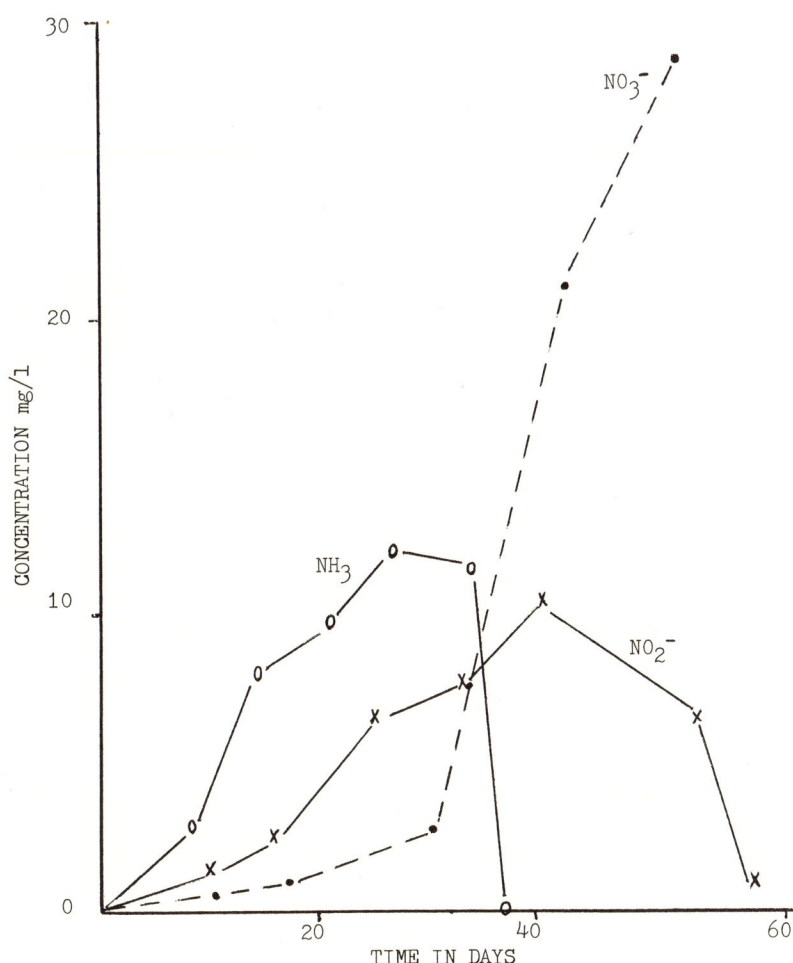

Fig. 1. Typical sequence of changes of inorganic nitrogen during the run-in period of a culture system using a biological filter at 12°C. The concentrations of the various forms of nitrogen at any given time depend on such factors as initial loading of animals, feeding, and temperature. The final level of nitrate will depend on the activity of denitrifying bacteria and algal growths. (From Roff, 1972).

CHEMICAL ANALYSIS OF INSTANT OCEAN SYNTHETIC SEA SALTS

ION	SYMBOL	PARTS PER MILLION
Chloride	Cl	18400
Sodium	Na	10200
Sulfate	SO_4	2500
Magnesium	Mg	1200
(Nitrate*)	(NO_3)	(730)
Potassium	K	370
Calcium	Ca	370
Bicarbonate	HCO_3	140
Boric Acid	H_3BO_3	25
Bromide	Br	20
Strontium	Sr	8
Silicate	SiO_3	3
Phosphate	PO_4	1
Manganese	Mn	1
Molybdate	MoO_4	0.7
Thiosulfate	S_2O_3	0.4
Lithium	Li	0.2
Rubidium	Rb	0.1
Iodide	I	0.07
Ethylenediamine-tetraacetic Acid	EDTA	0.05
Aluminum	Al	0.04
Zink	Zn	0.02
Vanadium	V	0.02
Cobalt	Co	0.01
Iron	Fe	0.01
Copper	Cu	0.003

Fig. 2. Chemical analysis of INSTANT OCEAN Synthetic Sea Salts at 15°C and a specific gravity of 1.025. Nitrate* which is NOT part of the formulation can accumulate in recirculating aquariums to levels where it becomes a major ion (Honig, 1934).

pH and Alkalinity

In the open oceans, the average pH values range from 7.5 to 8.4 (Sverdrup, Johnson and Fleming, 1942). In any given area, however, pH fluctuations are very slight. The same is not true in recirculating systems -- there is usually a continual decline of pH (Atz, 1964b; Hirayama, 1970).

This lowering of pH can be attributed to many factors -- the effect of accumulated nitrate ions, the respiration of the animals and bacteria, and the oxidation of metabolic wastes (Atz, 1964b). Low pH levels have been shown to have an adverse effect on the captive marine animals (Paccaud, 1962). Kuwatani and Nishii (1969a) showed that the growth of the Japanese pearl oyster (<u>Pinctada fucata</u>) was inhibited at a pH of 7.89. One hundred percent mortality was noted after 26 days at a pH of 7.36.

Phosphates and Other Inorganic Ions

The decomposition of metabolic wastes, uneaten food and dead plants or animals leads to an increase of phosphate ions. Like nitrates, phosphates have not been found to be very toxic to marine animals. Their accumulation does, however, affect culture conditions. Phosphates can precipitate as calcium salts on the filter beds reducing the buffering surfaces of the gravel and thus causing a drop in pH (Goldizen, 1970). Calcium and magnesium ions may also form insoluable precipitates with phosphate in the culture medium, thus reducing the alkalinity and adding to the acidification of the water (Saeki, 1962).

The breakdown of metabolic wastes also contributes to the increase of calcium, potassium, and sulphate ions (Saeki, 1958; Kelley, 1965). The exact influence of these ions on marine life has not been determined, but their accumulation does appreciably alter the captive environment.

Organic Material

Wallace and Wilson (1969) state that "Most dissolved organic compounds occur in seawater (open oceans) at concentrations in the parts per billion range, with the sum of such material rarely exceeding a few parts per million." The same is not true in recirculating systems. In addition to increases in inorganic compounds and ions, the decomposition of waste products, uneaten food, etc., also leads to increased levels of organic matter (Wilson and Greggs, 1941; Oliver, 1957). In recirculating systems that have been in operation for some length of time, the water often acquires a yellowish tint. This color is due to a build-up of organic material, collectively termed "Gelbstoff" (Kalle, 1966). Brockway (1950), Kawamto (1961), Atz (1964a) and Kuwatani (1969c) indicate that the increased levels of organic compounds are related to growth inhibitions in captive aquatic animals.

The basic differences between the waters of the open oceans and the water in the recirculating system have been summarized by Kelley (1965) as follows:

1. Nitrogen compounds appear in great abundance, initially as toxic ammonia, then as nitrite, and finally as nitrate.
2. The alkaline reserve of the system decreases with an associated drop in pH.
3. The concentration of phosphate, sulfate, calcium and potassium ions increases.
4. The total organic content of the water increases.
5. Bacteria increase numerically but with sharp loss of species diversity.

WATER MANAGEMENT TECHNIQUES

Fortunately, techniques are available to counter-act the above described changes in recirculated seawater systems.

By equipping the recirculating system with good aeration and an efficient biological filter (either of the undergravel type or as a separate component), (King and Kelley, 1971), the two most serious problems -- ammonia toxicity and loss of pH -- can be eliminated (Saeki, 1958; Kelley and Moreno, 1961; Kelley, 1963; Atz, 1964b; Spotte, 1970; Miklosz, 1971).

Nitrifying bacteria are found on the walls of the system and, when biological filters are used, throughout the filter substrate (Saeki, 1948; Hirayama, 1966a). Kawai, Yoshida, and Kiwata (1965) showed that the density of nitrifying bacteria was inversely proportional to the depth of the filter and that nitrifying activity more than 5 cm below the surface was 50% less that at the surface.

This observation led to the idea that greater filtering efficiency could be gained by increasing the surface area (as opposed to depth) of the filter. It should be noted, however, that in Kawai's studies sand (diameter of 0.7 mm) was used as the filtrant and as pointed out by Kelley (1963) and Spotte (1970), current configurations call for a filtrant size of 2-5 mm. Using a filtrant with this size, Siddall (1972) has shown that nitrifying bacteria are found in equal densities throughout the filter. This discovery should lead to more efficient utilization of space in the construction of filters for recirculating systems.

Kawai, Yoshida, and Kimata (1964) found that nitrifiers comprise only ten percent of the total aerobic bacterial population in the biological filters of recirculating systems. This ratio results when nitrifying bacteria are established in recirculating systems by exposing the culture water and filter to excretory products of "hardy" organisms, such as Homarus americanus (Spotte, 1970; Roff, 1972). Certain metabolites that are excreted by these animals inhibit the growth of the nitrifying bacteria. Siddall (1972) "seeded" a new filter with a small amount of filtrant from an old established recirculating system and then circulated an ammonia-enriched (160-320 mg/L NH_4Cl) non-organic synthetic seawater medium through the system for fifteen days. With this method he was able to increase the nitrifying activity of the filter by more than 250%.

The high concentration of accumulated nitrates resulting from this procedure necessitates the draining and refilling of the system with new seawater medium before any live marine organisms are introduced. During an extended test of 200 days, Siddall (1972, pers. comm.) noted no significant loss in nitrification rates even after animals were added to systems and their filters were exposed to the wide variety of organic compounds typical of old culture water.

Although further experiments on methods of increasing the nitrification rates of biological filters must be made, in theory it appears that specific biological filters (i.e. for ammonia, nitrite, etc.) can be established and maintained and that such filters will at least double the carrying capacity of the system.

For pH control the composition of the filtrant material is important -- the sand (silicon) or quartz gravels that are often used in fresh-water culture will not work in seawater. Calcium carbonate (calcite) is not soluble at the pH of seawater and therefore is not effective. In contrast, natural carbonates that contain a small percentage (a minimum of 4%) of magnesium, (e.g. dolmite, mollusc shells, or coral gravel) are soluble at the pH of seawater and will aid in maintaining optimum alkalinity and pH levels (Goldizen, 1970).

Because phosphates precipitate as calcium salts onto the filter bed, periodic stirring or tumbling of the filtrant is necessary (Goldizen, 1970; Siddall, 1972).

Regular partial changes of water are effective in controlling, by dilution, the undesirable accumulation of ions such as nitrates, phosphates, sulfates, etc. as well as various organics (Kuwatani and

Nishii, 1969b). The rate of exchange may vary depending on the organism cultured, its stage of development, the number cultured, etc., but for the wide variety of specimens kept in public aquariums, changes of 25% of the total volume per month have proved adequate (Kelley, 1963).

In addition to diluting potentially toxic substances, partial water changes replace necessary trace elements (Spangenberg, 1967) and help in pH control.

Culturing benthic algae, either micro or macroscopic, in recirculating systems also shows promise in at least partially controlling the accumulation of nitrates, phosphates, and organics. Using a separate autotrophic filter inoculated with non-nitrogen fixing species of Oscillatoria and Chlorococcum, Siddall (1972) reduced the nitrate level in a recirculating system by 68%. Shelborne et.al. (1963) using Enteromorpha and Sorokin (1971) using Chorella have shown the effectiveness of algal cultures in the maintenance of pH. James (1972 pers. comm.) suggests Enteromorpha, Ulva, Cladophora and Percursaria for use in autotrophic filter/buffer systems.

Activated carbon, ion exchange, and foam towers or protein skimmers are other devices currently being used in the water management of recirculating systems. All of these techniques work to a certain degree but there is a limit to their effectiveness. They have drawbacks.

Protein skimmers are extremely effective in removing dissolved proteins, amino acids, and other highly surface-active compounds, but they are relatively ineffective in removing carbohydrates (Goldizen, 1970). Moreover, according to Huckstedt (1969) the iodine that is present in the culture water as iodide is removed. Wallace and Wilson (1969) state that iron-bearing precipitates (ferric hydroxide and phosphates) and other trace elements are also removed by protein skimming.

Partial water changes, activated carbon, and protein skimming all assist in controlling excessive bacterial levels by depleting the amount of available nutrients. Further control of bacterial populations, viruses, and other parasites can be achieved with ultra-violet sterilization and ozone. From the standpoint of safety, only ultra-violet radiation should be considered.

Although ozone is effective in destroying bacteria, it is also toxic to humans (Milnes, 1961; Versagi, 1970). Recommended treatment levels in aquatic systems range from 0.1 - 0.3 mg/L/hr. At these concentrations not only are harmful bacteria killed, but so are the beneficial nitrifiers. It is important to note that concentrations as low as 0.25 mg/L effect red blood cells (Versagi, 1970).

In 1955 the American Conference of Governmental Industrial Hygenists established 0.1 mg/vol/eight hours as the maximum allowable concentration. This is just perceptible to the sense of smell. Most ozone generators used with recirculating systems are operated above this safety level.

Eagelton and Herald (1968) and Herald et. al. (1970) have shown the effectiveness of ultraviolet sterilization (2600 AU) in controlling bacteria in recirculating systems. Ultraviolet sterilization can keep bacteria levels at or below 200/ml which is in the range that Harvey (1955) reported for unpolluted seawater near shore.

DESIGN PARAMETERS FOR RECIRCULATING SYSTEMS

Goldizen (1970) presented the following criteria for the design of recirculating systems:

1. 500 L of medium per kg of animal.
2. A 0.1 m^3 filter bed consisting of 2 - 5 mm grains of a magnesium-containing calcareous filtrant.
3. A flow rate of 80 L per minute through each square meter of filter bed surface.
4. Replacement of 25% of the culture medium per month.

Systems of similar design will maintain 5 kg of the sea urchin, Strongylocentrotus purpuratus, or 7 kg of the American lobster, Homarus americanus.

For more complete control in culture work three components are recommended as additions to the above system:

5. An activated carbon filter.
6. An autotrophic filter/buffer.
7. An ultraviolet sterilizer.

With increasing coastal pollution -- ranging from oil spills to "red tides" -- not only individual investigators but whole marine laboratories are switching from "open flow" systems to recirculating ones. The newly dedicated South Carolina Marine Resources Center is perhaps a sign of the future. The Center, located near historic Fort Sumpter on scenic (but polluted) Charleston Bay, was designed as a recirculating system and uses a synthetic sea salt medium.

REFERENCES

Atz, James W., 1964a. Self-inhibition by captive fishes through water in which they live, Aquasphere (Boston) 2(1): 11-13.

___ 1964b. Some principles and practices of water management for marine aquariums, Seawater Systems for Experimental Aquariums, Bureau of Sport Fisheries and Wildlife, Report 63: 3-13.

Brockway, Donald R., 1950. Metabolic products and their effects, Progressive Fish Culturist, 12(3): 127-129.

Burrows, Roger E., 1964. Effects of accumulated excretory products on hatchery reared salmonids, Bureau of Sport Fisheries and Wildlife, report 66: 1-11.

Collier, A. & K. Marvin, 1953. Stabilization of the phosphate ratio of seawater by freezing, U.S. Fish and Wildlife Bulletin #79, 54: 69-76.

Collier, Albert W., 1953. The significance of organic compounds in seawater, Transactions of the Eighteenth North American Wildlife Conference, pp 463-477.

DeGraaf, 1964. Maintenance problems in large public aquaria, Archives Neerlandaises de Zoologic, 16(1): 143.

Eagleton, L. & E. Herald, 1968. Ultraviolet sterilization for marine aquaria, Salt Water Aquarium, 4(6): 158-164.

Emerson, David N., 1969. Influence of salinity on ammonia excretion rates and tissue constituents of euryhaline invertebrates. Comp. Biochem. Physiol., 29(3): 1115-1133.

Goldizen, Vernon C., 1970. Management of closed-system marine aquariums, Helgolander wiss. Meeresunters, 20: 637-641.

Harvey, W.H., 1955. The Chemistry and Fertility of Sea Waters, Cambridge Press, Cambridge, V.K., 224 pp.

Herald, E.S., R. Dempster, and M. Hunt, 1970. Ultraviolet sterilization of aquarium water, Aquarium Design Criteria (Special Edition Drum and Croaker): pp 57-71.

Hirayama, Kazutsuga, 1966a. Studies on water control by filtration through sand bed in a marine aquarium with closed system - III Relation of filtering velocity and depth of sand layer to purification of breeding water, Bulletin of the Japanese Society of Scientific Fisheries, 32(1): 11-19.

___ 1966b. Studies on water control -- IV Rated pollution of water by fish, and the possible number of weight of fish kept in an aquarium, Bulletin of the Japanese Society of Scientific Fisheries, 32(1): 20-27.

___ 1966c. Influence of nitrate accumulated in culturing water on Octopus vulgaris, Bulletin of the Japanese Society of Scientific Fisheries, 32(2): 105-111.

___ 1970. Studies on water control -- VI Acidification of Aquarium Water, Bulletin of the Japanese Society of Scientific Fisheries, 36(1): 26-34.

Honig, C., 1934. Nitrates in aquarium water, Journal of the Marine Biological Association of the United Kingdom, 19(2): 723-725.

Huckstedt, G., 1969. The pro and con of aquarium technology: Ozonizer, UV-radiating device, and protein skimmer -- critically considered, Aquarium Magazine 8: 325-327.

James, D.E., Botany Department, Carolina Biological Supply Co., Burlington, North Carolina. (Conference attendee).

Kalle, K., 1966. The problem of gelbstoff in the sea, Oceanography and Marine Biology Annual Review, 4: 91-104.

Kawai, A., Y. Yoshida and M. Kimata, 1964. Biochemical studies on the bacteria in aquariums with circulating system -- I Changes of the qualities of breeding water and bacterial population of the aquarium during fish cultivation. Bulletin of the Japanese Society of Scientific Fisheries, 30(1): 55.

___ 1965. II Nitrifying activity of the filter sand, Bulletin of the Japanese Society of Scientific Fisheries, 31(L): 65.

Kawamto, N.Y., 1961. The influence of excretory substances of fishes on their own growth, Progressive Fish Culturist, 23(2): 70-75.

Kelley, William E., 1963. Ideal configurations for semi-closed circulating aquarium systems, program 50th anniversary meeting American Society of Ichthyologist and Herpetologists.

___ 1965. A notable advance in marine biology, Wards Bulletin, 5(30): 1-2.

Kelley, William E. and Daniel H. Moreno, 1961. A filtering system for individual tanks, Bulletin of Aquatic Biology, 2(21): 117-118.

King, John M. and William E. Kelley, 1971. Airlift pumps and the efficiency of undergravel filters, Sea Scope, 2(3): 5-8.

Kuwatani, Y. and Tamotsu Nishii, 1969. Effects of pH of culture water on the growth of the Japanese pearl oyster, Bulletin of the Society of Scientific Fisheries, 35(4): 342-350.

___ and Fumio Isogai, 1969a. Effects of nitrate of culture water on the growth of the Japanese pearl oyster, Bulletin of the National Research Laboratory, 14: 1735-1747.

___ 1969b. Effects of the number of animals and the frequency of exchange of water on the growth of the Japanese pearl oysters in tank culture, Bulletin of the National Research Laboratory, 14: 1748-1764.

Miklosz, John C., 1971. Biological filtration, The Marine Aquarist, 1(5): 22-29.

Milnes, D., 1961. Ozone - a handbook of physical and technical data, Ozone Research and Equipment Corp., Phoenix, Arizona, pp 8.

Oliver, J.H., 1957. The chemical composition of seawater in the aquarium, Proc. of the Zoological Soc. of London, 129(1): 137-145.

Paccaud, Andre, 1962. Considerations generales sur la tenue des animax marins, specialement des poissons (des coraux), en eau der mer artificielle, Bulletin de Institute Oceanographique, Foundation Albert Ier Prince De Monaco, B(16): 39-48.

Roff, J., 1972. The nitrogen cycle in nature and culture, SEA SCOPE, 2(4): 1, 7-8.

Saeki, Aritsune, 1958. Studies on fish culture in filtered closed-circulation aquaria, Bull. of the Japanese Soc. of Scientific Fisheries, 23(11): 684-695. Translated 1964 by the Directorate of Scientific Information Services, Canada.

———— 1962. The composition and some chemical control of the seawater of the closed circulating aquarium, Bull. of the Marine Biol. Sta. of Asamushi, XI(2): 99-104.

Sandt, D.G., 1968. Closed marine system I theory, unpublished report, Natural Resources Institute, University of Maryland.

Shelbourne, J., J. Riley and G. Thacker, 1963. Marine fish in culture in Britain. I Plaice rearing in closed circulation at Lowestoff, 1957-1960, Jour. du Conseit International pour I' Exploration de la Mer, 28(1): 50-69.

Siddall, S., 1972. Studies of closed marine culture systems, Unpublished report submitted as Honor's Thesis to Department of Biology. Case Western Reserve University, Cleveland, Ohio.

Smith, H.W., 1929. The excretion of ammonia and urea by the gills of fish, Jour. of Biol. Chem., 81: 727-742.

Sorokin, C., 1971. The capacity of growing algal cells to affect the pH and the buffering properties of the medium, Plant and Cell Physiology, 12: 979-987.

Spangenberg, D.G., 1967. Iodine induction of metamorphosis in Aurelia, Journ. of Exper. Zool., 165(3): 441-450.

Spotte, Stephen H., 1970. Fish and Invertebrate Culture, Water Management in Closed Systems, Wiley-Interscience, New York, pp 160.

Sverdrup, H., Martin Johnson, and Richard Fleming, 1942. The Oceans: Their Physics, Chemistry and General Biology, Prentice-Hall, New York, pp 1087.

Versagi, F., 1970. Ozone and electronic air cleaners, Air Conditioning, Heating and Refrigeration News, Birmingham, Michigan, pp 11.

Waksman, S.A., M. Hotchkiss and C. Cory, 1933. Marine bacteria and their role in the cycle of life in the sea, II Bacteria concerned with the cycle of nitrogen in the sea, Biol. Bull. LXV(2): 137-166.

Wallace, G. and D. Wilson. Foam separation as a tool in Chemical Oceanography, N.R.L. Report 6958: 1-15.

Wilson, Louise P., and Mary A. Greggs, 1941. Analysis of seawater in a closed aquarium, Trans. of the N.Y. Academy of Sci., 3(8): 218-221.

MAINTENANCE OF SOME MARINE FILTER FEEDERS ON BEEF HEART EXTRACT

Helen M. McCammon

Environmental Protection Agency

Boston, Massachusetts

When considering eating habits of marine filter feeding invertebrates, the classical model of energy transfer through various trophic levels is not always valid. Filter feeders can feed directly on many different nutrients ranging from dissolved organics to large zooplankton. Furthermore, adaptations of feeding and digestive structures along with experimental evidence indicates that different classes of nutrients are utilized more efficiently by different filter feeders. For example, after some pre-oral screening, pelecypods ingest a variety of living and inert microscopic and sub-microscopic particles. Breakdown of the particles and extraction of the nutrients takes place in a complexly structured digestive system which varies in its morphology in different families probably as a result of adaptive plasticity (Morton, 1960). On the other hand, the pogonophores have evolved into a phylum with no digestive system; these invertebrates derive much of their energy from dissolved nutrients which are absorbed directly from solution along the entire body surface (Southward & Southward, 1970; Little and Gupta, 1968). Other filter feeders fall intermediate between these two groups in their digestive and food gathering adaptations.

Culturing of filter feeders such as many pelecypod and ectoproct species can be accomplished without too much difficulty if abundant stocks of phytoplankton are available. Mixtures of phytoplankton and larger zooplankton can provide adequate food supply for other species of bivalves and barnacles. However, certain groups of filter feeders cannot subsist long on plankton alone and need additional nutrients or a different diet from one that is ordinarily supplied in cultures. Beef heart extract with a vitamin supplement has been successfully used for maintaining forms which require dissolved nutrients either as an additional or a primary source of energy.

It is often advantageous to study the morphology as well as the natural habitat when first attempting to culture little known invertebrates in the laboratory. This leads to a better understanding of the culturing needs of the particular animals under consideration and the possibility of more rapid success in the laboratory environment.

Through laboratory studies of several species from the class Articulata, phylum Brachiopoda, a feeding procedure consisting of beef heart extract was evolved. The brachiopod morphology gives an indication that these invertebrates do not have the capacity to subsist on the very same nutrients that the pelecypods and many other filter feeders can, but they can do very well when a proper type of nutrient is provided.

COMPARATIVE MORPHOLOGY OF ARTICULATE BRACHIOPODS

Articulate brachiopods are more commonly recognized as fossils rather than living forms for they have a long geologic history and were dominant marine invertebrates during the Paleozoic era. In more recent geologic history, the articulates have been over-shadowed by the molluscs in dominance. However, in a few areas of the world today, the articulate brachiopods abound in size and numbers equal to those in the geologic past.

Although externally the articulate brachiopods superficially resemble the pelecypods in having two valves, internally the morphology is quite different (fig. 1). The brachiopods possess a lophophore as do all other members of the lophophorate phyla. The lophophore is generally used to create water currents for food gathering, sorting of particles and respiration. In the articulate brachiopods, the volume occupied by the lophophore is much greater than in the other invertebrates. The plectolophous lophophore which is the most complexly coiled type, is present in living species found in greatest abundance and which reach the greatest size. All lophophorate groups have a structurally simple digestive system in contrast to the highly complex molluscan digestive system (fig. 2). But the digestive system of the articulate brachiopods differs from other lophophorates and other large filter feeders in that it does not have a long intestine with an anal opening, but an intestine which is short and blind-ended. The reduced digestive system occupies a small portion of the posterior area of the valves with the end of the gut hanging free alongside the pedicle muscle. This

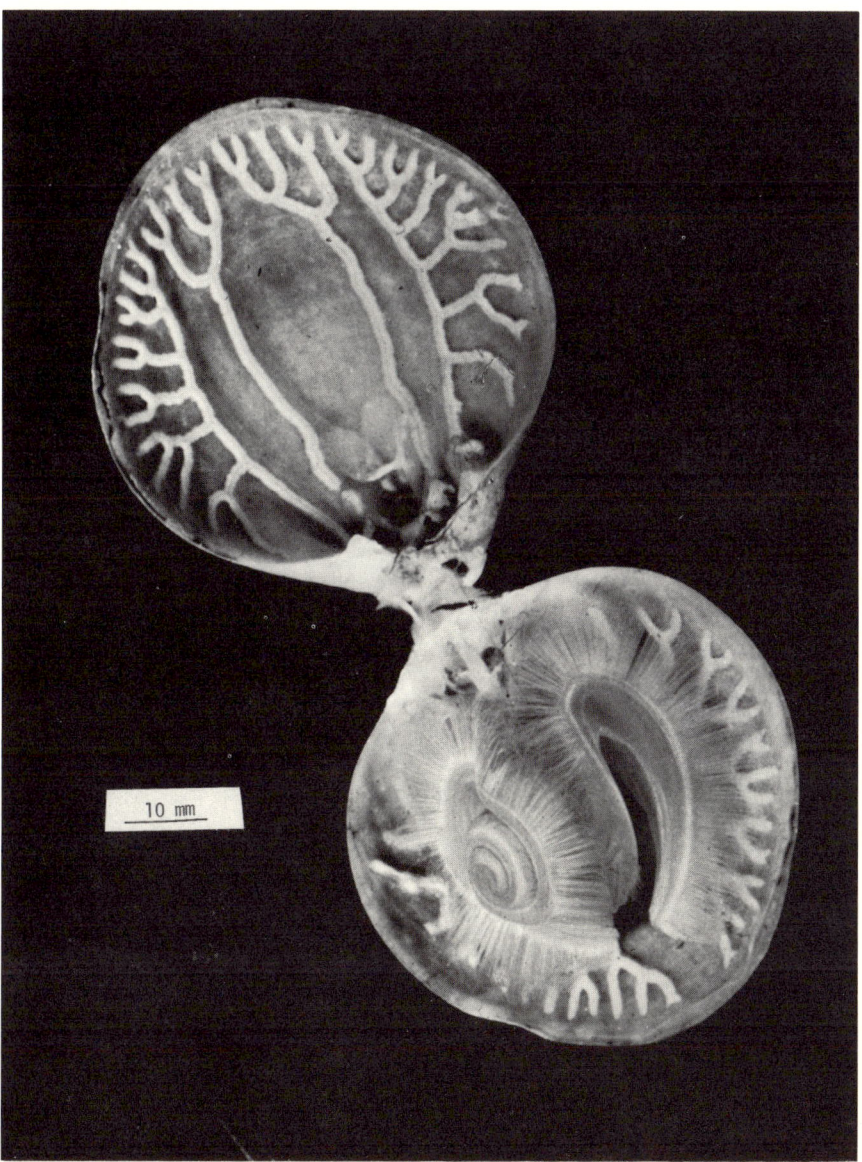

Fig. 1. Interior of the articulate brachiopod Magellania venosa. The pedicle valve (upper half) has the mantle canals filled with gonadal tissue. The brachial valve contains the large lophophore. The plectolophous lophophore is characterized by the double lateral arm and a medial planispiral coil. The planispiral coil has been pulled out in the illustration to show the configuration of the coiling.

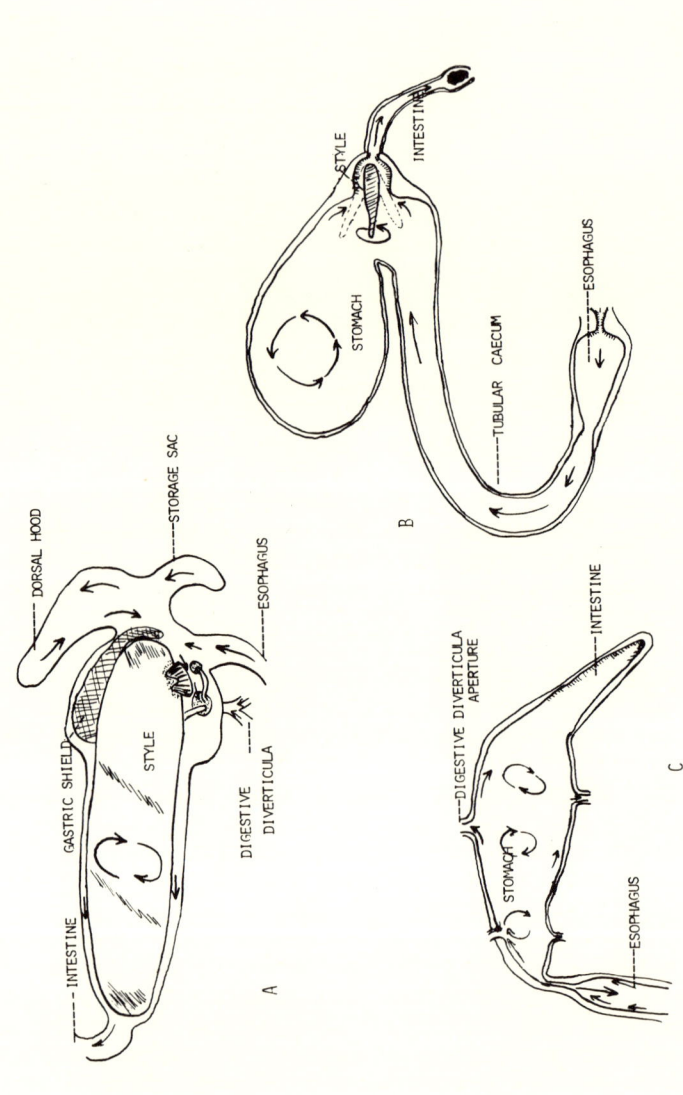

Fig. 2. Comparative anatomy of stomachs of filter feeders (adapted and modified from Morton 1960). A. Stomach of pelecypod, Tellina. B. Gut of ectoproct Membranipora. C. Gut of articulate brachiopod, Macandrevia.

evolved adaptation does not lend itself to indiscriminant ingestion of large quantities of both nutritive and non-nutritive particles and the resultant need for efficient elimination of copious waste material.

Comparison of the filtering and digestive systems of the pelecypods and lophophorates suggests a past evolutionary history of little food sorting before ingestion by the pelecypods, a somewhat more efficient sorting by ancestors of living ectoprocts, phoronids and inarticulate brachiopods, but a very efficient pre-oral sorting by the articulate brachiopods (Morton, 1960). This comparison also suggests that the hinged brachiopods are probably more sensitive to the diet available in the environment.

The greater food sensitivity of the articulates in contrast to bivalves became apparent to me when initial attempts to maintain the species Terebratulina septentrionalis (Couthouy) in the aquarium solely on plankton ended in failure. The species Magellania venosa (Solander), Waltonia inconspicua (Sowerby) and Notosaria nigricans (Sowerby) were maintained for many months on vitamins, glucose and various amino acids. When phytoplankton was added, irritation and rapid rejection with later mortality was observed in M. venosa with less, but nevertheless, negative reaction in the two intertidal species. The phytoplankton introduced at different times were Phaeodactylum sp., Dunaliella sp., and Isochrysis sp. Although there was a need for a more complete food to maintain vigor in the animals, phytoplankton did not appear to be the answer.

The idea for a simple satisfactory liquid diet for the brachiopods developed after a visit to the Vancouver Public Aquarium in British Columbia. After feeding chopped liver and other red meat to the fish, personnel at the Aquarium added meat juices that collected in the trays to the tanks containing filter feeders. On addition of the juice, a quick positive response was obtained from most of the mobile invertebrates, both filter feeders and others. Thus meat juice appeared to be a possible acceptable food source for the discriminating brachiopods, particularly beef heart juice which would be relatively inexpensive in comparison to its great nutrient value.

Because there was not a ready source of beef juice, a supply was created by processing meat through a blender. After initial trials, the procedure has been altered and adjusted so that now not only twelve species of articulate braciopods have been maintained in aquaria for over two years, but crinoids, small solitary tunicates, and many sponges appear to thrive with addition of the beef heart extract.

AQUARIUM REQUIREMENTS

Although there are standard techniques for introducing and maintaining invertebrates in a closed seawater system (King, chapter 1; James, 1972; Valenti, 1968), procedures which are considered pertinent to the beef extract feeding program are outlined below.

The aquarium system found to be the most successful with this method of feeding is one with a substrate filter and air-lift circulation. Because of the light load maintained in the tanks, no additional filters have been necessary. Artificial seawater (Instant Ocean Synthetic Sea Salts, Eastlake, Ohio) is used in preference to natural seawater to minimize introduction of unknown chemical and biological factors. The commercial sea salt is dissolved in distilled water when good quality tap water is not available.

The invertebrates fed beef heart have all been cold water forms. Thus the temperature of the aquarium is maintained at the lower temperature range in which the specimens ordinarily live. Before introducing the specimens into the tank, they are placed in a small volume of aerated seawater containing antibiotics at a concentration of 10000 units/liter penicillin "G", 0.1 g/liter streptomycin, and 0.02 mg/liter chloromycetin, with water at the same temperature as the maintenance tank. The specimens are kept in the antibiotic solution about 1 hour.

No food is given for the first week after introduction of the invertebrates into the main tank except for vitamins B_{12}, thiamine, and biotin which are added on about the third day and once a week thereafter. Feeding of the beef heart filtrate is started after the invertebrates have become acclimatized to the closed system.

BEEF HEART FILTRATE

A beef heart is cut up into 3 to 4 cm cubes with the adipose tissue removed so it doesn't clog the filters. The cubes are individually wrapped and frozen for use as needed. To make the extract, a beef cube is sliced up and placed in a blender half full of either seawater or distilled water and blended at high speed for several minutes. Then the fat at the surface is skimmed off. The liquified beef is passed through cheesecloth to catch the sinew, then through coarse paper toweling and finally through filter paper (Fig. 3). The filtrate is poured directly into the aquarium.

Initially, after introduction of the specimens, a small amount of the extract is added to the tanks twice a week with a gradual increase of filtrate to maintenance level. Quantity of extract at maintenance level is dependant upon the size of the aquarium and the

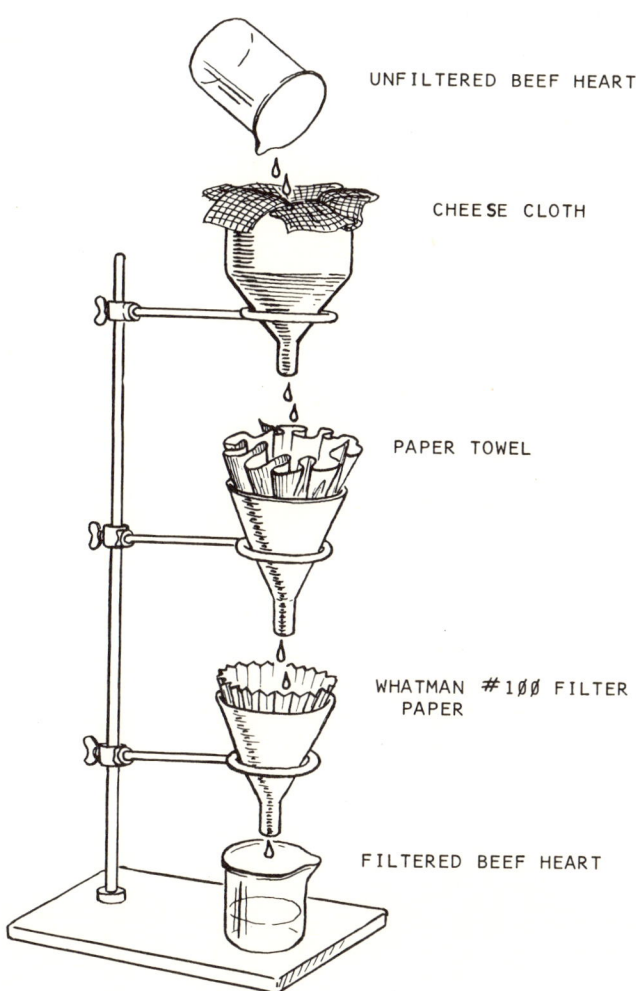

Fig. 3. Apparatus set up used for processing of beef heart.

number of specimens in it. A 3 to 4 cm cube of beef has been sufficient to feed 400 brachiopods in each of three 100 liter tanks for each feeding.

The beef filtrate can also be used with great success as a supplemental food for filter feeders and other invertebrates that prefer a more solid diet. Barnacles, coelenterates, pelecypods, ectoprocts, copepods and filtering worms exhibit more vigor and robustness with the addition of beef filtrate along with their regular diet.

Although liver is also of high nutritional value and may be used, the odor after blending is often objectionable to persons processing the extract. Commercially purchased beef extract (Baltimore Biological Lab., Baltimore, Md.) can be used if beef heart is not immediately available, but it has not been used over a long term basis for the invertebrates do not appear as vigorous and healthy as with the freshly processed filtrate.

Keeping the aquarium in darkness, with a black cover over the transparent surfaces, prevents nuisance algal growth in the tanks. Algae tend to foul the surfaces in the aquarium and seem to discourage vigorous growth of marine filter feeders which are normally found on undersides of rocks and in dark crevasses.

DISEASE PROBLEMS

The beef filtrate and vitamin feeding regime can be carried on indefinitely but periodic observation of the behavior of the invertebrates is necessary to detect onset of any diseases in the closed system. Brachiopods and tunicates are particularly prone to fungal infection if adequate circulation is not present. The infection appears as blackened areas having a noxious odor. At the first sign of this infection, feeding is stopped and a commercial fungicide is added directly into the water. Treatment is continued until all trace of fungus has disappeared. Fungus infection can appear within the first month after introducing specimens into tanks, or it may appear after temperatures have, for some reason reached a level higher than normal in the aquarium.

Bacterial infection has been recognized when the invertebrates appear sluggish and the shelled forms feel slippery. For correction of the problem individuals can be placed for a short period into an antibiotic solution and the water in the tank sterilized with ultraviolet light. Feeding is discontinued until healthy conditions return.

Generally, however, when the beef heart diet is used, problems with infection are minimal or non-existent once the specimens have been established in the aquarium for there is no accumulation of uneaten food. Removal of algal growth is not necessary if the tanks are kept dark.

NUTRIENT UTILIZATION

The manner of assimilation of the beef heart by the filter feeders is not known. Heterotrophic organisms could be an important intermediary in the utilization of the beef heart solution in the closed system. The role of bacteria in the food chain has been documented for many types of filter feeders and shown to be significant (Jørgensen, 1966). But recent experiments indicate that bacteria have little effect on uptake of certain nutrients by brachiopods (McCammon & Reynolds, in press). Figure 4 summarizes the results obtained by McCammon and Reynolds of uptake of 0.1 μc/ml ^{14}C-glucose by the braciopod Terebratalia transversa (Sowerby) over a 24 hour period under both aseptic and non-aseptic conditions. The specimens kept under aseptic conditions had a somewhat higher uptake of glucose after 24 hours suggesting that perhaps the bacteria were themselves diminishing the nutrient pool for the brachiopods.

The beef extract when added to the water illicits a positive chemotactic response from many invertebrates. Crinoids wave their arms and move rapidly about in the tanks, contracted sea anemonies will expand their tentacles and brachiopods will open their valves if they are closed. Whether the invertebrates are actually ingesting the beef molecules or are only being stimulated by it can be debated. However, MacGinitie (1945) was able to show that large molecules were caught on mucus nets and were indeed ingested by echiuroid worms.

Recently, many experiments have shown that invertebrates can absorb small molecules directly into the body (Stephens, 1968). But the percent contribution of this uptake to total energy input is generally believed to be small in soft bodied forms with normal digestive systems.

In contrast, pogonophores, which do not possess a digestive system, appear to derive much of their energy from absorption of both small and moderately sized molecules directly from the surrounding liquid (Southward & Southward, 1970, 1972; Little & Gupta, 1968, 1969).

Experimental studies have shown that the lophophore in articulate brachiopods can concentrate organic molecules such as glucose and assimilate this material for transport through the body via the

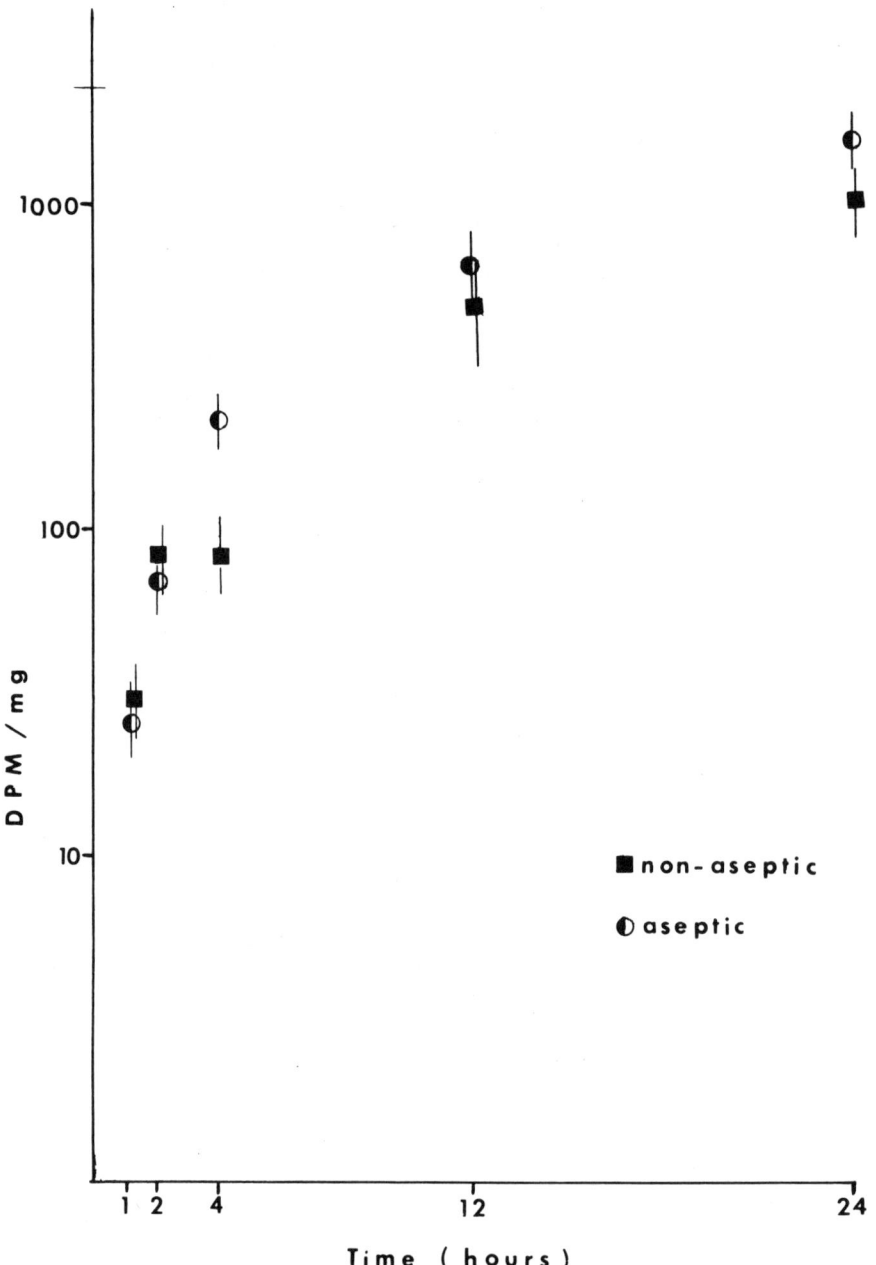

Fig. 4. Uptake of ^{14}C-glucose by <u>Terebratalia transversa</u> under aseptic and non-aseptic conditions. Each point represents the mean ($\pm$ standard error) of six individuals.

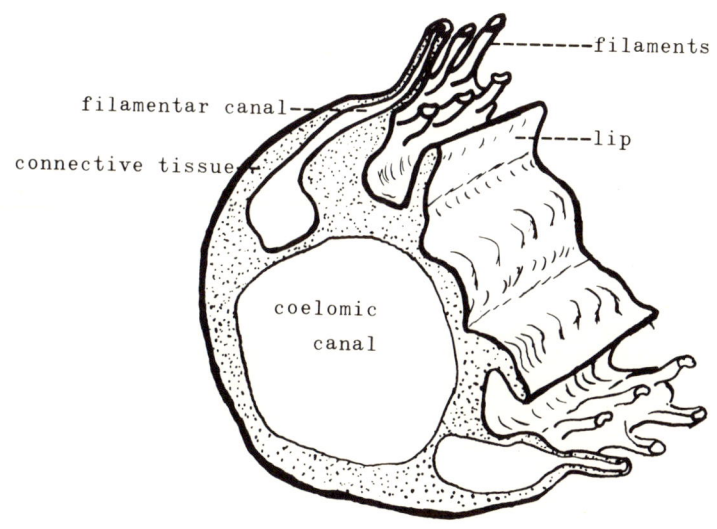

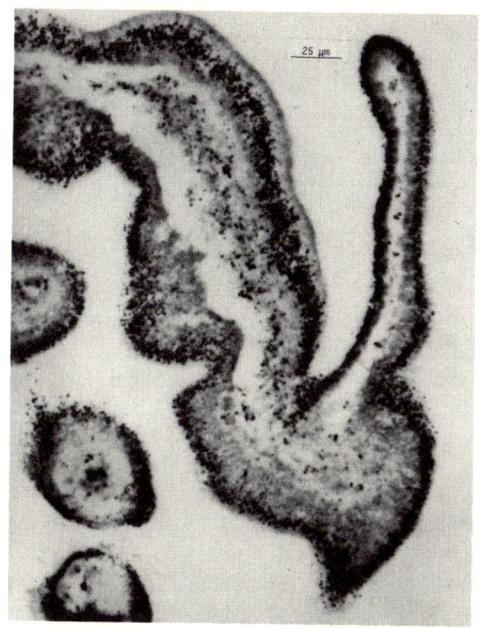

Fig. 5A. Generalized diagram of section of side arm of plectolophous lophophore (after Williams and Rowell, 1965). B. Autoradiograph from side arm of lophophore of <u>Terebratalia transversa</u> after one hour immersion in 0.1 uc/ml solution of ^{14}C-glucose (3 week exposure to emulsion: toluidine blue stain).

coelum before any of the nutrient reaches the digestive system
(McCammon & Reynolds, in press). Figure 5A is a diagramatic representation of a cross section through the lophophore. Figure 5B
is an autoradiograph made from a section through the lophophore,
one hour after administration of ^{14}C-glucose into the water. The
presence of ^{14}C label is indicated in the tissue by blackened spots
where the radioactive particle exposed the NTB-2 (Kodak) emulsion
that was used to stain the slide.

Figure 5B shows that a heavy concentration of labelled material occurred on the exposed ciliated surfaces of the lophophore
and filaments. A moderate concentration of label was present in
the connective tissue with heavy local concentration in the brachial canal. A series of autoradiographs was made from tissue in
brachiopods exposed to progressively longer intervals in the radioactive solution indicates that the nutrients entered along the
lophophoral and mantle surfaces and was distributed throughout the
body by the coelomic fluid before any labelled material was detected in the lumen of the stomach.

Beef heart is particularly nutritious with high concentrations
of vitamins and proteins. It may be that some of the particles
are broken into component molecules through mucoid action along the
exposed ciliated surfaces of the invertebrates. Although these
beef derived molecules are considerably larger than glucose molecules many may be assimilated directly along the lophophoral surface of the brachiopods and not necessarily through a secondary
pathway of heterotrophic organisms.

SUMMARY

Culturing marine filter feeders is not always successful with
only a phytoplankton diet. Many filtering organisms survive much
better on dissolved food or a combination of both particulate and
dissolved food, depending on their physiologic and anatomic adaptations.

Beef heart filtrate is a simple, readily available food source
for certain filter feeders such as articulate brachiopods, crinoids,
small tunicates and sponges. It is also an effective supplemental
food for filter feeders preferring a more solid diet such as barnacles, various coelenterates, ectoprocts, worms and pelecypods.
Specimens maintained on such a diet can live for many years in a
closed system.

ACKNOWLEDGEMENTS

It is difficult to thank everyone who has helped me in my culturing attempts over the years, for many people stand out equally in their kind assistance and advice they have given. But without the financial support of the National Science Foundation, it would not have been possible to embark on the study of filter feeders from various parts of the world. Support was obtained from NSF Grants GA 255, GA 675, GB 8227, and GB 20067.

REFERENCES

James, D.E., 1972. Maintaining the marine aquarium. Carolina Tips, 35: 5-7.

Jørgensen, C.B., 1966. Biology of suspension feeding. Pergamon Press, New York, 357 pp.

Little, C. & B.L. Gupta, 1968. Pogonophora: Uptake of dissolved nutrients. Nature, 218: 873-874.

___ 1969. Studies of Pogonophora. III Uptake of nutrients. Journ. Exp. Biol., 51: 759-773.

MacGinitie, G.E., 1945. The size of the mesh openings in mucous feeding nets of marine animals. Biol. Bull. 88: 107-111.

Morton, J.E., 1960. The functions of the gut in ciliary feeders. Biol. Reviews, 35: 92-139.

Southward, A.J. & E.C. Southward, 1970. Observations on the role of dissolved organic compounds in the nutrition of benthic invertebrates. Experiments on three species of Pogonophora. Sarsia, 45: 69-96.

___ 1972. Observations on the role of dissolved organic compounds in the nutrition of benthic invertebrates. II Uptake by other animals living in the same habitat as pogonophores, and by some littoral Polychaeta. Sarsia, 48: 61-70.

Stephens, G.C., 1968. Dissolved organic matter as a potential source of nutrition of marine organisms. Am. Zool. 8: 95-106.

Valenti, R.J., 1968. The salt water aquarium manual. Aquarium Stock Co., New York, 162 pp.

Williams, A., & A.J. Rowell, 1965. Brachiopod Anatomy. In: Treatise on invertebrate paleontology; R.C. Moore, Editor. Part H. Brachiopoda. 6-57 p., Lawrence, Kansas; University Kansas Press & Geol. Soc. America.

CULTURE OF PHYTOPLANKTON FOR FEEDING MARINE INVERTEBRATES

Robert R.L. Guillard

Biology Department, Woods Hole Oceanographic Institution

Woods Hole, Massachusetts

These pages describe relatively simple and reliable methods for the culture of marine phytoplankton species useful for feeding marine invertebrates. The methods suffice for the most fastidious algae now routinely cultivable, and simplifications indicated for less demanding species are easily made; for example, omission of silicate for plants other than diatoms. Certain modifications of techniques, ancillary methods, and precautions will be treated briefly because questions often arise concerning them, but documentation will be minimal and hopefully restricted to publications readily available.

The methods apply to the production of sterile cultures of considerable density in volumes up to 18 liters, in commercially available 5 gallon borosilicate glass carboys. Apparatus for growing unialgal (not bacteria-free) cultures in volumes up to 160 liters has been described (Ukeles, 1971; Siegelman and Guillard, 1971). Such cultures, started from sterile carboy cultures, can be used in turn to inoculate larger outdoor tank or pond cultures. This is another problem entirely. All non-sterile cultures suffer the disadvantage that they may become contaminated with bacteria that injure or destroy the animals being fed. Larger non-sterile cultures appear to be more susceptible to invasion by competing algae and to predation by animals. It is true that carefully managed tank algal cultures as used by many hatcheries are quite reliable, but the only algal cultures suitable for critical studies are both bacteria-free and harvested while the algae are sill actively growing; that is, before maximum cell density is attained.

The methods to be described take an algal culture to the point of usefulness for experiment, for maintenance of relatively small animal populations, or for inoculation of larger unialgal cultures.

The brief and lucid book by Fogg (1965) provides an excellent background for the algal culturist, and the terse article on laboratory cultures by Myers in the treatise by Lewin (1962) outlines many important concepts.

SPECIES OF ALGAE EMPLOYED

Inasmuch as animals from fish to protozoa have been fed on algal cultures, it is not surprising that algal species from all major taxonomic groups are involved. The scattered literature and the book by Jørgensen (1966) reveal that most studies of marine animals concern molluscs or arthropods. Most of the successful algal species belong to one of six algal classes. (There are some 15 algal classes according to the system of Christensen. The book on algae by Morris (1967) obtainable in paperback, discusses these classes and describes or mentions many of the algae that are important as food species.) The six classes referred to, and representative useful species, are:

Class Bacillariophyceae, the diatoms: Skeletonema costatum, Thalassiosira pseudonana (ex Cyclotella nana), Thalassiosira fluviatilis, Phaeodactylum tricornutum, Chaetoceros calcitrans, Chaetoceros simplex, Ditylum brightwellii.

Class Haptophyceae, golden brown flagellates (similar to the class Chrysophyceae mentioned next): Isochrysis galbana, Dicrateria inornata, Crisosphaera carterae, Coccolithus huxleyi. Probably Monochrysis lutheri belongs here.

Class Prasinophyceae, a group of greenish-colored algae (recently removed from the green algae proper): Pyramimonas grossii, Tetraselmis (Platymonas) suecica, several other species of this genus, and Micromonas pusilla.

Class Chlorophyceae, the green algae sensu strictu: Dunaliella tertiolecta, Chlorella autotrophica, Chlorococcum sp. Nannochloris atomus, Chlamydomonas coccoides, Brachiomonas submarina.

Class Chrysophyceae (sensu strictu), golden brown flagellates: possibly certain Monochrysis species.

Class Cryptophyceae, a group of naked flagellates: Chroomonas salina (often put into the genus Cryptomonas or the genus Rhodomonas).

Starter cultures are ordinarily obtained from investigators working on particular animals or from algal culture collections that are maintained at various institutions as part of research effort. A list of sources is given in the Handbook of Phycological Methods (Stein, 1973). At least one commercial biological supply house, Carolina Biological Supply Co., furnishes some useful species.

Some care should be exercised when obtaining inoculating cultures. Not all cultures bearing the same genus and species names are equivalent. Quite aside from instances of outright misidentification of species, there may be differences between different isolates from nature; also, prolonged subculture and reisolation of clones in various laboratories may yield variant strains. The clonal designation of the inoculum should be ascertained (and used in publication), and the history of the strain traced back to the original isolation.

A plea can be made at this point to investigators working with animals in natural situations that they arrange to isolate the dominant local microalgae whenever animals are growing extremely well. Overall, this will get good food species into culture faster than screening random isolates. The methods involved are beyond the scope of these pages, but they are described in the handbook referred to above (Stein, 1973).

PROVISION OF A SUITABLE PHYSICAL ENVIRONMENT FOR CULTURES

This includes the provision of light, of temperature control, of air for pH control and to supply carbon dioxide (which is in effect a nutritional aspect), and finally, includes discussion of the kinds of culture vessels and associated materials that are needed. It is best to consider the glassware first, because its geometry imposes conditions on all the rest. Culture glassware in turn depends upon the system of maintaining stock cultures and of subsequent food production. These interacting factors are conveniently discussed together.

Culture Vessels and Routine

Stock cultures are kept in standard 125 ml Erlenmeyer flasks or in one of three types of culture tubes: 20 x 120 mm or 20 x 150 mm screw-capped, or 18 x 150 mm without lip (the question of stoppers is considered later). Auxiliary stocks for starting larger cultures are kept in 125 ml standard or screw-capped Erlenmeyer flasks. Intermediate-scale production for feeding or inoculation of carboys is in 2 liter Erlenmeyer flasks or the efficient 2.8

liter Fernbach flasks. Carboys of 3.5 or 5 gallon capacity are used for the largest cultures; choice may be dictated by size of the autoclave or culture shelves or by physical strength of the phytoplankton farmer.

Approximate dimensions of glassware are tabulated below:

<u>Tubes</u>: diameter - 18 mm (3/4 in.) to 20 mm (13/16 in.)
length -- 120 mm (4.75 in.) to 150 mm (5.9 in.)
Allow another 25 mm (1 in.) if cotton stoppers are used.

<u>125 ml Erlenmeyer flasks</u>: diameter - 65 mm (2.5 in.)
height --- 11 cm (4.25 in.)
Allow another 4 cm (1.5 in.) for the plug.

<u>2 liter Erlenmeyer flasks</u>: diameter - 17 cm (6.5 in.)
height --- 31 cm (12.3 in.)
Allow 10 cm (4 in.) for cotton stoppers.
Allow 15 cm (6 in.) if aerator tubes are used.

<u>2.8 liter Fernbach flasks</u>: diameter - 21 cm (8.25 in.)
height --- 23 cm (9 in.)
Allow additional height as for 2 liter flasks above.

<u>3.5 gallon carboys</u>: diameter - 24-25.5 cm (9.4-10.1 in.)
<u>(11 liters)</u> height --- 40.5-45 cm (16-17.4 in.)
Allow another 15 cm (6 in.) for the aeration system.

<u>5 gallon carboys</u>: diameter - 28.5-30 cm (11.3-11.8 in.)
<u>(18 liters)</u> height --- 51-52 cm (20-20.5 in.)
Allow another 15 cm (6 in.) for the aeration system.

These physical dimensions will determine the arrangement of the culture facility, or conversely, the facility can influence the choice of glassware. Algal production efficiency in terms of power and space consumption can be maximized by proper allocation of shelf space to vessels of different heights, by appropriate spacing of shelves, and by arrangement of light banks to meet the light intensity requirements of different cultures.

In summary of the overall process of food production -- not of specifics of technique, which will be discussed later -- the system calls for stock cultures, and if necessary, sub-stocks in 125 ml flasks or tubes, or both; for bigger cultures in 2 or 2.8 l vessels; and for carboy cultures if necessary. Safety of the stock cultures is also an important consideration. Each type of vessel is placed so as to maximize production.

Provision of Light

Fluorescent tubes are now the source of choice for routine production and much experimental work. Tubes are available commercially that have quite different spectral emissions. Tubes with certain of these spectral characteristics were developed to match the absorption spectrum of green plants, or to match the appearance of daylight to the human eye. However, there is no definite evidence that such tubes are better for growing marine algae, most of which in fact are golden-brown in color and well adapted to absorbing the bluish-green light that has maximum penetration in clear seawater. Manufacturers use different names to describe the spectral properties of their bulbs. Descriptive literature is available. The most commonly used bulbs have spectral properties like those called "cool-white" by the Sylvania Co. A successful combination is equal numbers of "daylight" and "natural" bulbs (same terminology).

The intensity and duration of illumination are both important. It is surprisingly difficult to measure either the total light or the photosynthetically active light received by a culture, and confusing to describe it when measured, because of the different kinds of units that can be employed. Because the commonest and certainly least expensive type of measuring device is a hand-held foot-candle meter ("light meter") such as the General Electric Co. Type 213, we shall use illumination (human eye response) units, rather than energy units; however, in deference to the metric system we use the lux (meter-candle) rather than the foot-candle. The conversion, fortunately, involves only multiplying by about ten-more exactly:

$$1 \text{ foot-candle} = 10.76 \text{ lux (meter-candles)}$$

$$1 \text{ lux} = 0.0929 \text{ foot-candles}$$

If energy-measuring devices are used, or given in references, the following conversion factors will be found useful; they apply to cool-white fluorescent bulbs, (μw designates microwatts):

$$1000 \text{ }\mu w/cm^2 = 1.43 \times 10^{-2} \text{ cal}/cm^2/\text{min or langleys/min}$$
$$= \text{approximately 280 foot-candles}$$
$$= \text{approximately 3020 lux}$$

For other types of fluorescent bulbs the conversion factors may differ by + or - 20%.

Maddux (1966) has described a small integrating light meter that can measure total radiation inside a vessel; this is remarked in hopes that more workers will use such a device. Most devices measure the illumination perpendicular to a flat surface on the meter. Any effort to sum the light received from all directions is an approximation. To serve as a general guide, some light measurements were made in a typical growing chamber having four 42 in. "cool-white" tubes suspended below perforated stainless steel shelves (about 50% of the surface perforated). The space reserved for carboys is illuminated from above by four tubes and from behind by four additional tubes. The walls of this particular chamber were gray stippled enamel, not particularly reflective.

1. Shelf 35 cm (14 in.) beneath four F42 T 12 CW (Sylvania Co.) tubes; light also came up through the shelf perforations from the tubes below: intensity was 4000 lux from above, 1500 lux from below, and 1500-2000 lux measured horizontally. This shelf space was used for stock cultures and Fernbach flasks. Half this light intensity will do for stock cultures, but not for the Fernbach cultures. If the shelf was moved to 25 cm (10 in.) which is about as low as it can be for culture tubes or for 125 ml flasks, the intensity from above was about 6000 lux.

2. Shelf 43 cm (17 in.) below the tubes, otherwise as above; illumination was not measurably different. This shelf is suitable for large Erlenmeyer flasks.

3. Space for carboys on the floor, below the lowest shelf; measurements made at the position of the center of a carboy: illumination from above (45 cm or 18 in. from the tubes) 6000 lux; towards the rear tubes, 10,000 lux; from below, 2500 lux; from the interior of the room 1500 lux. Either the rear or top lights alone will do until the cultures become quite dense.

Not all marine phytoplankters will tolerate continuous illumination, but most species commonly used as foods will. For most such plants there is a level of illumination at which growth rate will be increased by either higher intensity or longer exposure or both. This effect ultimately saturates and there is obviously no use supplying more light energy than will result in increased yield. This will depend on the volume of the culture. It is much more difficult to saturate a dense carboy culture than a stock culture. As a guide, note that a stock culture of <u>Thalassiosira pseudonana</u> can divide over three times a day in continuous light of about 3500 lux, and in 14 hours of illumination a day at ca. 4500 lux (vertical illumination). The light conditions as described above for carboys will give excellent growth in 14 hr./day.

Photoperiod is set by means of electric time clocks like those used commercially for poultry houses. These are reliable and inexpensive.

It should not be forgotten that daylight will suffice for cultures. Stock cultures will do well illuminated by a window provided that they do not receive direct sunlight. Carboy cultures of many species will even tolerate full sunlight if they are dense, but they must be shaded in earlier stages. It is in general difficult to establish routine algal production using natural light. The chief problems are temperature control and -- the weather.

A final point; not all algae will grow under artificial illumination, but these species are for obvious reasons not part of the current armamentarium of the marine culturist.

Provision of Temperature Control

Room temperature is not really satisfactory for marine algae even if they are grown without artificial illumination. Virtually all marine or estuarine species, even if from the tropics, die at temperatures a little above $35^{\circ}C$ ($95^{\circ}F$); in temperate climates, especially in metropolitan areas, room temperatures exceeding these (above $38^{\circ}C$) can be expected. Temperature control is mandatory if artificial lighting is used.

The commonest arrangement and certainly the most useful is a walk-in commercial culture box ("environment chamber") or an air-conditioned room with provision for lights. There are many kinds of smaller culture boxes, commercial or home-made, that do not permit the investigator to enter them bodily. Some of these will accept carboy cultures. If they are big enough for the production desired they may offer advantages over walk-in boxes or rooms. All these devices are in effect small air-conditioned rooms (like a commercial refrigerator). Most use air-cooled compressors, but the bigger ones may require water-cooled compressors if the supply of cool air is insufficient. This can be an important point, not only for the comfort of the investigator who must live in proximity to his culture box, but for the safety of the phytoplankters. The compressor is most likely to fail under conditions of heat stress when it is needed most.

Water baths offer better but unnecessarily fine temperature control and are in general less convenient. However, a cheap and reliable source of cold water can be exploited for temperature control, especially for carboy cultures. Depth of immersion in running water is a simple control mechanism for cultures well stirred by aeration.

For all air-conditioned chambers and for water baths with temperature control by heating and cooling units, it is necessary to have auxiliary safety controls. These are not always provided on commercial chambers. The auxiliary temperature thermoregulators and associated relays (if needed) should be arranged as follows:

1. One regulator should open when the temperature in the culture chamber (or bath) reaches a predetermined value above the setting of the main thermoregulator. (For example, if a box is set to run at $20° \pm 2°C$, the safety regulator may be set to open at $25°C$).

2. The thermoregulator should remain open until reset manually.

3. On opening it should inactivate <u>all</u> circuits associated with the culture box, this includes lights, heaters, fans, stirrers, pumps, and the cooling compressor. The last is important, because a compressor may fail and act as a heater, thus heating a chamber when cooling is called for by the thermoregulator.

4. On opening, the safety thermoregulator should set off an audible alarm.

5. Another thermoregulator should open when the temperature in the culture chamber reaches a predetermined value <u>below</u> the setting of the main thermoregulator. It should be further arranged to act as in 2-4 above.

In some laboratories the safety thermoregulators open if the main power is interrupted momentarily, even if the temperature in the chamber does not go beyond the safety limits. While this is not a desirable feature, it does little harm in general.

The above precautions are not trivial. The better the temperature control of a chamber, the more heat it can transfer quickly, and the more damage it can do if the controls fail. A good quality commercial chamber can freeze seawater stock cultures solid overnight, or attain a temperature of $85°C$.

Thermoregulators using bimetallic strips, mercury contacts, or thermistors plus solid state circuitry are needlessly sensitive or expensive. Commercial thermoregulators using a liquid-filled bellows and having the sensor in the form of a thin tube, coil, or bulb are good to + or - $2°F$ ($\pm 1.1°C$) and are very reliable. The relays in the thermoregulators are often rated high enough to carry the load of a small compressor, but long experience has shown that it is best to use a sturdy relay in addition. Overdesign is long-term economy here. Some investigators prefer to put the sensing element in water or in a glass flask or tube exposed to light so that it detects the temperature to which the algae are actually exposed, not just the chamber air temperature. This technique damps out small temperature oscillations and prevents too frequent operation of the compressor (at least in some designs).

A number of obvious considerations can help a growth chamber to "fail safe". Location is one. A thermally well-buffered place is desirable; to put a culture box in a hot cellar between the boiler and the autoclave is to invite trouble. Fluorescent lamp ballasts liberate about half the heat of operation; they can be re-wired to be located outside the chamber, near its roof for heat dissipation, preferably readily accessible for replacement. This technique seems to be detested by designers, who prefer to leave the ballasts within and increase the size of the compressor, which is no help at all.

Virtually all important food algae will grow somewhere between $16-24°C$; if one were forced to pick one temperature as the best compromise, it might well be $20°C$.

Aeration of Cultures

Aeration serves of course to keep algae in suspension, but its main purpose is to supply the carbon needed for plant growth. It also serves to stabilize the pH (in seawater, to keep it from getting too high too fast) through the interaction of the carbonate system with pH. A most readable account of this system is given in the first part of Chapter I and the first part of Chapter X in the excellent little book by Harvey (1955). An example shows the need for aeration. A seawater of 35 °/oo salinity near room temperature contains 25 mg/liter of inorganic carbon (about 2.08 mM). A reasonably dense culture of a small diatom often used for food, Thalassiosira pseudonana (clone 3H) has about 3×10^6 cells/ml. In a liter of this culture, 16 mg of carbon, or 64% of that initially present in the seawater, has been converted to organic matter (the diatoms). There is not enough carbon left in the medium for another cell division. Air contains about 0.03% CO_2 by volume; it thus has 0.164 mg/liter of inorganic carbon, 1/155 of the concentration found in seawater. To provide for only one cell division during a day in a 15 liter carboy culture like the one mentioned above would require the complete absorption of all the CO_2 in 1463 liters of air, meaning continuous pumping at a rate of 1 liter/min as a minimum. This will give some estimate of the volume of air that it may be necessary to supply. Not all of the CO_2 will be absorbed by the liquid, of course. (This can be helped by use of glass frits or commercial aquarium aerating stones, some of which will tolerate autoclaving. Note also that absorption is helped by the high pH in dense cultures.)

The effect of aeration in holding down the pH increase in growing cultures is beneficial. However, air (0.03% CO_2) cannot control pH completely in dense cultures. Increasing the CO_2 content of the aerating gas makes pH control possible and also opens the possibility of increasing the total inorganic carbon content of a medium by simultaneously adding bicarbonate. This is discussed by Spencer (1966; see his Tables 1 and 4). Denser algal crops can thus be grown. However, the gain may be offset by the additional apparatus needed to mix gases, and in any event, it has not been established by experiment that such dense cultures as might be grown are free of injurious side effects. (For example, harmful algal metabolites might accumulate in greater proportion than the cell density.)

Carboy cultures always require aeration. Fernbach flask cultures of 1 liter or less grow well if shaken by hand at least once a day, as do half-liter cultures in 2 liter Erlenmeyer flasks, but 2 liter cultures in Fernbach flasks and 1.5 liter cultures in Erlenmeyer flasks need aeration. Stock cultures should never be aerated. Diffusion of CO_2 is adequate for the populations attained in the 40-50 ml of medium in the 125 ml Erlenmeyer flasks. Tube cultures are generally of 10 or 15 ml volume, and are not aerated. It is likely they become CO_2 limited when partly grown.

Aeration should always be gentle during the first day or two after inoculation, then increased in rate as the culture grows.

Aeration devices are discussed with the question of stoppers, in the next section. The choice of air pumps depends upon the volume of air required and the continuity of demand. Commercial brass needle valves are used to control the flow rate and can be arranged in a manifold in the culture facility for simplicity of operation.

Various Materials

Some supplies common to the algal culture facility are listed, with comments. (a) Pasteur pipettes, 9 in., for making transfers of stock cultures; (b) glass test tubes, 30 x 4 cm to hold sterile Pasteur pipettes -- they should be cotton plugged; (c) assorted measuring pipettes, to 10 ml, preferably with expanded mouthpiece end to hold cotton plug; (d) a few 50 ml volumetric pipettes, wide bore (or with ends cut) for inoculating intermediate size cultures; (e) pipette sterilizing cans, and brown paper for wrapping volumetric pipettes; (f) plastic cylinders or cut-off plastic bottles, to hold used pipettes in water -- facilitates cleaning; (g) pipette rinser, or cylinder; (h) latex tubing and glass tubing for aerators (1/16 in. wall, 3/16 bore latex is convenient with 5 mm and 6 mm/glass tubing); (i) screw clamps for aerators; (j) silicone stoppers for aerators (optional method); (k) cotton for stoppers; (l) cheesecloth for stoppers; (m) autoclavable plastic calcium carbonate tubes to hold cotton for sterilization of the air stream; (n) aluminum foil, cut into squares, as dust covers for cultures; (o) membrane filter apparatus, with glass fiber filter discs and also membrane filters if sterilization by filtration is indicated; (p) autoclave: in this connection, note that pressure cookers will do for small vessels, and there are small cylindrical autoclaves available that will just hold one 5 gallon carboy.

The culture vessels themselves have been discussed before. It may be mentioned that transparent polycarbonate 125 ml flasks and Fernbach flasks are available. These appear to be free of detectable toxic effects, though they must leach to some extent. They are not fragile, and will stand up under routine use for at least a few years. They have not been tested for longer.

For a large collection of algae it will pay to get glass dispensers for media. These dispense 10 to 15 ml for tubes and 40 or 50 ml for flasks. The kind that is by far the best is unfortunately somewhat fragile, expensive, and hard to find; it has a glass rod plunger and valve seat, and operates by tipping.

Stoppers and Aerators: Unthreaded culture tubes should be closed with cotton plugs as is customary. Threaded tubes can be closed with the screw caps if some precautions are taken. Teflon lined caps are preferred and any caps used should be autoclaved several times in changes of seawater to leach out phenalics (the water will come out brown). The materials that leach into tubes are toxic and render medium in the tubes useless for isolating algae, because only one algal cell is added, but stock cultures are inoculated with a few drops of a dense culture, so that the toxins are diluted. Workers who feel uneasy about this practice autoclave the medium in threaded tubes covered with glass vials or with soft plastic unthreaded caps, and autoclave the screw caps in petri plates, then change caps at the time of inoculating the tubes. This is inexpensive and easy. Cotton plugs allow adequate gas exchange. Screw caps should be tightened then loosened one quarter turn at most for gas exchange. They seem relatively safe; many cultures in peptone enriched medium (1 gm/l) have been kept for years without obvious contamination. The real disadvantage of screw caps is that if the culture vessel tips over, sterility is probably broken, but if a cotton stoppered vessel tips, one has only to sterilize a new stopper and exchange it for the wet one.

Flasks for stock cultures should be cotton stoppered but subcultures may be kept in screw capped flasks if desired. Unaerated Fernbach and large Erlenmeyer flask cultures are also cotton stoppered.

A word is in order about making cotton stoppers. One neat way of making plugs for 125 ml flasks is as follows: cut or pluck a swatch of cotton about 11 cm (4.5 in.) square and 8 mm (3/16 in.) thick. Then roll it, or fold the edges over and roll it -- the cotton fibers running along the axis of the cylinder of the roll. Before folding the roll in half, pluck a tuft of cotton from its middle region, at the fold, so that the fold, which will enter the flask first, is a bit smaller than the body of the plug. Center the fold of the plug in a square of cheesecloth about the same size as the original piece of cotton, and fold the cloth around the plug, trimming excessively long ends. A plug should be just tight enough so that a 125 ml flask of medium can be picked up by it, and should protrude enough so that it can be held by the little finger and heel of the hand, leaving the other fingers free to handle the pipette. Plugs so made can be used dozens of times; if wet with medium, they can be rinsed in distilled water and dried. They should be discarded when the cotton begins to give off dust. Larger plugs are made in much the same way. Plugs can accommodate aeration tubes, siphons, and even tubes to be used as inoculation ports, if desired. Use several layers of cheesecloth, and punch holes through it and the cotton swatch for the tubings. Space the tubings as desired with cotton and tie with white string, then fold up the cheesecloth-covered swatch and tie.

It is usual to put covers of aluminum foil over the stoppers of stock culture flasks as additional dust protection. Squares of foil are pressed gently down over the plugs then squeezed around the neck of the flask; the foil should be loose enough so that it comes off easily without pulling the plug when picked up by one corner. The foil is not replaced after the stock culture flask is opened to inoculate its daughter culture. Foil should not be left on flasks for more than a few weeks in warm humid climates, because fungi will grow in the cotton.

Aerators are assembled from 6 mm glass tubing, latex tubing, and plastic $CaCl_2$ tubes stuffed with cotton to filter bacteria and other contaminants from the air stream. If the air supply has oil or other contamination, another $CaCl_2$ tube should be put in the line before the sterilizing one; this tube should contain charcoal pellets or a mixture of charcoal and cotton to remove contaminants. It need not be sterile. (Commercial charcoal filters are available.)

An efficient apparatus for sealing large flasks and carboys can be made of suitable diameter silicone stoppers with holes

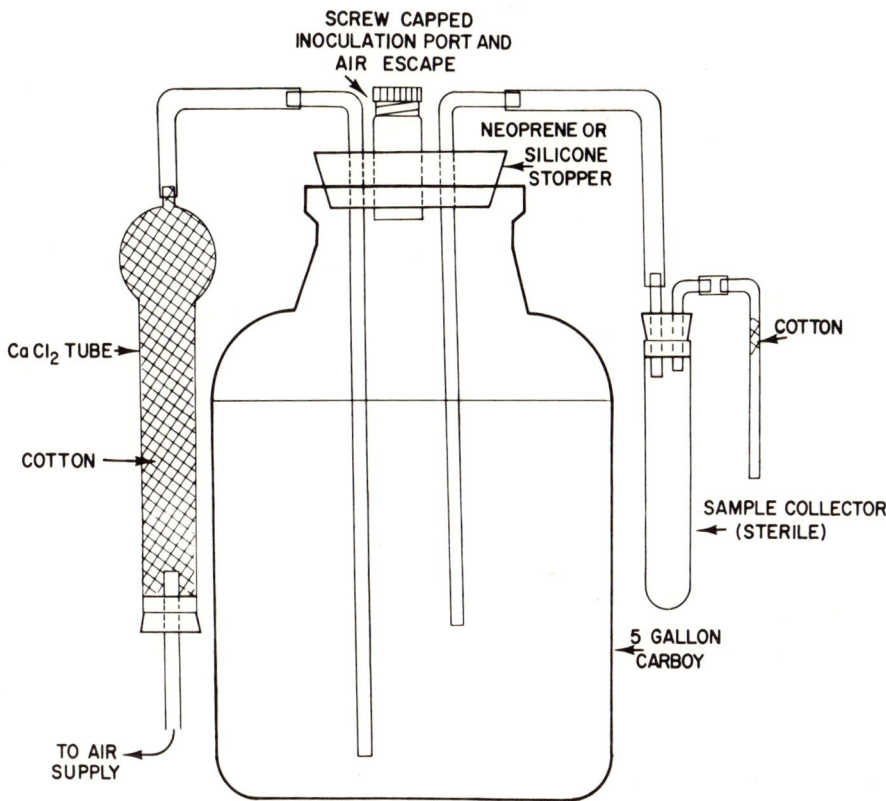

Fig. 1. Diagram of Apparatus for Culturing Algae.

bored for the aerator, sampling tube if desired, and inoculating tube. The inoculating tube serves also as the air exhaust. It is conveniently made by cutting the bottom off a 20 x 150 mm screw-capped tube. The silicone plug becomes slippery when wet and must be tied down on the carboy. It is easy to make a stainless steel wire collar with loops, that remains on the carboy permanently and allows for securing the aerator assembly with string.

When large cultures are autoclaved, superheating of the medium causes violent bubbling that can wet the stoppers and even blow them (and some of the medium) out of the flasks. Slow cooling helps this, but is not good for the medium in other respects. While the stoppers can be held down by brute force, it is easier to autoclave the large cultures with loose cotton plugs or big plastic beakers over them, so that exchange of gas is easy. The aerator assemblies can be autoclaved at the same time and exchanged for the plugs or beakers while the vessels of medium are still hot, or can be autoclaved separately wrapped in brown paper and put on the carboys when convenient. In the event aerators or siphons are autoclaved in the cultures, bend the latex tubing near where it joins the glass tube and hold with screw clamps (not pinch clamps) in order to prevent siphoning of the culture medium under pressure; medium can be lost and the air sterilizing plug is sure to be spoiled. Neoprene stoppers can be used for aerator assemblies but these should then be autoclaved separately, not on the medium.

Fernbach cultures can be sampled (for counting) fairly easily through the inoculating port or by pulling the plug, but either process may be difficult for carboy cultures. In this case provide a glass siphon tube in the carboy, joined to latex or plastic tubing outside. This tubing is joined by a piece of 5 mm glass tube about 7 cm long to a #1 two-hole stopper; through the other hole of the stopper is passed another piece of 5 mm glass tube, bent at about right angles. This bent tube is stuffed with a piece of cotton (as is a pipette). The stopper is now inserted into a 16-18 mm x 150 mm test tube. In effect one of them has a two-stage siphon, with the test tube acting as a reservoir. Suction applied (by pump or mouth) to the bent glass tube will draw liquid from the culture into the test tube; when suction is released the siphon is broken provided the culture tube is above the level of liquid in the carboy. The test tube with sample can then be removed and replaced aseptically with another sterile empty tube (see Fig. 1).

ROUTINE FOR MAINTENANCE AND PRODUCTION OF ALGAE

Reliability of both maintenance and production demands a routine, the details of which depend upon the species of algae,

PHYTOPLANKTON FOR FEEDING MARINE INVERTEBRATES

the physical circumstances, and the demand. Some comments are given on typical operating procedures.

Stock cultures in 125 ml flasks are generally transferred once a week; flagellates in tubes will usually go 2 weeks, and green algae in tubes even longer. The important point is to transfer before half the life span of a culture has passed. For example, an alga that will survive in culture about 15 days must be transferred by 7 days. In case of accident -- a batch of toxic medium, dropping a tray of flasks, etc., -- the parent culture (two weeks old) will still be able to start a new stock.

If a culture collection is small it is easiest to schedule all transfers according to the needs of the most fragile species, generally once a week. It may pay to schedule weekly, fortnightly, and monthly transfers in a big collection. In a typical routine, medium is sterilized on a Monday or Tuesday and transfers made the following day; this allows leeway for accidents, holidays, etc., away from weekends. The 7 rather than 6 or 8 day schedule follows sociological rather than phycological rhythm, but fits well.

Stocks in tubes or 125 ml flasks are inoculated from a Pasteur pipette with 1/2 to 2 ml of culture, depending on the species. Diatoms generally need little inoculum, green algae sometimes more, and dinoflagellates usually more to avoid lag. Size of the inoculum can be adjusted so that the crop in the flask will peak at 5-7 days.

Stock cultures are inviolate, they are opened only once, to inoculate the daughter stock culture. At this same time any auxiliary 50 ml cultures for experiment or for routine inoculation of larger food cultures can be started. Use the parent (week-old) culture if it is necessary to start cultures during the week, not the current stock culture.

It is very important to keep the old ("parent") stock cultures in a different location from the current stock -- in a different room -- so that both cultures will not be lost by power or compressor failure, etc. in the main culture facility, whatever that may be. Arrange, if possible, for a slightly lower temperature and definitely lower light intensity (not exceeding 1000 lux) for the old stock cultures. Repeated experience has shown that failure to keep stocks in two places is the most frequent cause of loss. Even a desk top in dim daylight can be made to serve.

Auxiliary 125 ml flask cultures started at transfer time are grown to near maximum density for inoculating 1-2 liter cultures; 10-50 ml of inoculum per 2 liter culture is used for various species. A 10 ml inoculum of a fast growing diatom (ca. 3 divisions per day) will generate a 2 liter culture ready to feed in three days. Flagellates, particularly dinoflagellates, will generally do better with larger inoculum. A schedule for inoculation of the 1-2 liter cultures is easily established, and medium can be sterilized with that for the stock cultures, or separately.

Carboys are inoculated with 200-500 ml of culture, certain plants, notably dinoflagellates, doing better with larger inocula. The 100 ml volumetric pipette with the end cut off, or a siphon tube, can be used to make the inoculation. Algal crop production is more efficient in terms of its quality and of space and effort required if the carboy cultures are inoculated heavily and grown rapidly to final density. Note that there is little point in using carboy sized cultures at all if 2 liters or less of culture is needed per day, provided the algae grow reasonably rapidly.

A requirement of several carboys a week puts a heavy burden on the autoclave. Tests with a commercial "Autoclave Maximum" thermometer show that it takes about 45 minutes at full pressure for the temperature at the center of the carboy holding 18 liters of seawater to attain $120°C$. The full cycle takes about two hours. This can cause autoclave scheduling problems on the days on which the stock culture media are being autoclaved; these should not be exposed to such a long period of high temperature. (Some further notes on sterilizing follow the section on media.)

Attention is again drawn to the point that in moderate to large scale production one can effect savings by providing light and aeration at only the levels needed as the cultures grow -- often simply by moving a half-grown culture into the position vacated by one harvested.

Harvesting algal cultures is simple, because it is usual to supply the food to the animals in the required quantity -- known from the literature, presumably -- simply by pouring an appropriate amount of the algal culture into the animal tanks. The volume to feed is usually determined from measurement of the algal population in the culture by counting cells in a hemacytometer or other counting device. Alternatively, the wet packed cell volume of 10 ml of culture is often measured by centrifuging at a relative centrifugal force of ca. 1000 in Hopkins vaccine centrifuge tubes. These and other methods are described in detail in articles in the treatise edited by Stein (1973), which should be consulted.

It is efficient from the point of view of production and important nutritionally to harvest each culture during the right period of its life. Begin harvesting during the late phase of logarithmic growth, when the density is high enough so that a reasonable volume will suffice to feed the animals, and deplete the culture early in the stationary phase, before the cells become obviously unhealthy and metabolites begin to accumulate. For experimental purposes, when constancy of nutrition is important, it would be better to use cells harvested from early log phase growth, because it has been found that the nutritional status of the algae changes before the growth rate is altered. This, however, is a problem that calls for more refined techniques.

The attractive possibility of continuous sterile culture is unfortunately not now technically feasible, and the practice of refilling a harvested carboy with fresh sterile medium in contraindicted. The labor gained is not worth the damage from contamination.

Test algal cultures for sterility occasionally by inoculating up to 0.5 ml of one into ca. 5 ml of "f/2" (See Table 1) plus 0.1% (W/V) of Bactopeptone in a tube. Allow at least 15 days in darkness for bacterial growth to be evidenced, but examine occasionally during this time because some bacteria grow for a short time then die. Tests can also be made on the same medium solidified with 1% agar in Petri plates. Several cultures can be tested on the same plate. There are much more complex and somewhat more effective test media in the literature, but relatively little is gained for the extra effort. The most important point is to keep the nutrient level low, not exceeding 1 g/l total.

The algal cultures should be examined microscopically occasionally with bright field or phase contrast illumination.

The glassware cleaning routine should be simple and reliable. If the glassware is used only for growing algae it need not be "chemically clean," but it should be freed of the organic material often left by algae and of the film of precipitate that usually forms because of the alkaline conditions in cultures. Scrubbing with detergent is usually enough, but may leave some of the film of precipitate. In many laboratories the glassware is routinely treated momentarily with mild acid first, then dipped into detergent solution, rinsed with tap water, and finally rinsed with distilled or deionized water if the local tap water leaves a film. The mild acid is usually 20-30% commercial HCl, though H_2SO_4 should do as well. A row of small flasks, emptied of contents, can be lined up and 60-75 ml of the acid poured into the first. This is swirled and the contents poured into the second, while the first is set under the tap to fill. This is repeated; the acid can be saved if not too diluted with seawater. The flasks are then emptied and washed with detergent, using the brush only as necessary. Big flasks and carboys are treated with 200-300 ml of the acid. If HCl is used the vessels should be filled with water to drive out the acid fumes. Aerator and siphon assemblies can usually be flushed intact if the cotton air sterilizing assemblies in the $CaCl_2$ tubes are removed. Pipe cleaners may aid in cleaning the aerator tubing, and glass frits will require occasional soaking in concentrated nitric acid. The cotton in the air stream sterilizers should be replaced occasionally, especially if it gets wet.

MEDIA

Three types of media are described; enriched natural seawater, artificial seawaters, and freshwater. Enriched natural seawater is, for economic reasons, the only one practical for large scale production. Artificial seawater is necessary for experimental work where natural seawater supply is temporarily cut off or unfit for use due to contamination. The freshwater medium is of course necessary for growing any freshwater algae that may be useful as foods. Also, inasmuch as it amounts to an enriched natural river water, it is useful for salinity studies in which one wishes to simulate conditions in an estuary. Enriched natural or artificial seawater can be mixed with the freshwater medium to give mixtures having similar plant nutrient characteristics, but different salinities. Note that if it is desired to grow euryhaline algae in seawater diluted to match the salinity required by some animal, then it is not in general necessary to add the nutrients for the freshwater medium. One can simply enrich the diluted seawater as if it were ordinary seawater unless the salinity is reduced to perhaps 3 °/oo or less.

Enriched Seawater Media

These are based on enrichment "f" of Guillard and Ryther (1962). A number of very similar enrichments are described in the literature and many are surely at least as good as the ones given here. These are specified merely because the author has had most experience with them and they have been widely used in many parts of the world.

The composition of the basic enrichment "f/2" is given in Table 1, that of the medium with added ammonium ion ("h/2") in Table 2, and that of the modification using citrate as the chelator ("f/2-Beta") in Table 3.

Table 4 has directions for making working stock solutions of the major nutrients (nitrate, phosphate, silicate, ammonium). Table 5 has directions for the primary stock solutions of the trace metals, used for making the working trace metal stocks. Table 6 describes the trace metal EDTA working stock solution. Table 7 describes the trace metal citrate working stock. Table 8 has directions for the vitamin primary and working stock solution.

Note that for all working stocks except the vitamin stock one uses 1 ml of stock solution per liter of final medium; of the vitamin working stock only 1/2 ml is used per liter of final medium.

It is customary to filter the seawater, using fine paper, glass fiber, or membrane filters of large porosity (over $1\,\mu$). It is often desirable to let the seawater settle for a day or so, then decant. Oceanic seawater may be clear enough to use without filtration.

Seawater can be stored in glass or plastic carboys. Plastic ones will leach when new and should be soaked for several days with dilute HCl then water. Activated charcoal will remove most organic contaminants from seawater and can be used to detoxify water stored in doubtful containers. A few grams of activated charcoal should be put into a carboy set on a magnetic stirrer (or shaken) for about a day; the water is then filtered and stored in a non-contaminating container.

It may be convenient to omit certain nutrients at times. As mentioned in Tables 1 and 2, ammonium ion is generally supplied only when absolutely needed, and silicate can be omitted unless diatoms are involved. By no means all useful species of algae need vitamins, though some without a requirement are stimulated by them. Vitamin B_{12} is one most often needed, with thiamine following and biotin a poor third (only a few flagellates need it). However, it is just as well to supply all three vitamins if any are required at all, because the cost is slight, while the trouble of making an auxiliary solution if needed later is considerable.

Inasmuch as soil extract is necessary for, or at least beneficial to, certain species that may be of interest, a procedure is described here. It is essentially as given by Gross (1932).

> Approximately equal weights of garden soil (not recently fertilized) and distilled water are autoclaved and allowed to stand for several days. The supernatant is removed by decanting, filtering, or centrifuging the preparation. The extract (supernatant) is re-autoclaved if necessary and stored sterile in the cold. Use 10-50 ml per liter of final culture medium. We have successfully used seawater instead of distilled water to make the extract.
>
> A method for making a concentrated alkaline extract (use only ca. 2.5 ml/l) is described by Provasoli et al. (1957, page 411).

TABLE 1

Composition of Enrichment "f/2"

Major Nutrients (stock solutions are described in Table 4)

$NaNO_3$	75 mg (883 μM)
$NaH_2PO_4 \cdot H_2O$	5 mg (36.3 μM)
$Na_2SiO_3 \cdot 9H_2O$*	15-30 mg (1.5-3 mg Si or 54-107 μM)

Trace metals (working stock solution is given in Table 6)

$Na_2 \cdot EDTA$†	4.36 mg (<u>ca</u> 11.7 μM)
$FeCl_3 \cdot 6H_2O$†	3.15 mg (0.65 mg Fe or <u>ca</u> 11.7 μM)
$CuSO_4 \cdot 5H_2O$	0.01 mg (2.5 μg Cu or <u>ca</u> 0.04 μM)
$ZnSO_4 \cdot 7H_2O$	0.022 mg (5 μg Zn or <u>ca</u> 0.08 μM)
$CoCl_2 \cdot 6H_2O$	0.01 mg (2.5 μg Co or <u>ca</u> 0.05 μM)
$MnCl_2 \cdot 4H_2O$	0.18 mg (0.05 mg Mn or <u>ca</u> 0.9 μM)
$Na_2MoO_4 \cdot 2H_2O$	0.006 mg (2.5 μg Mo or <u>ca</u> 0.03 μM)

Vitamins (working stock solution is given in Table 8)

Thiamin·HCl	0.1 mg
Biotin	0.5 μg
B_{12}	0.5 μg

Seawater ‡	to one liter

Notes: *Silicate may be omitted for organisms other than diatoms. The concentration used for diatom cultures may be adjusted according to the silicate concentration of the seawater used and the need for avoiding precipitation.

Recently it has been found (by K.C. Haines and the author) that seawater enriched as "f/2" but without added Si will dissolve a great deal of silicate (up to 290 μM) from 125 ml borosilicate glass flasks, on autoclaving. This does not occur in polycarbonate flasks, of course, and less will dissolve if the medium is autoclaved in large vessels. Diatom growth is completely satisfied by the silicate provided by the small flasks. However, until this phenomenon has been investigated further, we recommend that silicate be added to all "f/2" media if diatoms are being grown in it.

†The quantities of $Na_2 \cdot EDTA$ (disodium salt of ethylenediamine-tetra-acetic acid) and $FeCl_3 \cdot 6H_2O$ given here yield approximately the same amounts of iron and EDTA as results from using 5 mg/l of Ferric sequestrene (13% iron), which is the material specified in the original formulation of medium "f". "Ferric Sequestrene" is the sodium iron salt of EDTA produced by Geigy Industrial Chemicals, Saw Mill River Road, Ardsley, N.Y. 10502.

‡The salinity of the seawater used should be adjusted according to the needs of individual organisms. Most species will tolerate 28-34 °/oo. Certain species isolated from the open ocean will not grow well below this salinity range, but most species from neritic and estuarine environments can be maintained at 20 °/oo or less. See text regarding filtration of seawater.

TABLE 2

Composition of Enrichment "h/2"

Medium "f/2"	1 liter (See table 1)
NH_4Cl*	26.5 mg (0.5 mM) (Stock solution described in Table 4)

*NH_4Cl is generally used only for those species that cannot grow on nitrate. Note that while many species will not tolerate even 0.1 mM (Guillard, 1963), many others that are commonly used as food will grow well on "h/2" even though they do not need ammonium ion. Thus "h/2" can sometimes be used for all algae in a collection if any need ammonium ion.

We have found that when 0.5-1.0 mM NH_4Cl is added to seawater and autoclaved in 50 ml lots in 125 ml flasks, 25-30% of the ammonium ion is lost. Therefore, if the amount of ammonium ion is critical, sterile NH_4Cl should be added aseptically to medium autoclaved separately. The NH_4Cl stock solution can be autoclaved because of its low pH.

TABLE 3
Enrichment "f/2-Beta"

This medium differs from "f/2" in that citrate is used as the chelator, in a 10:1 ratio, and the iron content is slightly lower than that of "f/2". Iron is supplied as the citrate.*

Major nutrients: as in "f/2" (see Table 1)

Vitamins: as in "f/2" (see Table 1)

Trace metals: (Working stock solution is given in Table 7)

$FeC_6H_5O_7 \cdot 5H_2O$ 3 mg   (0.5 mg Fe or 8.9 µM; 8.9 µM citrate)

$C_6H_8O_7 \cdot H_2O$ 16.8 mg   (80.1 µM citrate)

all other trace metals: as in "f/2"

*Droop (1961) has made the important point that if a trace metal solution is to serve for media made with a wide range of salinities, then citrate has advantages over EDTA. Citrate has lower affinities for Ca and Mg than does EDTA. Hence, changes in the concentrations of Ca and Mg exert less influence on the availability of iron and the other heavy metals when citrate is used than when EDTA is used.

Enrichment "f/2-Beta" is offered in case it is useful in this context, and as an alternative if EDTA is unavailable or must be avoided for some reason.

If the medium with added ammonium ion (Table 2) is used in conjunction with the "Beta" trace metal chelate, the designation "h/2-Beta" is indicated.

Note the remarks concerning pH in Table 7.

TABLE 4

Stock Solutions for Major Elements

(n% (W/V) means n grams brought to a volume of 100 ml with distilled water.)

The following are made 10^3 x more concentrated than in the final medium. Use 1 ml per liter of seawater to obtain medium "f/2", "h/2", or "f/2-Beta".

TABLE 4 continued:

Material	Stock Solution, % (W/V)
$NaNO_3$	7.5
$NaH_2PO_4 \cdot H_2O$	0.5
NH_4Cl	2.65
$Na_2SiO_3 \cdot 9H_2O$	3 (heat to dissolve if necessary)

TABLE 5

Primary Stock Solutions for Trace Elements

(n% W/V means n grams brought to a volume of 100 ml with distilled water.)

Salt	Primary Stock, % W/V
$CuSO_4 \cdot 5H_2O$	0.98
$ZnSO_4 \cdot 7H_2O$	2.2
or $ZnCl_2$	1.05
$CoCl_2 \cdot 6H_2O$	1.0
$MnCl_2 \cdot 4H_2O$	18
$Na_2MoO_4 \cdot 2H_2O$	0.63

It is convenient to make **primary** **stocks** of individual trace metals, with the elements 10^6 x more concentrated than in the final media "f/2" etc. A formulation is suggested above. Note that the sulfate salts are best because precipitation is minimized.

TABLE 6

Trace Metal Working Stock Solutions, EDTA Chelated

A. Trace metal stock solution, using "Ferric Sequestrene" as iron and chelator source (see footnote to Table 1).

Dissolve 5 g ferric sequestrene in ca. 900 ml of distilled water, add 1 ml of each trace metal primary stock, (Table 5) and bring to 1 liter. pH is ca. 4.5.

Use 1 ml/l of this stock solution to make medium "f/2" or "h/2".

B. Trace metal stock solution, using ferric chloride and di-sodium EDTA. Dissolve 3.15 g $FeCl \cdot 6H_2O$ and 4.36 g Na_2 EDTA in ca. 900 ml of distilled water; add 1 ml of each trace metal <u>primary stock</u>, (Table 5) and bring to one liter. pH is ca. 2.0. The solution remains clear if left at pH 2.0. If titrated to ca. pH 4.5 (taking ca. 7 ml of N·NaOH), a precipitate will form, resembling that in the solution made with ferric sequestrene.

Use 1 ml/l of this stock solution to make medium "f/2" or "h/2". Even if the trace metal stock solution is left at pH 2.0, the effect on the pH of the seawater medium is negligible unless a large amount is used. (Eight times the usual amount (8 ml/l) lowered the pH of "f/2" from 7.8 to 7.2).

TABLE 7

Trace Metal Working Stock Solution, Citrate Chelated

Dissolve 16.8 g citric acid ($C_6H_8O_7 \cdot H_2O$) in ca. 900 ml distilled water. Add 3.0 g ferric citrate ($FeC_6H_5O_7 \cdot 5H_2O$) and 1 ml of each trace metal <u>primary</u> stock (Table 5). Bring to one liter. Heat or autoclave briefly to hasten complex formation (liquid clarifies). pH is about 2.3. Insofar as practical protect this trace metal solution from light and oxygen. Dispense in tubes or ampoules and keep sterile in darkness.

Use 1 ml/l of this stock solution to make medium "f/2-Beta" or "h/2-Beta" (Table 3).

Note that the pH of this stock solution is ca. 2.1 and 1 ml/l lowers the pH of seawater to ca. 7.2, so the effect cannot be ignored, but it is either acceptable to or actually favorable to the algae that have been grown on this medium.

TABLE 8

Vitamin Primary Stock Solutions and Vitamin Working Stock Solutions

<u>Primary stock solutions</u>: Biotin is obtained in crystalline form; allow for about 4% water of crystallization. A <u>primary stock</u> solution is made containing 0.1 mg/ml by weighing about 10 mg and adding distilled water, 9.6 ml for each mg of biotin. Make the solution slightly acid if it is to be autoclaved. Keep the solution sterile and frozen.

Vitamin B_{12} is similarly obtained as crystals and 11% should be allowed for water of crystallization. Weigh (or buy in weighed amounts) and make a <u>primary stock</u> solution having 1 mg/ml. Acidify the solution if it is to be autoclaved, and keep sterile and frozen.

Primary stocks can be put up in ampoules or small screw-capped test tubes.

<u>Vitamin working stock solution:</u> To make the vitamin stock solution bring 1.0 ml of biotin primary stock and 0.1 ml of B-12 primary stock to 100 ml and add 20 mg of thiamine HCl. No primary stock of thiamine is needed.

The vitamin stock solution is dispensed in 1, 2, or 5 ml lots in ampoules or in 10 ml lots in screw-capped test tubes, autoclaved, then stored in a refrigerator. Use 1/2 ml per liter of final medium, adding before autoclaving. One-tenth this concentration is probably always adequate. If an ampoule is opened and the contents not completely used, freeze the remainder.

TABLE 9

ASPM Base (see p. 56)

Compound	Amount per Liter of Distilled Water, Grams	Concentration Millimolar
NaCl	23.38	400
KCl	0.75	10
$MgSO_4 \cdot 7H_2O$	4.93	20
$MgCl_2 \cdot 6H_2O$	4.07	20
$CaCl_2$	1.11	10
$NaHCO_3$	0.17	2
H_3BO_3	0.25	0.4

The above is enriched as "f/2", "h/2", etc. as described in Tables 1, 2 or 3.

pH of the final medium should be about 7.5. The buffer tris is often used, for those algae that will tolerate it. (Glycylglycine is better tolerated but prohibitively expensive, and it is utilized by bacteria). Tris is added up to 5 mM. A stock solution is made as follows:

Tris (hydroxymethyl) amino-methane is added to distilled water (it will not dissolve) and titrated with concentrated HCl to make 25% solution of pH 7.1-7.3. (About 29.3 ml of conc. HCl will bring 50 g of Tris to ca. pH 7.2 in a final volume of 200 ml). 1.21 ml of this solution contains 1 mM, hence it is used at 1-5 ml/liter; add to the medium before autoclaving.

TABLE 10

Clark's Artificial Seawater Medium (see p. 56)

(Grams per liter of medium)

NaCl	26.73		$SrCl_2 \cdot 6H_2O$	4	mg
$MgCl_2 \cdot 6H_2O$	4.82		RbCl	28	μg
$MgSO_4$	3.25		$ZnSO_4 \cdot 7H_2O$	2.2	mg
$CaCl_2 \cdot 6H_2O$	2.28		$MnCl_2 \cdot 4H_2O$	810	μg
KCl	720	mg	$Na_2MoO_4 \cdot 2H_2O$	130	μg
$NaHCO_3$	190	mg	$CoCl_2 \cdot 6H_2O$	23	μg
KBr	67	mg	$CuCl_2 \cdot 2H_2O$	11	μg
H_3BO_3	58	mg	$NaNO_3$	10	mg
$Na_2SiO_3 \cdot 9H_2O$	7	mg	NaFeEDTA	10	mg
K_2HPO_4	1	mg	TRIS	500	mg
$AlCl_3 \cdot 6H_2O$	440	mg	Thiamine-HCl	200	μg
NH_4Cl	6	mg	Vitamin B_{12}	1	μg
LiCl	800	μg	Biotin	1	μg
			KI	6.6	μg

Distilled water to make one liter

(Notes next page)

Clark adjusted the pH to 7.4 before use, but it has been found satisfactory for many species when used at pH 8.2, which is the pH without adjustment. Tris may be omitted. See Table 9.

The vitamin solution used in other media given in this paper (Table 8) yields the concentrations required for this medium if used at 1 ml/l instead of 1/2 ml/l. The medium has proved satisfactory with 1/2 ml/l of vitamins.

TABLE 11

Composition of Medium WC (see p. 57)

Nutrient Elements	Stock Solution, g/l Use 1 ml/l Final Medium	Final Medium mg/l	μM	Elements in Final Medium	mg/l	μM
$CaCl_2 \cdot 2H_2O$	36.76	36.76	250	Ca	10.02	250
				Cl	17.73	500
$MgSO_4 \cdot 7H_2O$	36.97	36.97	150	Mg	3.65	150
				S	4.81	150
				SO_4	14.40	150
$NaHCO_3$	12.60	12.60	150	Na	----	150
				CO_3	9.00	150
K_2HPO_4	8.71	8.71	50	K	3.91	100
				P	1.55	50
$NaNO_3$	85.01	85.01	1000	N	14.01	1000
				Na	-----	1000
$Na_2SiO_3 \cdot 9H_2O$	28.42	28.42	100	Si	2.81	100
				Na	----	200
				Total Na	31.05	1350

<u>Trace Elements</u>: Per liter, use one ml of one of the trace element solutions described in Table 6 or 7 plus 1 ml of the boron stock solution given in Table 12. (The latter provides 6 mg/l of boric acid or 1.25 mg B per liter).

Check the pH if the more acid trace metal solutions are used and adjust as necessary. The formulations at pH 4.5 with no further adjustment yield culture media with pH of 6.95 - 7.1 after autoclaving and equilibrating. This has proved satisfactory for most algae.

<u>Vitamins</u>: Use as in medium "f/2" - Tables 1 and 8

<u>Ammonium</u>: Add as for medium "h/2" - Tables 2 and 4, for algae that require ammonium.

<u>Buffers</u>: The best is glycylglycine, at 3.79 mM (500 mg/l). Tris may also be used, with caution. See notes to Table 9. Most algae grow without added buffer. The bicarbonate content of the medium can be increased. (Note that seawater has 2 mM bicarbonate).

TABLE 12

Boron Stock Solution

H_3BO_3 0.6% (W/V) (0.6 g per 100 ml distilled water)

Four ml per liter provides as much B as is present in seawater, 4.6 mg B or 24 mg H_3BO_3 per liter. Boron is not present in the trace metal mix (Tables 6 and 7) because this element is abundant in seawater. However, if seawater is diluted tenfold or more, the possibility of a boron deficiency exists, at least for some diatoms. In any event, boron is presumably required for the freshwater media (though it is quite possible that enough is provided by borosilicate glass).

Artificial Seawaters

There are many formulations for artificial seawaters in the literature. Some were designed to simulate natural seawater, others deliberately simplified and designed for growing algae. For the present purposes, a basic difficulty is that the simpler media will not suffice for some of the more fastidious pelagic algae, and in addition usually favor one group of algae or another, while the most widely applicable medium of which the author is aware is distressingly difficult to make.

The source paper is by Provasoli et al., (1957), with additional details furnished by McLachlan (1964). The article by McLachlan in the treatise edited by Stein (1973) should be consulted.

Here we suggest two formulations. One, called ASPM base, consists of the major components of seawater according to McLachlan's modification of the well-known ASP medium. This ASPM base is treated as if it were seawater, and enriched with the major nutrients, trace elements, and vitamins of enrichment "f/2" or "h/2". Table 9 gives the composition of ASPM base and further directions for enrichment.

The other formulation is the much more complex one of Clark (1966), which is given in its entirety in Table 10, together with some notes. In its complexity and the number of elements involved this medium does not differ much from others published by the authors cited above, but it is the one with which most experience has been had in this laboratory. (The nitrogen and phosphorus can be increased to the "f/2" levels.)

Commercial Artificial Seawaters

A number of commercial preparations are now available. These differ somewhat in particulars, and the user should get the manufacturers' data, particularly that related to chelators and trace metals. Most such preparations that have been tried have simply been enriched as if they were natural seawater (e.g. as "f/2"), and most are reasonably good for certain algae. Five preparations are listed (alphabetically) below. There are certainly more.

Dayno Sea Salt
    Dayno Sales Co.
    678 Washington Street
    Lynn, Massachusetts 01901

Instant Ocean
    Aquarium Systems Inc.
    33208 Lakeland Blvd.
    Eastlake, Ohio 44094

Rila Marine Mix
    Rila Products
    Teaneck, New Jersey

Sea Salt Mix (unnamed)
    Marine Research Associates
    Box 7
    Westport Point, Massachusetts 02791

Utility Seven-Seas Mix #156
    Utility Chemical Co.
    Manufacturing Chemists
    145 Peel Street
    Patterson, New Jersey

Freshwater Medium

Freshwater media, like marine ones, may be designed from two points of view. A medium may either simulate the composition of some natural water, in which case it should be adequate for all inhabitants of that water, or else it may be specially designed around the nutritional or physical requirements of a particular species or group of species. The second type of medium amounts to a selective enrichment, and cannot be expected to suffice for all inhabitants of a particular natural environment. These considerations are explored by Provasoli and Pintner (1960), who give several formulations, and give references to the literature. Hutchinson (1957) discusses the composition of natural lake waters.

The medium "WC" given in Table 11 is a modification of Gerloff, Fitzgerald and Skoog's (1950) version of the famous Chu #10. Medium WC has grown many common freshwater algae and some rather particular ones. Many algae grow in it with no buffer. Note that as it stands it has a bit more nitrate and phosphate than "f/2" does, but this can be adjusted. The adjustment will change the total solids content a little, the ion ratios, and the Na content.

Notes on Sterilization

Autoclaving medium is still the most reliable routine way of sterilizing it on the carboy scale of operation, though filtration techniques are now excellent. Fifteen minutes at ca. 15 lbs/in^2 is adequate for volumes up to 2 liters, but carboys should be left at full pressure for about 45 minutes. If there is any doubt, check with a maximum recording thermometer or with the commercially available heat-resistant bacterial cultures designed to "prove" autoclave function. Remove vessels from the autoclave as soon as possible after autoclaving and arrange a system so that they cool quickly. Small flasks will cool quickly if set on a metal shelved cart or surface. Bigger vessels can even be set in cold water.

A precipitate forms in medium "f/2" upon autoclaving. It is not generally injurious to algae. If gathered on a Millipore filter it is reddish in color and contains most of the added phosphorous. The quantity of precipitate depends upon the salinity of the seawater used, its phosphate (and possibly silicate) concentration, and upon the volume of medium autoclaved. The latter dependence is presumably because the medium when autoclaved in large batches remains hot longer and comes more slowly into equilibrium with atmospheric carbon dioxide. The quantity of precipitate can be reduced by reducing the salinity, by autoclaving in small lots (e.g., 50 ml in 125 ml Erlenmeyer flasks), or best by adding sterile phosphate, silicate, and trace metals aseptically after medium has been autoclaved, cooled, and aerated or shaken. This is the preferred method of preparing sterile medium in carboys. Medium in 1 or 2 liter lots can usually be cooled quickly enough in running tap water to minimize precipitation. Inasmuch as vitamin solutions are stored sterile, it is no problem to add the vitamins after autoclaving carboy lots of medium or of seawater to be made up to medium.

An alternative to adding nutrients aseptically after autoclaving is to lower the pH of the medium to ca. 4.8 with H_2SO_4 before autoclaving, then to add the same number of equivalents of either NaOH or $NaHCO_3$ after the water has cooled.

This work was supported by NSF Grant GA 33288.
Contribution #3233 from the Woods Hole Oceanographic Institution.

REFERENCES

Clark, R.C., Jr. 1966. Saturated hydrocarbons in marine plants and sediments. M.S. Thesis, Massachusetts Institute of Technology, June 1966. (Dept. of Geology and Geophysics), 96 pp.

Droop, M.R. 1961. Some chemical considerations in the design of synthetic media for marine algae. Botanica Marina II, Fasc. 3/4, 231-246.

Fogg, G.E. 1965. Algal Cultures and Phytoplankton Ecology. University of Wisconsin Press. 126 pp.

Gerloff, G.C., G.P. Fitzgerald, and F. Skoog. 1950. The isolation, purification and nutrient solution requirements of blue-green algae. pp. 27-44. In: The Culturing of Algae, C.F. Kettering Foundation, J. Brunel, G.W. Prescott, and L.H. Tiffany, Editors. 114 pp.

Gross, F. 1932. Notes on the culture of some marine plankton organisms. J. Mar. Biol. Assoc. U.K. 21: 753-768.

Guillard, R.R.L. 1963. Organic sources of nitrogen for marine centric diatoms. Ch. 9 pp. 93-104. In: Symposium on Marine Microbiology, C.H. Oppenheimer, Ed. C.C. Thomas, Springfield, Illinois. 769 pp.

Guillard, R.R.L. and J.H. Ryther. 1962. Studies on marine planktonic diatoms I. Cyclotella nana Hustedt and Detonula confervacae (Cleve) Gran. Can. J. Microbiol. 8: 229-239.

Harvey, H.W. 1955. The Chemistry and Fertility of Sea Waters. Cambridge University Press. 224 pp.

Hutchinson, G.E. 1957. A Treatis on Limnology. Vol. I, John Wiley, N.Y. 1015 pp.

Jørgensen, C.B. 1966. Biology of Suspension Feeders. Pergamon Press. 357 pp.

Maddux, W.S. 1966. A 4π light meter. Limnol. Oceanogr. 11: 136-137.

McLachlan, J. 1964. Some considerations of the growth of marine algae in artificial media. Can. J. Microbiol. 10: 769-782.

Morris, I. 1967. An Introduction to the Algae. Hutchinson University Library, London. 189 pp.

Meyers, J. 1962. Laboratory cultures. Ch. 39, In: Physiology and Biochemistry of Algae, R.A. Lewin, Ed. Academic Press, 929 pp.

Provasoli, L. and I.J. Pintner. 1960. Artificial media for freshwater algae: problems and suggestions. 84-96, In: The Ecology of Algae, Special Pub. No. 2, Pymatuning Laboratory of Field Biology, University Pittsburgh, C.A. Tryon, Jr. and R.T. Hartman, Eds. 96 pp.

Provasoli, L., J.J.A. McLaughlin and M. Droop. 1957. The development of artificial media for marine algae. Arch. Mikrobiol. 25: 392-428.

Siegelman, H.W. and R.R.L. Guillard. 1971. Large-scale culture of algae. Methods in Enzymology 23: 110-115.

Spencer, C.P. 1966. Theoretical aspects of the control of pH in natural seawater and synthetic culture media for marine algae. Botanica Marine 9: 81-99.

Stein, J.R. 1973. Handbook of Phycological Methods. Cambridge University Press, 448 pp.

Ukeles, R. 1971. Nutritional requirements in shellfish culture. In: Proceedings of the Conference on Artificial Propagation of Commercially Valuable Shellfish, Oct. 22-23, 1969. College of Marine Studies, University of Delaware, Newark, Delaware.

BACTERIAL PATHOGENS ASSOCIATED WITH CULTURED BIVALVE MOLLUSK LARVAE

Haskell S. Tubiash

U.S. Department of Commerce, National Oceanic and
Atmospheric Administration, National Marine Fisheries
Service, Oxford Laboratory, Oxford, Maryland

Whenever man has established a husbandry, be it plant or animal, he has inadvertently, but concomitantly also established communities of competitors, predators, and parasites. The culture of marine invertebrates has been no exception. Like the rest of us, marine invertebrates spend their lives in environments permeated with bacteria and other microorganisms -- the vast majority of which are either entirely harmless, or may indeed, serve as the first link in the food chain. However, our concern here is with the small minority of microbial pathogens, since it would be presumptuous if not impossible to cover all bacteria associated with marine larvae; and this discussion will be limited to some experiences with bacterial pathogens of larval juvenile bivalve mollusks.

A decade ago advances in methods for induced spawning of bivalve mollusks and for algal food production had established the feasibility for artificial culture of commercially valuable shellfish species (Loosanoff & Davis, 1963). However, research groups and pilot-plant shellfish hatcheries were hindered by a disturbingly high incidence of fatal epizootics among larval and juvenile hatchery stock, paralleling the experiences of those who pioneered in the husbandry of plants and higher animals.

At about this time, Paul Chanley, Harry Davis, Herb Hidu, Warren Landers and I were working with Victor Loosanoff at the Bureau of Commercial Fisheries Biological Laboratory in Milford, Connecticut, where the mission was principally dependent on successful hatchery production of bivalve mollusks. From time to time the laboratory suffered inexplicable, catastrophic overnight losses of larval stock. Close microscopic examination of affected live cultures often showed shimmering, vibrating particles around the margins of many larvae, like swarming bees. Subsequently this phenomenon was recognized as the active swarming of bacteria (Fig. 1).

MATERIAL, METHODS AND RESULTS

Single larvae from infected cultures of hard clams, Mercenaria mercenaria and American oysters, Crassostrea virginica, were collected in capillary pipettes, washed in sterile seawater, ground in a tissue grinder, and cultured in a conventional seawater nutrient medium. Isolates were selected from agar plates, grown on agar slants, and washed suspensions were utilized to challenge clam and oyster larvae for determining pathogenicity. Approximately 10,000 three to seven day old larvae in 400 ml of seawater were seeded with about 5×10^8 test bacteria. The larval cultures were incubated at 28°C and examined for mortality at twenty four and forty eight hours. Suitable controls were included. Pathogen invariably caused disease and very high mortality within twenty four hours, while saprophyte-seeded cultures could not be differentiated from normal controls after forty eight hours. In this manner an array of pathogens were isolated from diseased hard clams and oyster larvae from the Milford Laboratory and from an outbreak in juvenile hard clams at a Quinby, Virginia hatchery. (Fig. 2; Colony of Vibrio anguillarum, freshly isolated.) (Fig. 3; Electron Photomicrograph of V. anguillarum.)

Host ranges were determined in a similar manner. Larvae of the hard clam, American oyster, European oyster (Ostrea edulis), bay scallop (Aequipecten irradians), and shipworm (Teredo navalis), were used as hosts for challenge by Strain 17, a pathogen of high virulence. Mortality curves were constructed by sampling and counting viable and dead larvae hourly for twenty four hours.

Course of the Disease

The course of the disease, which we named bacillary necrosis (Tubiash, Chanley & Leifson, 1965), is swift and dramatic. Within four to five hours early signs of infection are a reduction of mobility and a tendency for an appreciable number of larvae to lie quiescent with both foot and velum extended. Five to six hours after seeding "swarms" of bacteria originating from discrete foci appear on the margins of some of the larvae. The swarming phenomenon is a pathognomonic sign of bacillary necrosis, although at this point infected larvae may seem otherwise normal. The swarming becomes progressively more intense and diffuse, resembling the swarming of bees in a migrating hive. By eight hours, death, ac-

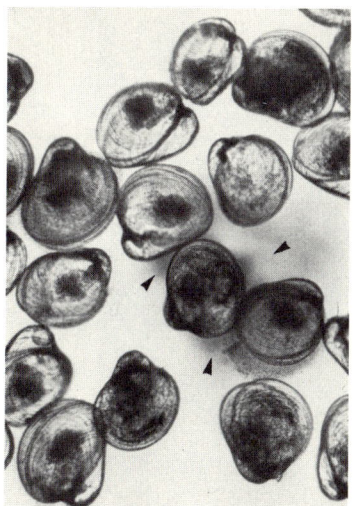

Fig. 1. Bacteria swarming aroung infected oyster larvae (arrows) resemble a shimmering haze. Many of the embryonic oysters have already been killed.

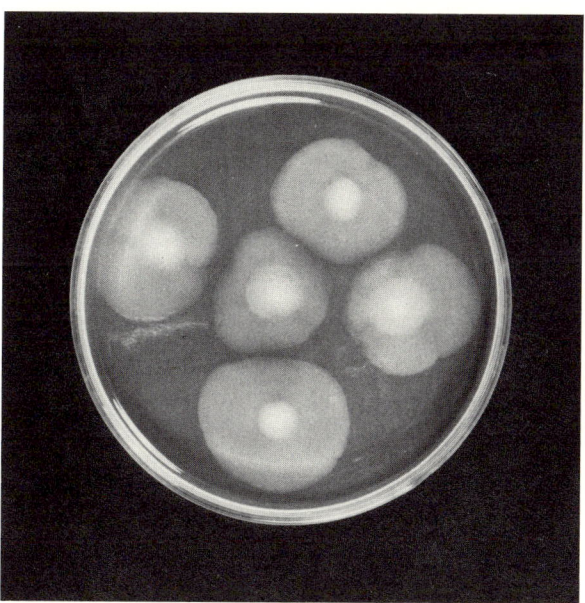

Fig. 2. Day-old colonies of <u>Vibrio</u> <u>anguillarum</u> freshly isolated from a case of bacillary necrosis in larval hard clams.

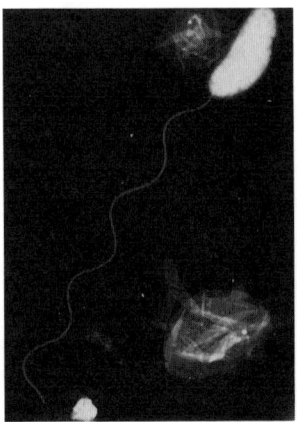

Fig. 3. Electron photomicrograph from V. anguillarum colonies in figure 2. The bacterium is about 1.5 microns long exclusive of the flagellum.

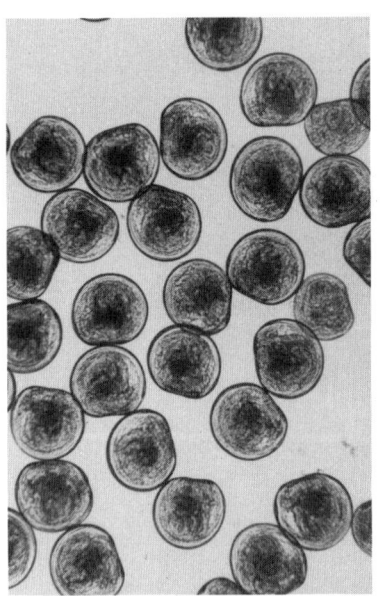 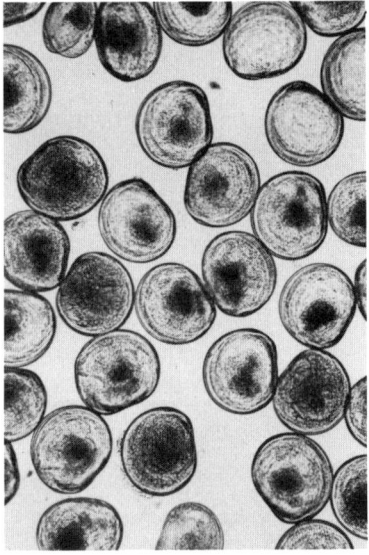

Fig. 4a. Straight-hinged European oyster larvae immediately prior to challenge with V. anguillarum. Shells are filled with healthy, intact tissue.

Fig. 4b. Larvae from same culture 11 hours post-challenge. Most of the larvae are dead and shells are either empty or contain necrotized tissue. Larvae are about 200 microns in diameter.

companied by granular necrosis is widespread. Under low-power magnification, the larval tissues appear to scintillate from the activity of invading bacteria and, in a heavily seeded culture, mortality is often complete within eighteen hours. Loosely attached or detached portions of the velum frequently continue their ciliary action after all other tissues are destroyed. (Fig. 4a and b, O. edulis challenge.) (Fig. 5, M. mercenaria curve.)

Host Range

The pathogenic strains obtained from oyster larvae were initially neither as virulent nor as motile as those of clam origin, but each was interchangeably pathogenic for larvae of homologous or heterologous species of all lamellibranchs challenged. (Table 1; Host Range.)

TABLE 1

Susceptibility of Selected Invertebrate Larvae to Bacillary Necrosis

SUSCEPTIBLE LARVAE	RESISTANT FORMS
Mercenaria mercenaria	Busycon sp. (Gastropod)
Crassostrea virginica	Artemia nauplii (Crustacea)
Ostrea edulis	
Aequipecten irradians	
Teredo navalis	
Crepidula fornicata (Gastropod)	

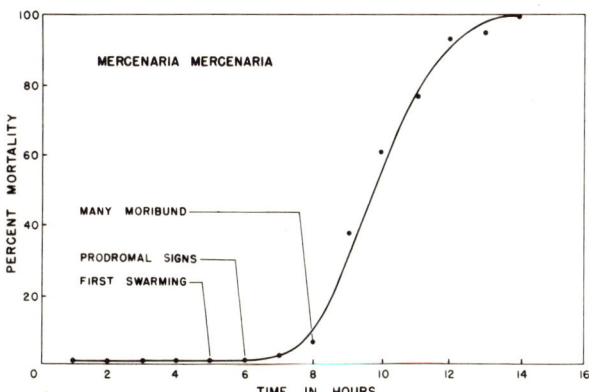

Fig. 5. Mortality curve of hard clam larvae after challenge with V. anguillarum, indicating course of the disease.

Characteristics and Taxonomy

All the pathogens proved to be gram-negative motile, flagellated rods, members of the order Pseudomonadales, resembling the majority of marine and aquatic forms. In order to permit rapid diagnosis and designation of the etiologic agents responsible for new disease episodes, diagnostic antisera were prepared from rabbits. The antigens consisted of live organisms grown twenty four hours on slants, washed, and resuspended in seawater to a density of approximately 1.5×10^9 organisms/ml. Rabbits were hyperimmunized by four subcutaneous doses administered weekly. A week after the final injection they were bled by cardiac puncture. The serum was separated, inactivated, and tube-tested for agglutinating titer against homologous and heterologous strains of pathogens.

All rabbit antisera agglutinated homologous antigens to satisfactory titers and were tested for cross-agglutination against heterologous antigens. In this manner five serological strains of pathogens were typed and labelled: Milford A, J and O and Virginia C and D. Type A included four Milford and two Virginia isolates and Types C and D each included two Virginia isolates. Type J consisted entirely of pathogens isolated from disease episodes in seed clams (M.mercenaria), while Type O was from oyster spat.

By the use of computerized numerical taxonomy 200 characteristics were utilized for the identification of these bacteria and several larval isolated pathogens from Chesapeake Bay. With these determinative methods the bacteria responsible for bacillary necrosis were identified as Vibrio anguillarum, Vibrio alginolyticus, and a yet unnamed Vibrio sp. (Tubiash, Colwell, and Sakazaki, 1970).

Histopathology

Sections of seven day old oyster larvae immediately after challenge with pathogen #17 (V. anguillarum) and twelve hours after exposure were prepared by Dr. Ruben Cares at Kings Park State Hospital. The cellular invasion, destruction, and necrosis could be readily seen. Observations of the necrotic areas confirmed a massive invasion and proliferation of bacterial cells in these areas, forming the basis for the name bacillary necrosis. These pathogens liquify gelatin rapidly and the proteolytic enzymes are probably instrumental in the rapid and dramatic necrosis. (Fig. 6; Normal Larva Section); (Fig. 7; Necrotic Larva Section); (Fig. 8; Bacteria in Shell).

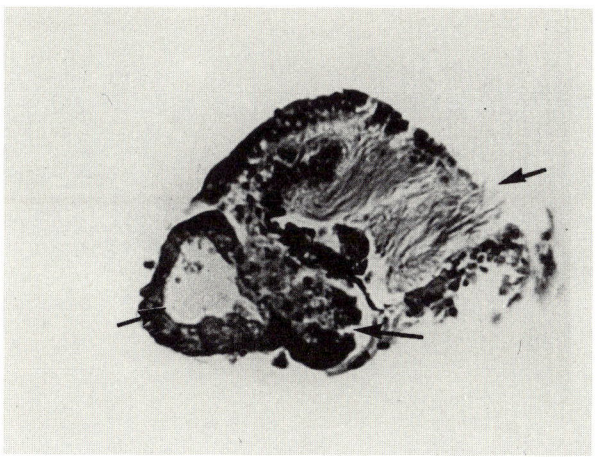

Fig. 6. Longitudinal section of normal seven day old hard clam larvae stained with H & E. Arrows indicate ciliated velum, esophagus and stomach.

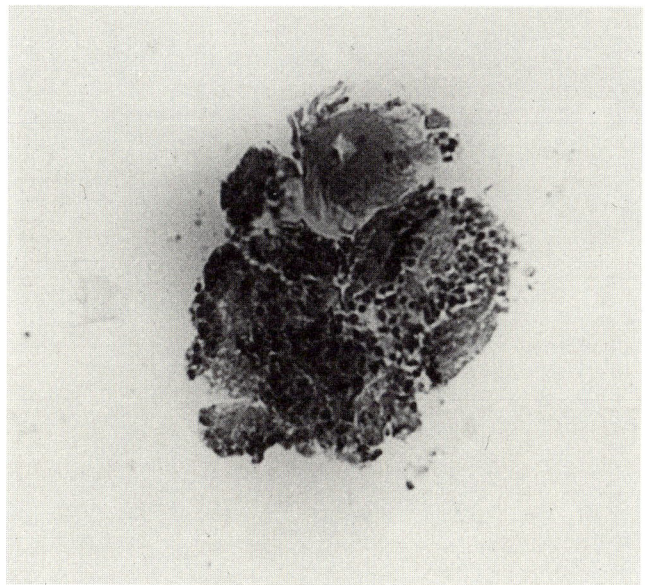

Fig. 7. Similar larva from culture eighteen hours after challenge with V. anguillarum. Loss of cellular integrity is caused by the necrotizing bacteria. H&E.

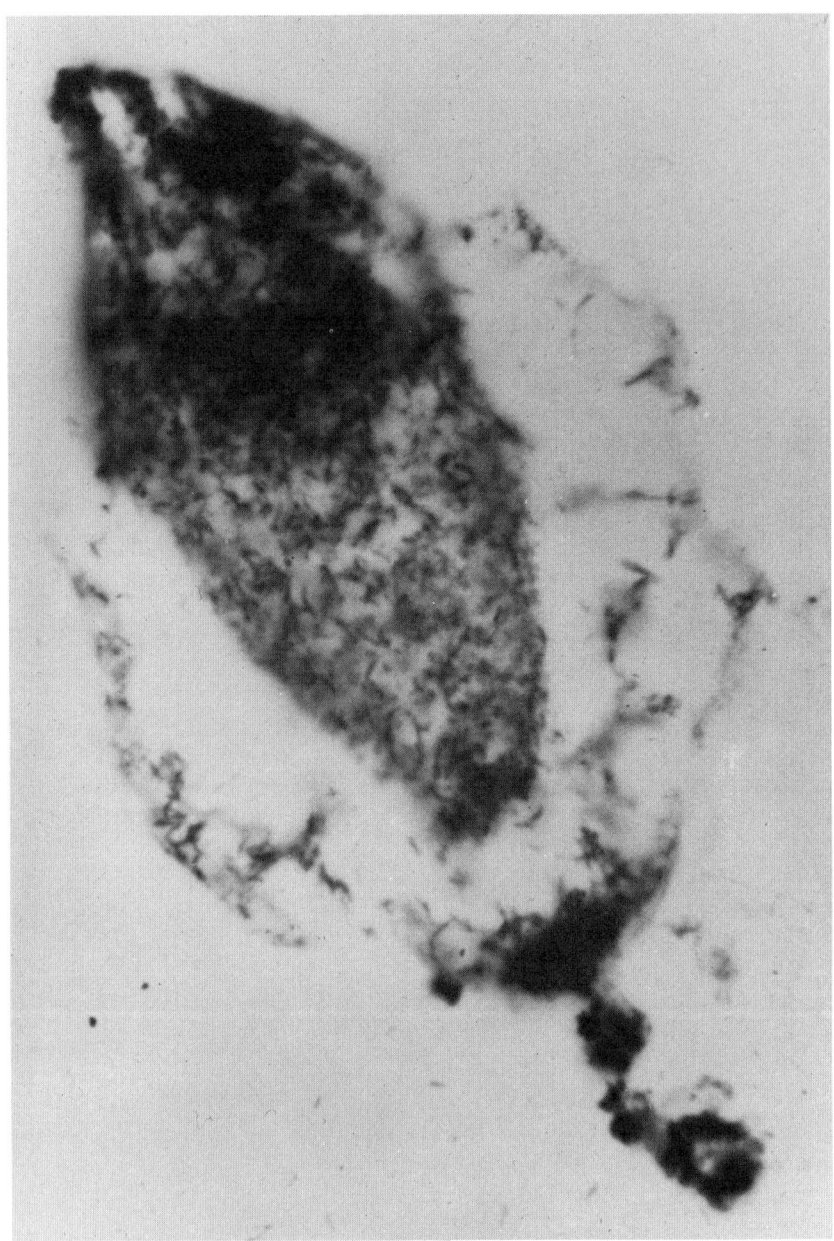

Figure 8. Section through larval shells converted to bacterial culture twenty four hours after seeding with V. anguillarum. Giemsa stain.

Screening of Antibacterials

In order to establish prophylactic or therapeutic regimes, promising antibiotics were selected using "Multidisk" sensitivity disks (Colab) on agar plates spread with twenty four hour broth cultures of larval pathogens. Antibiotics tested included chloramphenicol, kanamycin, neomycin, penicillin, polymyxin B, dihydrostreptomycin, and tetracycline. Those antibiotics whose antibacterial spectra appeared promising as indicated by inhibitory zones, were further tested in vivo for prophylactic and therapeutic value, using drug levels from 5 - 200 ppm in 400 ml cultures of clam larvae, which were then seeded with a virulent pathogen. Percent mortality was determined after forty eight hours' incubation at $28^{\circ}C$. Many antibiotics and other antibacterials are highly toxic to shellfish larvae and controls for drug toxicity were included in all assays.

"Combistrep" (Pfizer) a formulation of 25% streptomycin and dihydrostreptomycin has proved very effective against challenge by these vibrios. Fifty to 100 ppm of the active antibiotic was useful experimentally, for both prophylactic and therapeutic application. Most surprisingly, this veterinary preparation, which was marketed in liquid form for poultry treatment has been found to enhance the growth of larval cultures (Hidu and Tubiash, 1963). Chloramphenicol was equally as effective against pathogen #17 but, to our knowledge, is not available in such a convenient, inexpensive veterinary formulation. Pfizer no longer markets Combistrep, but similar veterinary proprietary preparations such as VetStrep and BiStrep are available.

DISCUSSION

In larval culture as elsewhere an ounce of prevention is indeed worth a pound of cure. While antibiotics may be used to save the patients, or an appreciable fraction of them the key word in preventative medicine be it human, veterinary, or food technology is SANITATION. Everything that effects the larvae is waterborne and the way to avoid difficulties with bacillary necrosis or any other disease is zealously to guard and control the physical, chemical, and biological quality of the water. Every hatchery and laboratory has its unique problems and must establish its own controls, standards, and prophylactic measures, whether they be filtration, ultraviolet irradiation, ozonation, or antibiotic treatment -- with the latter to be used when all else fails. Sanitation and quality of the food cultures are probably more important than sanitation of larval cultures.

Several years ago in connection with high oyster mortalities in Chesapeake Bay we did a comparative study of oyster-associated bacteria in a disease-free and enzootic area (Lovelace, Tubiash and Colwell, 1968). Bacterial analyses were done on mud, water, and oyster tissues. About the same time Murchelano and Bishop (1969) were doing a study on bacteria associated with oyster larvae at the Milford Laboratory, and it might be of interest to compare the two results, bearing in mind that they originate from disparate sources and areas (Table 2). In both cases the predominant genera were Vibrio, Pseudomonas, Achromobacter and Cytophaga/Flavobacterium). For whatever it may mean, vibrios were twice as plentiful in samples from Marumsco Bar, the high morality area, as they were in samples from disease-free Eastern Bay.

While the etiologic agent of the oyster mortalities in Delaware and Chesapeake Bays proved to be a protozoan, Minchinia nelsoni, vibrios have long been associated with pathology in fin fishes, and from all indications they are plentiful in both healthy and diseased bivalve mollusks and their environments in highly productive areas. It might be pertinent to speculate on their natural role in the recruitment of commercial mollusks. Oyster setting occurs in Chesapeake Bay from mid-June through August and in Long Island Sound from mid-July through August, concomitant with increasing air and water temperatures (Loosanoff, 1966; Shaw, 1969). It is therefore possible that, during midsummer, when conditions favor both molluscan spawning and bacterial proliferation, natural epizootics of bacillary necrosis may limit the recruitment of commercially valuable bivalve mollusks.

On the other hand, bacteria probably serve a more universal, basic and benign role in molluscan ontogeny as an embryonic food source.

TABLE 2

Generic % Distribution of Oyster-Associated Bacteria

TAXONOMIC GROUP	CHESAPEAKE BAY MARUMSCO BAR	CHESAPEAKE BAY EASTERN BAY	L.I. SOUND HATCHERY LARVAE
Vibrio spp.	46	20	19
Pseudomonas spp.	22	9	49
Achromobacter spp.	13	17	8
Corynebacterium spp.	5	7	-
Cytophaga/Flavobacterium spp.	3	30	15
Micrococcus/Bacillus	-	6	8
Enterics	7	3	-
Others	3	8	2
Isolates Tested	125	104	176

REFERENCES

Hidu, H. and H.S. Tubiash, 1963. A bacterial basis for the growth of antibiotic-treated bivalve larvae. Proc. Nat'l. Shellfish Assoc. 54: 25-39.

Loosanoff, V.L. 1966. Time and intensity of setting of the oyster, Crassostrea virginica, in Long Island Sound. Biolog. Bull. 130: 211-227.

Loosanoff, V.L. and H.C. Davis. 1963. Rearing of bivalve mollusks, p. 1-128. In: F.S. Russel (ed.), Advances in Marine Biology, Vol. 1, Academic Press, Inc. New York, N.Y.

Lovelace, T.E., H. Tubiash, and R.R. Colwell. 1968. Quantitative and qualitative commercial bacterial flora of Crassostrea virginica Chesapeake Bay. Proc. Nat'l. Shellfish Assoc. 58: 82-87.

Murchelano, R.A. and J.L. Bishop. 1969. Bacteriological study of laboratory-reared juvenile American oysters (Crassostrea virginica). Journ. Invertebr. Pathol. 14: 321-327.

Shaw, W.N. 1969. Oyster setting in two adjacent tributaries in Chesapeake Bay. Transact. Ame. Fish. Soc. 98: 309-314.

Tubiash, H.S., P.E. Chanley, and E. Leifson. 1965. Bacillary necrosis, a disease of larval and juvenile bivalve mollusks. I. Etiology and epizootiology. Journ. Bacteriol. 90: 1036-1044.

Tubiash, H.S., R.R. Colwell, and R. Sakazaki. 1970. Marine vibrios associated with bacillary necrosis, a disease of larval and juvenile bivalve mollusks. Journ. Bacteriol. 103: 272-273.

MARINE MICROBIOLOGY: SOME PRACTICAL ASPECTS FOR AQUACULTURE

Joseph M. Cassin, *Patricia E. Cassin, Elsa Brunn,

Kenneth Frenke, Michael Priano, Heidi G. Wetherall,

Neil T. Wetherall

Adelphi University Institute of Marine Science, Garden City, New York; *Queensborough Community College, City University of New York, Bayside, New York

Any in situ culture of marine invertebrates (clams, oysters, prawns, etc.) should consider resident bacterial flora. Resident flora will indicate: 1) The degree and type of heterotrophic activity at the decomposition level of the trophic structure; 2) the degree and type of fecal intrusion on water quality.

In marine ecosystems, bacterial activity contributes to gross productivity and nutrient recycling. Takahashi and Ichimura (1968) have shown that autotrophic bacteria contribute 3 to 5% gross productivity in some aquatic systems. Heterotrophic bacteria are important in inorganic nutrient regeneration, degrading detritus and complex organic molecules and pathogenicity.

Marine heterotrophic bacteria are mainly gram-negative motile highly pleomorphic rods. Groups most commonly encountered are Pseudomonas, Achromobacter, Flavobacter, Micrococcus, and Corynebacterium. As in clinical microbiology, the marine microbiologist is confronted with a plethora of physiological quirks, which demand sampling, culture, and enumeration ingenuity on the part of the investigator. Marine and estuarine bacteria have varying tolerances to temperature, salinity, and pressure.

Ingram (1962) has thoroughly reviewed psychrophily, mesophily, and thermophily. Psychrophily has been the subject of many papers (Burton and Morita, 1963; Morita and Burton, 1963; Morita and Haight, 1964). Malcolm (1969) demonstrated the obligate psychrophily of Micrococcus cryophila to be derived from temperature sensitivity of two amino acid activating enzymes and the cognate species of tRNA. Glutamyl-trans-synthetase and prolyl-tRNA synthetase are molecular determinants of the temperature sensitivity.

Larson (1962) has reviewed the salinity requirements of microorganisms. Slightly halophilic organisms tolerated 2-5 °/oo salinity; moderately, 5-20 °/oo; and extremely, 20-30 °/oo. The predominance of gram-negative bacteria in marine biotopes has been related to composition of cell wall. Halophiles are characterized by an absence of muramic acid and a greater proportion of proteins and nucleic acids (Brown, 1964). Stanley and Morita (1968), Holmes and Halverson (1965), Ishida (1970), and Dundas (1970) have demonstrated salt requiring enzymes in bacteria of marine origin. Possibly differences in ionic requirements and cell wall composition contribute to rapid die-off of terrestial forms in the marine biotope.

Zobell and Johnson (1949) and Zobell and Morita (1957) have presented classical studies on barophily in some bacteria recovered from marine sediments.

It is not the purpose of this presentation to explore in depth these traits of marine bacteria; they do give rise to inherent difficulties in sampling, culture, enumeration, and taxonomic methods. The marine microbiologist seeks to answer these questions:

1. How many organisms are present per unit volume?
2. Who are they (taxonomy)?
3. What role do they perform in the trophic structure?

We will therefore consider sampling, cultural, taxonomic, and physiological parameters.

SAMPLING

An excellent discussion of aquatic sampling methods may be found in Rodina's (1972) <u>Methods in Aquatic Microbiology recently</u> translated by Colwell and Zambruski.

The Nansen bottle has had wide use in sampling physical and chemical parameters. Sorokin (1962) compared bacteriological samples using sterile and non-sterile bottles. Vessel sterilization with heat or alcohol returned samples with counts 296 to 200,000 less than non-sterilized bottles. The apparatus is lowered into the sea as an open cylinder. If organic slimes or detritus are present, recovery may appear higher than actually exists <u>in situ</u>.

The Niskin sampler (Niskin, 1962) consists of an anodized aluminum and stainless steel frame. Water is collected in 1 to 5 liter sterilized polyethylene bags. The sampler is not adversely affected by hydrostatic pressure which prevents use of glass sampling apparati at certain depths.

The rubber bulb apparatus has been successfully used by Murchelano and Brown (1970) in monitoring bacterial flora over shellfish beds in Long Island Sound and embayments. A sterile collapsed bulb with a closed hollow glass rod is fixed to oceanographic wire. A series of bulbs may be fixed to desired depths. A messenger breaks the glass tube, and a sterile sample is obtained. This apparatus is useful in shallow water. At greater depths, hydrostatic pressure may prevent expansion of the bulb. The volume of sample is limited by size of the rubber bulb.

In recent studies of heterotrophic, coliform, and Streptococcus Group D bacterial populations in the New York Bight, we tested a deep sea pump system designed by Weyl and Hardy (1972), of Marine Sciences Research Center, State University of New York, Stony Brook. Simultaneous chemical, physical, and biological samples may be taken. The pump is lowered to the desired depth, to a maximum of 50 meters. A flushing time of 3 to 8 minutes insures proper sample composition. The system is satisfactory for bacterial sampling only when using selective, diagnostic media. We have demonstrated the efficiency of the system for monitoring fecal indicators (Frenke, 1972; Brunn, 1973).

Sediment sampling is especially important in determining residual populations of human fecal contaminants. Sediments are usually sampled with a coring or grab apparatus. Sterile samples of a specific volume may be removed for enumeration and diagnosis. For deep sea work, the Shipek Sampler (Hydroproducts, Inc., San Diego, Calif.) has proven most useful in sediment bacterial enumeration in the New York Bight and Great South Bay. An undisturbed sediment sample is obtained. As the sample is raised, water is excluded from the sample, thereby preventing aquatic contamination.

CULTURE AND ENUMERATION

As with sampling procedures, culture and enumeration methods should be easy to perform and reproducible. Jannasch and Jones (1959) have evaluated five different culture methods (agar plate, silica gel, extinction dilution (MPN), membrane filtration macrocolony, and membrane filtration microcolony), and two direct microscopic methods (membrane filtration and Cholodny method) of enumerating marine bacteria populations. Direct counts yielded numbers 13 to 9700 times higher than culture methods. However, no distinction was made between viable and dead or inactive cells. Microcolony and MPN values were 30-35 times higher than macrocolony counts. Differences between methods were attributed to aggregation or bacteria, surface tension, selectivity of media, inactive or dead cells, or accumulated debris.

Nuzzi (1969) has compared pour plate, spread plate, and direct count methods. The pour plate method gave the most replicable counts. Modified mud extract agar gave a greater number of colonies in 21 days than Zobell's Marine Media 2119E (Difco).

One of us (Brunn, 1973) recently evaluated several culture media used to enumerate marine heterotrophic bacteria. Five pour plate and two membrane filtration methods were compared. Pour plate media compared were Zobell's Marine Media 2119E (Difco) (ZA), Mud extract agar (MEA), Yeast extract agar with aged seawater (SWYE), nutrient agar with aged seawater (NASW), and nutrient agar with double distilled water (NADW). Two membrane filtration methods were compared using total count broth (Millipore Corp.) (MFTGE), and Zobell's 2119E as a broth (MFZ). Figure 1 gives colony counts averaged from 10, 1, and 0.1 ml inocula sampled from the same source in Great South Bay, and incubated at $12^{\circ}C$, $21^{\circ}C$, and $37^{\circ}C$ up to 216 hours. Greatest reproducibility was found at 168 hours at $12^{\circ}C$ and $21^{\circ}C$ using ZA medium. Although MEA gave slightly higher numbers at 216 hours of incubation, ZA gave more reproducible results using shorter incubation times. Forty-eight to 72 hours incubation in ZA allows ample time for marine bacteria to reach stationary growth phase. Thus this medium seems to be ideal for recovering maximum numbers of bacteria in minimum time. Mucoid overgrowth was observed in all media at $37^{\circ}C$. The membrane filtration methods yielded more erratic results.

TAXONOMY

Taxonomy should indicate physiology and ecology of the microbial flora as well as phylogenetic relationships. Because of inherent differences in somewhat arbitrary classification schemes, it is difficult to compare and evaluate data from various sources on marine isolates. For example, Scholes and Shewan (1964) have noted that three separate and distinct taxonomic keys are currently in use: Bergey's Manual (United States and British Commonwealth), Prevot's diagnostic key (Continent), and the Krassilnikov schema (Russia and Eastern Europe). The inherent pleomorphic traits of marine bacterial isolates greatly decreases the usefulness of these diagnostic keys. Table 1 compares marine bacterial flora recovered from various marine ecosystems. In most studies Pseudomonas comprises the greatest proportion of forms recovered. These data illustrate the great variability which exists in the literature. One wonders whether the variability lies in the diagnostic methods used, or whether the areas studied really do have different bacterial flora.

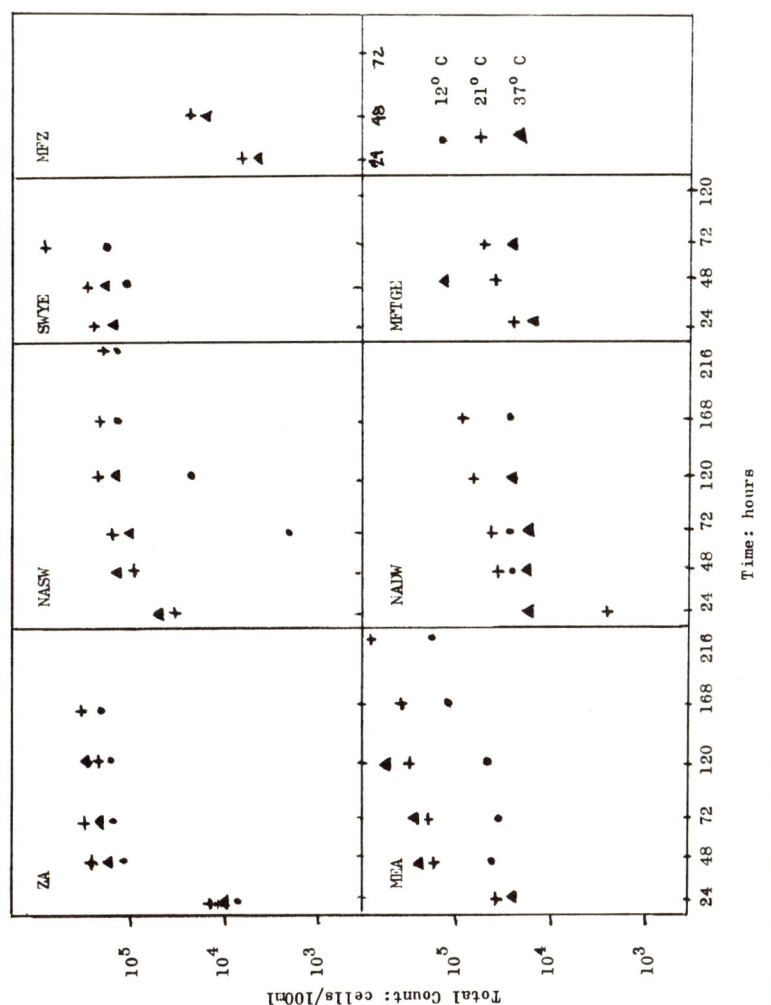

Fig. 1. Comparative growth of inocula from same source in various media. ZA, Zobell's 2119E (Difco); MEA, mud extract agar; NASW, nutrient agar + seawater; NADW, nutrient agar + distilled water; SWYE, aged seawater + 0.2% yeast extract and 1.0% agar; MFZ, Zobell's 2119E marine broth (Difco); MFTGE, total count broth (Millipore Corp). Incubation temperatures: • = 12°C; + = 21°C; ▲ = 37°C.

These studies are based on Shewan and associates' (1960) determinative scheme for identification of certain gram-negative rods. However, this scheme does not take into account many other physiological and biochemical traits which define the role of marine bacteria within the ecosystem. Colwell and Wiebe (1970) suggested use of "core" characteristics when classifying aerobic heterotrophic bacteria to reduce present variability in diagnostic methods (Table 2). The core characteristics may also be applied to numerical taxonomy, i.e., correlated characters identify taxa on the basis of overall similarities among isolates. This proposal has two merits: methodology could be standardized, and core traits would contribute much more ecologically significant information. We applied this concept in our studies of 140 isolates from the New York Bight. Results are summarized in Table 3. Data emphasize the physiological role of marine isolates in recycling materials rather than taxonomy. For example, 47.5% of isolates had an absolute requirement

TABLE 1

Percent composition of heterotrophic marine bacteria reported by various authors.

Genera	Komagawa Bay[1]	North Sea[2]	Goose Creek[3]	Algae[4]	L.I. Sound[5]
Vibrio	37.3	5.5	-	8.0	4.9
Pseudomonas	29.8	10.0	8.7	36.0	40.6
Achromobacter	21.3	22.0	47.8	16.0	28.6
Aeromonas	0.4	-	-	-	-
Photobacter	0.4	-	-	-	-
Flavobacter	5.5	7.5	39.2	24.0	23.1
Bacillus	0.4	-	-	-	0.1
Micrococcus	0.4	31.0	-	4.0	0.3
Coryneforms	2.3	12.0	-	-	-
Rhodospirillum	-	-	4.3	-	-
Miscellaneous	2.3	12.0	-	12.0	1.6

1. Simidu and Aiso, (1962)
2. Anderson, (1962)
3. Nuzzi, (1969)
4. Berland et al., (1969)
5. Murchelano and Brown, (1970)

for Na^+ and 86% of the isolates possessed lipolytic capabilities. Many isolates were able to exist and be physiologically active at low temperatures found within marine systems.

WATER QUALITY PARAMETERS

Monitoring of an aquaculture facility for fecal bacterial intrusion in water column and sediment is most important. The universal norm of fecal pollution is coliform diagnosis. It is assumed that fecal coliform presence indicates pathogens of human origin. Recent research experience in the New York Bight and waters contiguous to the South Shore of Long Island has emphasized the importance of also monitoring Lancefield's Streptococcus Group D. Coliforms indicate latent fecal intrusion, while the Streptococcus Group D indicates recent pollution. The former group is easily diagnosed in water and sediment by APHA standard methods (American

TABLE 2

Some suggested "core" characteristics for numerical taxonomy of bacteria.*

I. Cell Morphology: Size, shape, spore, motility, gram reaction, inclusions, arrangements, colony, pigments, liquid.

II. Physiology: Temperature, pH, ionic requirements.

III. Biochemistry: Carbohydrates, carbon sources, nitrogen, polysaccharides, lipolytic activity, oxidase, sulfur, phosphatase, antiobiotic sensitivity, IMVic.

*Colwell and Wiebe, 1970.

TABLE 3

Some "core" characteristics of 140 marine isolates from the New York Bight (Priano, 1974).

Core Trait	Percent of Total
Motility	68.5
Salt requirement	70.0
Na requirement	47.5
Temperature: 4°C	47.0
7°C	58.5
35°C	50.5
Glucose utilization	81.5
Na-citrate "	56.5
Amylolysis	63.5
Proteolysis	50.0
Lipolysis	86.0
NO_3-NO_2 utilization	43.5
NO_2-N_2 "	16.4
Cysteine "	2.8
Thiosulfate "	29.3
Mixed amino acids (9 groups) utilization	1.4 - 62.8
Glutamate utilization	12.0
Proline "	24.2
Methionine "	13.6

Public Health Association, 1971). Streptococcus Group D has been difficult to diagnose because of false positive reactions. One of us, Wetherall, (1972) has recently improved methods of fecal Streptococcus Group D recovery from estuarine waters and sediments. The selectivity of the Streptococcus presumptive KF broth (BBL) was improved by incubation at 45°C rather than 37°C. Bile esculine azide broth medium effectively confirmed the presence of Group D Streptococci in estuarine waters. Use of these conditions eliminated false positive reactions.

Investigation of fecal indicators in the New York Bight demonstrates the importance of studying sediments as well as the water column (Frenke, 1972). Figures 2 and 3 show the distribution of fecal coliforms in water and sediment for August 1971. A marked contrast exists between water and sediment values. Highest water value was an MPN of 500/100 ml, but a value of 1500/cc of sediment

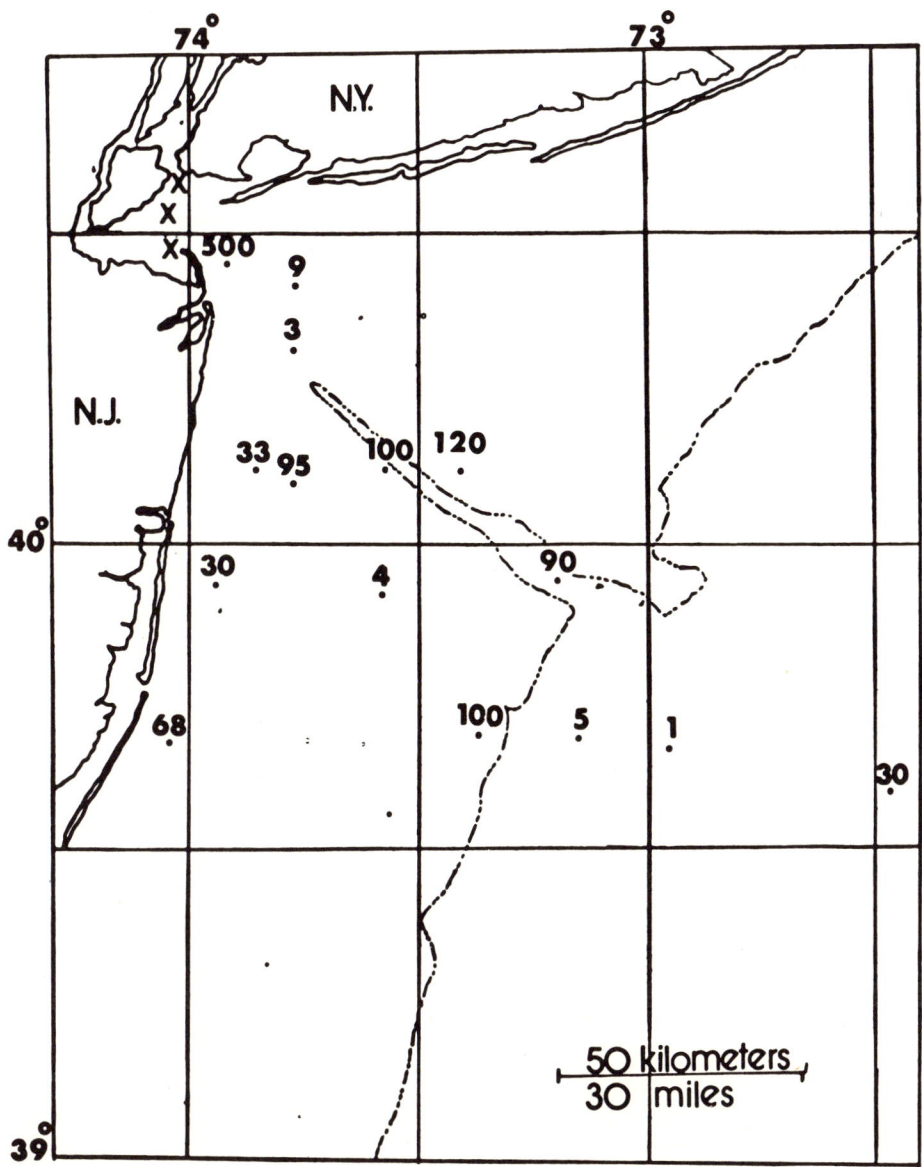

Fig. 2. Distribution of fecal coliforms in water column of the New York Bight, August 1971. X identifies samples with mucoid overgrowth. Dots represent stations sampled. Dots with no accompanying numbers indicate no fecal coliforms recovered (Frenke, 1972).

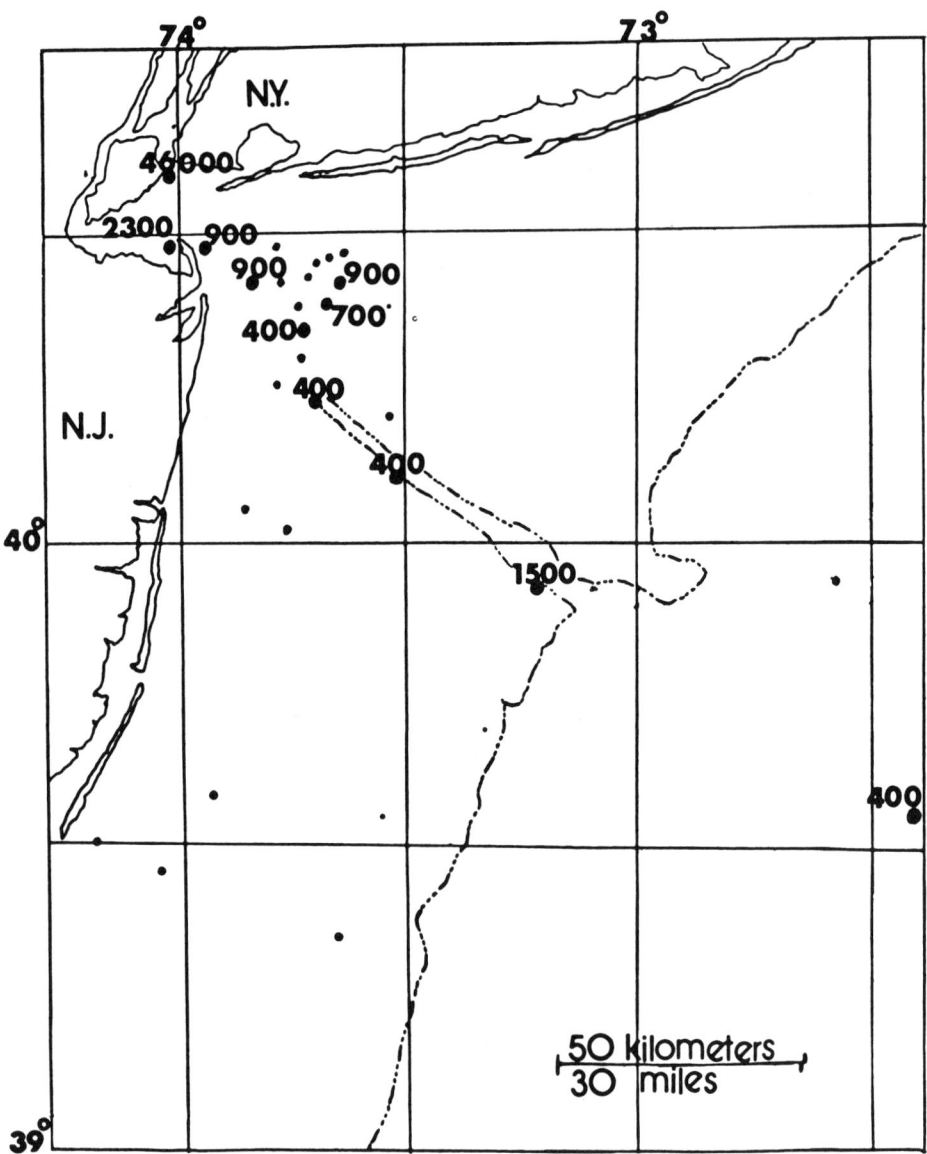

Fig. 3. Distribution of fecal coliforms in sediments of the New York Bight, August 1971. Dots represent stations sampled. Dots with no accompanying numbers indicate no fecal organisms recovered (Frenke, 1972).

was recovered in part of the Hudson Canyon. Some of these findings have been confirmed by other investigators (Wetherall, N.T., 1972; Wetherall, H.G., 1973).

We have also surveyed the sewage dumping ground area of the New York Bight for the distribution of Streptococcus Group D in water and sediments (Wetherall, H.G., 1973). Figure 4 shows the location of the sampling stations. Table 4 lists the coliform and Streptococcus MPN for water and sediments. Again there is a marked contrast between water and sediment values, especially with the Streptococcus data from the stations in the immediate area of sludge dumping.

It has usually been assumed that antibiotic factors present in seawater contribute to diminishing the numbers of terrestial bacteria. In a marine system under heavy ecological stress, such as the New York Bight, bacteria were removed from the water column, but apparently were not killed, since large numbers of viable organisms were recovered from sediments. One would question the validity of pollution level studies which rely solely on estimation of fecal contamination of the water column.

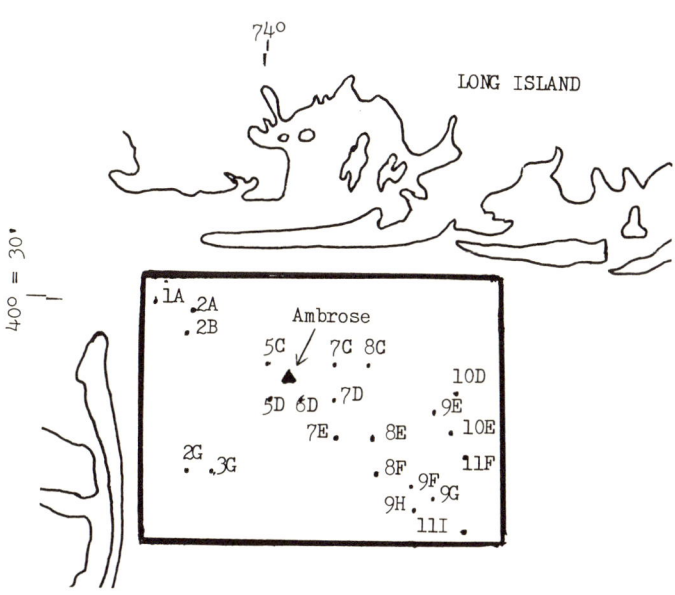

Fig. 4. Location of stations for coliform and Streptococcus Group D samples, August 1972 (Wetherall, H.G., 1973).

TABLE 4

Coliform and Streptococcus Group D MPN values/100 ml for water and sediment. R/V Venture cruise: 7/1 - 7/9, 1972.

Station	Coliforms				Streptococcus Group D	
	Water		Sediment		Water	Sediment
	Total	Fecal	Total	Fecal		
1A	1609	350	9	6.1	7.8	6
2A	1609	920	33	23	9	6
2B	>1609	1600	49	23	22	23
5C	93	9.1	43	0	0	0
7C	1609	1.8	33	1.8	0	0
8C	79	7.8	175	7.8	31	49
5D	23	23	0	0	0	0
6D	460	240	23	0	0	0
7D	1609	920	17	0	2	8
10D	33	2	>1609	>1609	0	>1609
7E	918	6.8	9	0	2	8
8E	0	0	1609	1.8	0	348
9E	21	6.8	430	430	0	130
10E	240	9.3	>1609	220	0	240
8F	9	0	109	1.8	0	1609
9F	5	0	175	33	4	9
11F	79	23	17	1.8	0	22
2G	56	43	94	7.8	0	0
3G	542	7.9	15	4	0	33
9G	22	9.3	0	0	7.2	0
9H	79	4.5	918	4.5	0	130
11I	63	4	240	7.8	7.5	11

ACKNOWLEDGEMENT

This research was supported in part by Adelphi University Institute of Marine Science; EPA Training Grant T 900-320, Dr. A.H. Brenowitz, Project Director; and Research Contract AU 376 from Research Foundation, State of New York.

REFERENCES

American Public Health Association. 1971. Standard Methods for the Examination of Water and Wastewater, 14th Edition.

Anderson, J.W.I. 1962. Heterotrophic bacteria in North Sea Water. Ph.D. Thesis, Glasgow University.

Berland, B.R., M.G. Bianchi, et S.Y. Maestrini. 1969. Étude des bactéries associées aux Algues marines en culture. I. Détermination préliminaire des espèces. Marine Biol. 2: 350-355.

Brown, A.D. 1964. Aspects of bacterial response to ionic environment. Bact. Ref. 28: 296-329.

Brunn, E. 1973. Distribution of Fecal Indicators in Great South Bay and Contiguous Waters. M.S. Thesis, Adelphi University. In preparation.

Burton, S.D. and R.Y. Morita. 1963. Denaturation and renovation of malic dehydrogenase in a cell-free extract from a marine psychrophile. Journ. Bact. 86: 1019-1024.

Colwell, R.R. and W.J. Wiebe. 1970. "Core" characteristics for use in classifying aerobic, heterotrophic bacteria by numerical taxonomy. Bull. Ga. Acad. Sci. 28: 165-185.

Dundas, I.E.D. 1970. Purification of ornithine carbamyltransferase from Halobacterium salinarum. Eur. Journ. Biochem. 16: 393-398.

Frenke, K. 1972. Distribution of fecal coliforms in sediments of the New York Bight. M.S. Thesis, Adelphi University.

Holmes, P.K. and H.O. Halverson. 1965. Purification of a salt-requiring enzyme from an obligately halophilic bacterium. Journ. Bact. 90: 312 ff.

Ingram, J.L. 1962. Temperature relationships. In: Bacteria, I.C. Gunsalus and R.Y. Stanier, eds. Academic Press. N.Y. 4: 265-295.

Ishida, Y. 1970. Growth patterns of halobacteria in media containing salts, sugars, and other compounds. Bull. Jap. Soc. Sci. Fish. 36: 481 ff.

Jannasch, H.W. and G.E. Jones. 1959. Bacterial populations in seawater as determined by different methods of enumeration. Limnol. Oceanog. 4: 128-139.

Larson, H. 1962. Halophilism. In: Bacteria, I.C. Gunsalus and R.Y. Stanier, eds. Academic Press. N.Y. 4: 297-336.

Malcolm, N.L. 1969. Molecular determinants of obligate psychrophily. Nature 221: 1031-1033.

Morita, R.Y. and S.D. Burton. 1963. Influence of moderate temperature on growth and malic dehydrogenase activity of a marine psychrophile. Journ. Bact. 86: 1025-1029.

Morita, R.Y. and R.D. Haight. 1964. Temperature effect on growth of obligate psychrophilic marine bacterium. Limnol. Oceanog. 9: 103-106.

Murchelano, R.A. and C. Brown. 1970. Heterotrophic bacteria in Long Island Sound. Marine Biol. 7: 1-6.

Niskin, S.J. 1962. A water sampler for microbiological studies. Deep Sea Res. 9: 501-503.

Nuzzi, R. 1969. A study of bacterial cycle in Goose Creek, N.Y. 1966-1968. Ph.D. Thesis, Fordham University.

Priano, M. 1974. Heterotrophic Bacteria in New York Bight and Long Island Sound. M.S. Thesis, Adelphi University.

Rodina, A.G. 1972. Methods in Aquatic Microbiology. R.R. Colwell and M.S. Zambruski, trans. University Park Press, Baltimore, Md.

Scholes, R.B. and J.M. Shewan. 1964. The present status of some aspects of marine microbiology. Adv. Mar. Biol. 2: 133-169.

Shewan, J.M., G. Hobbs, and W. Hodgkiss. 1960. A determinative scheme for the identification of certain genera of gram-negative bacteria, with special reference to the Pseudomonadaceae. Journ. App. Bact. 23: 379-390.

Simidu, U. and K. Aiso. 1962. Occurrence and distribution of heterotrophic bacteria in seawater from the Komagawa Bay. Bull. Jap. Soc. Sci. Fish. 28: 1133-1141.

Sorokin, Yu. I. 1962. Problems of sample selection for the study of marine microflora. Okeanol. 5: 888-897.

Stanley, S.O. and R.Y. Morita. 1968. Salinity effect on the maximal growth temperature of some bacteria isolated from marine environment. Journ. Bact. 95: 169-173.

Takahashi, M. and S. Ichimura. 1968. Vertical distribution and organic production of photosynthetic sulfur bacteria in Japanese lakes. Limnol. Oceanog. 13: 644-655.

Wetherall, H.G. 1973. Techniques and Studies on Fecal Pollution Indicators in Marine and Estuarine Sediments. M.S. Thesis, Adelphi University.

Wetherall, N.T. 1972. Recovery and Enumeration of Lancefield's Group D Streptococci from Estuarine and Marine Waters. M.S. Thesis, Adelphi University.

Weyl, P. and D. Hardy. 1972. Personal Communication.

Zobell, C.E. and F. Johnson. 1949. The influence of hydrostatic pressure on the growth and viability of terrestrial and marine bacteria. Journ. Bact. 57: 179-189.

Zobell, C.E. and Morita, R.Y. 1957. Barophilic bacteria in some deep sea sediments. Journ. Bact. 73: 563-568.

CULTURE OF SALT MARSH MICROORGANISMS AND MICROMETAZOA

John J. Lee and William A. Muller

Biology Department, City College of the City University
of New York; New York, New York
New York Institute of Technology, Wheatley Road
Old Westbury, New York

In the Northeastern United States there is an ever growing interest in the preservation and management of those remaining marine wetlands which have escaped "development" in one form or another. Even today many governmental units view the wetlands as the least expensive solution to many pressing social problems such as solid waste disposal, public housing, expansion of public transportation facilities, and valuable real estate. In spite of these needs the renaissance of interest in preservation of salt marshes stems from an awareness that salt marshes and associated shallow embayments are important links in the coastal marine food web and a recognition that domestic industrial and governmental sources of pollution have had a strong impact on the productivity of those marshes which remain.

Recent public hearings and written debates on environmental impact statements underscore the huge gap between sound impartial data on which management decisions can be based and best guesses that no environmental damage will be done. Even today, with highly sophisticated means of analysis, the prospect of understanding the intricacies of the functional relationships between organisms living in a single community is incredible. Yet studies of this sort are tempting since they seem to offer the most promise toward providing the information necessary for the development of comprehensive and realistic plans for the management and restoration of critical ecosystems.

Our interest in salt marsh communities as a whole has gradually evolved from a particular interest in some of the microscopic herbivores in the marsh. Although there is a great deal of knowledge and research in progress revolving around salt marshes and other shoal water marine habitats (Teal and Kanwisher 1966, 1970; Pomeroy 1959, 1963; Jeffries 1962; Jitts 1959; Odum and Smalley 1959; Pomeroy et al. 1972; Teal 1962; Stowe et al. 1971; Burkholder 1956; Burkholder and Bornsider 1957; Udell et al. 1969; Bradshaw 1968; Phleger 1960; etc.) very little is known of the intricacies of the microbial and micrometazoan communities or assemblages at the base of the detritus-marine food web. An assessment of their roles in the food web involves a knowledge of their distribution and population dynamics, their food sources, trophic dynamics and assimilation at each node in the web, special growth requirements, mineral cycling abilities and rates and their utilization by predators.

We are presently engaged in developing mathematical models of salt marsh epiphytic communities by employing the data we have already gathered in continuous flow systems and by testing the durability of the organisms under abiotic and biotic stresses. Toward this end we have developed many methods for the culturing of marine epiphytic microorganisms. Some of these creatures present special problems. The rather long life cycle of many foraminifera results in the deterioration of an agnotobiotic culture, by bacteria, algae, and ciliates, before the forams can multiply. Nematodes are quite prolific but are fragile and quickly die when they are transferred by conventional means. Then too, many of the diatoms have interesting nutritional needs that must be supplied by various differential media.

MATERIALS, METHODS, AND RESULTS

The aims of the study should be carefully defined before a sampling program is initiated. Agnotobiotic (crude mixtures of the desired organism(s) in the presence of unknown other organisms) cultures are the easiest to set up but they are notoriously undependable in the long run. Agnotobiotic cultures, however, are perfectly useful in life cycle and physiological studies where culture associates do not critically effect results.

Collection of organisms for setting up agnotobiotic cultures of marsh organisms is usually a very simple matter. Only a few precautions need to be observed. Organisms are harvested from the marsh into plastic buckets. If brass sieves are used to separate organisms, exposure of the organisms to the brass should be as brief as practical. We routinely use no. 4 (4.76 mm), no. 18 (1.0 mm) and no. 35 (0.5 mm) sieves to remove small clams, larger microarthropods, annelids and coarser sand grains. Many organisms such as nematodes and foraminifera are easily damaged or killed by rough or excessive processing. If these organisms and others are the aim of the collection, procedures must be very gentle. We shake or gently rub together macroalgae in pails of seawater to concentrate and isolate members of the epiphytic communities. The pails are allowed to stand for 15-20 minutes in order for the organisms to settle to the bottom. The upper 7/8 of the contents of the pail is then decanted. Collections are transferred from the plastic pails to wide mouth screw-capped polyethylene bottles or to plastic bags. Great care must be taken to avoid placing too much sample in a container. As a general rule we fill our sample containers approximately 20% in volume with an epiphytic or epibenthic collection. The sample is then overlain by seawater to 80% of the total container volume. Samples in the field are immediately transferred to an ice chest or submerged under the water at the collection site to avoid killing many organisms by the combined effect of elevated temperature, anoxia, and rapid bacterial growth.

As soon as possible on return to the laboratory, collections are placed in 18 x 12 x 2 1/2 in. Pyrex baking dishes (Corning 3170) or in Wheaton (155 x 55 mm) stacking culture dishes (Wheaton 41684). Since evaporation is a serious problem in culture maintenance the Pyrex baking dishes are used in pairs. One inverted dish serves as a lid for another. The dishes are held together by ordinary household freezer tape. Although there are many types of "finger bowls", "stacking aquaria" or stacking culture dishes" on the market we prefer the Wheaton (Wheaton Glass Co., Millville, N.J. 08332) because they fit together snugly (the bottom of one is the lid for the one below) and they have straight sides almost perpendicular to their bottoms. These design features permit the use of rubber dish seals (Fisher 8-759) which bind the stacks together and retard evaporation.

If small micrometazoa and microorganisms are the culture goal, larger competitive or predator organisms are removed by pipette or with forceps. Some organisms we have cultured agnotobiotically are also removed at this stage. Epiphytic and epibenthic foraminifera in natural collections spread out in finger bowls or baking dishes, climb to the surface of the sediment overnight and collect in small balls of algae which they draw together with their pseudopods. Some species of nematodes also migrate to the surface layers where they can be collected, whereas others migrate along the sides of the dishes in the water column above the sediment where they too can be collected.

Crude agnotobiotic cultures are generally incubated in moderate light ($\sim$250 $\mu w/cm^2$) at approximately the water temperature at the time of collection. Fluorescent light seems to be as good, in terms of light quality, as incandescent for the growth of most littoral benthic marine algae which are the basic food source in many cultures (Saks and Lee 1973). Some organisms will survive longer or grow better if the cultures are incubated at slightly lower (3-8°C) than field collection temperatures; others will not. Most cultures are incubated in Sherer Model CEL 4-4 environmental chambers which are programed for an 18 hr. light/6 hr. dark cycle. We prefer this model chamber because it has moveable shelves illuminated from below. The best medium for crude agnotobiotic cultures is Millipore filtered seawater from the collection site. Erdschreiber is a medium often employed for crude culture study. It is basically an enriched seawater medium composed of seawater 95 ml, soil extract 5 ml, $Na_2HPO_4 \cdot 7H_2O$ 0.02% (W/V). Erdschreiber promoted the growth of many species of algae but other species do not grow well in it at all (Lee et al. 1973). Some crude cultures grow well in seawater enriched only with 1-2% (V/V) soil extract. Soil extract is a tricky substance to prepare. The growth promoting properties of the extracts vary greatly with the source and type of soil. The extracts from some soils we have tried have been strongly toxic. Directions for making soil extract have been detailed previously (Lee et al. 1970).

When fresh cultures are set up from field collections we have found it helpful to decant the old medium and replace it with fresh culture fluid daily. Later as the cultures become established the interval between medium changes can be extended to weekly or monthly. It is not always practical or advantageous to grow the organisms with their food. Sometimes the food organisms require larger volumes of medium, different nutrients, specialized culture conditions or reproduce at rates incompatible with their intended predator. The waste products of some algae inhibit the growth of some microherbivores (Muller and Lee 1969).

In our experience crude (= raw) unselected agnotobiotic cultures are the least desirable type of laboratory culture. The reproductive rates of desired organisms in crude cultures are generally low and unpredictable. Blooms of the desired organisms occur, but they are rare. Gnotobiotic (all of the organisms present are known) or more limited agnotobiotic cultures are generally an immediate goal of every culture program. Depending upon one's experience and the facilities available, several avenues of approach can be attempted: 1) fortuitous methods; 2) tracer feeding and related inductive techniques; 3) reductive methods.

Fortuitous methods for the culture of new organisms are always worth the effort since they require the least amount of effort. As soon as possible after collection the desired species of foraminifera, nematodes, microcrustaceans, or other microherbivores are picked or separated from the other organisms and debris with the aid of glass needles, fine camel's hair brushes, forceps, or pipettes. If gnotobiotic cultures are desired the organisms can be washed free of unwanted associated organisms at this stage. Several different methods can be used depending upon the hardiness of the animals. Foraminifera can be brushed free of algae using glass needles or a camel's hair brush and then transferred to the wells of a sterile 9-hole spot plate (Pyrex 7220) kept within a microscope glove box (Germfree). Few nematodes and microcrustacea survive this kind of brushing treatment. Sterile seawater with or without mixtures of antibiotics (Table 1) are aseptically added to the other 8 wells. The animals are then transferred aseptically in the smallest practical volume of washing fluid from well to well. Approximately 20 serial transfers in the course of 4 hours will free most animals of externally associated microorganisms.

The species of foraminifera isolated in this way in our laboratory are Allogromia laticollaris and Rosalina leei which are grown in liquid Erdschreiber (~26%). Spiroloculina hyalina grows on Erdschreiber agar slopes with an overlay of 5 ml Millipore (HA 0.45 µM) filtered seawater (~26%). Other species (Ammonia beccarii, Quinqueloculina lata, Elphidium spp. Milliamina fusca, Trochammina inflata, Bolivina spp and Rotaliella spp) grow on Millipore filtered seawater (25%).

We have successfully used another technique for the aseptic washing of delicate nematodes. Rhabditis marina was obtained in monoxenic culture by washing the worm in a conical screw-capped centrifuge tube. The worm was picked from an agnotobiotic culture and transferred to a sterile centrifuged tube of seawater. The tube was gently agitated with the aid of a vortex mixer and then slowly centrifuged. Inspection of the tube under a stereomicroscope allowed us to determine when all the worms reached the bottom of the tube. Using aseptic technique the supernatant fluid was then slowly withdrawn and replaced with fresh sterile seawater. Monoxenic cultures were obtained in this case after 15 serial transfers.

After separation from the collection or after aseptic serial washing, if that has been done, the animals are inoculated into finger bowls, petri dishes, plastic tissue culture flasks, or test tubes with natural seawater, or Erdschreiber and mixtures of potential food organisms. The reproductive rate of most microherbivores is always higher in mixtures of food than with single species (Muller and Lee, 1969).

Studies on trophic dynamics have provided considerable feedback toward improving culture methods (Lee et al. 1966; Muller and Lee 1969). The principal salt marsh herbivores, foraminifera and nematodes, are selective feeders. Of several dozen diatoms and chlorophytes tested as food for foraminifera, only 4-5 species are eaten in significant amounts ($40\text{-}150 \times 10^{-5}$ mg/foram/day). One species, Spiroloculina hyalina, utilized only one diatom, a species of Amphora, and relies more heavily on marine bacteria as its principal food source.

Nematodes are also apparently very selective feeders. Tietjen et al. (1970) in a study of Rhabditis marina found that only 3 strains of marine bacteria and 2 chlorophytes (Nannochloris sp and

TABLE 1

Antibiotic Mixes (all values in µg/100 ml)

Components*	Mix 1	Mix 2	Mix 3	Mix 4	Mix 5	Mix 7	Mix 8
Chloromycetin		250	250		250		
Albamycin		200			400		
Coly-mycin		60			120		
Erythrocin				50		100	100
Dihydro-streptomycin SO$_4$	2000		2000				
Fungizone				50			50
Mycostatin	50			50			
Aerosporin			60				
Penicillin	1000						

*Aerosporin (Polymyxin B Sulfate), Burroughs Wellcome & Co., Tuckahoe, New York.
Albamycin (Novobiocin sodium), Mix-O-vial; n, n, -Dimethylacetamide (diluent) 10% v/v. The Upjohn Co., Kalamazoo, Michigan.
Chloromycetin (Chloramphenicol-(Sodium succinate)), Park Davis & Co., Detroit, Michigan.
Coly-mycin (Sodium colistimethate), Warner-Chilcott Laboratories Div., Morris Plains, New Jersey.
Erythrocin (Erythromycin lactobionate), Abbott Laboratories, North Chicago, Illinois.
Fungizone (Amphotericin B), E.R. Squibb & Sons, Inc., New York, New York.
Mycostatin (Squibb Nystatin), E.R. Squibb & Sons, Inc., New York, New York.

Dunaliella parva) were eaten in large numbers. Only the bacteria, Micrococcus sp, Flavobacterium sp, and Pseudomonas sp were digested. In later studies they showed that the Pseudomonas had an essential growth factor required by the host and which was not provided by the other 2 species tested. Over the years we have found that the following group of diatoms and chlorophytes are good food organisms for many species of salt marsh foraminifera, nematodes and microcrustacea: Nitzschia acicularis, Cylindrotheca closterium, Phaeodactylum tricornutum, Amphora spp, Fragilaria construens, Nitzschia brevirostris, Navicula diversistriata, Dunaliella prava, Dunaliella salina, Chlamydomonas spp, Nannochloris sp, Chlorococcus sp, Monochrysis sp, and Isochrysis sp. As a general rule we add $\sim 1 \times 10^4 - 1 \times 10^5$ algae/ml of culture fluid. To the inexperienced or casual observer it may seem that one alga should be as good as another. Growth curves we have obtained with foraminifera (Fig. 1 and 2, see also Muller and Lee 1969) and nematodes clearly demonstrate that this is not the case.

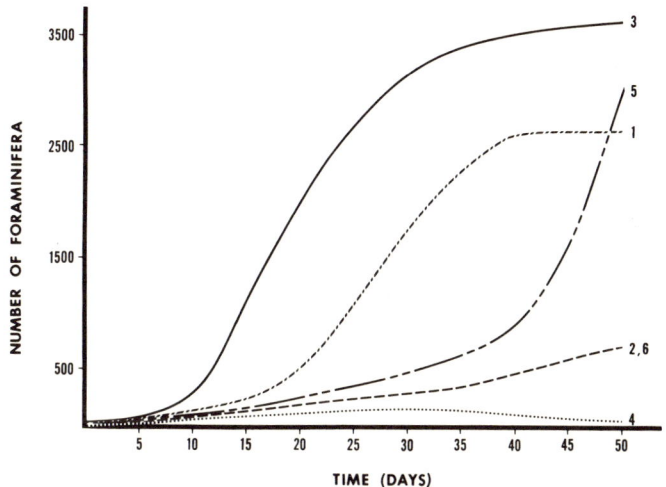

Fig. 1. Growth curves of foraminifera on various food organisms.
1) Allogromia laticollaris feeding on: Amphora sp; Phaeodactylum tricornutum.
2) Allogromia laticollaris feeding on: Chlorococcum sp; Nannochloris sp.
3) Spiroloculina hyalina feeding on: Amphora sp.
4) Spiroloculina hyalina feeding on: Chlorococcum sp; Nannochloris sp.
5) Quinqueloculina lata feeding on: Chlorococcum sp.
6) Quinqueloculina lata feeding on: Nitzschia acicularis; Amphora sp.

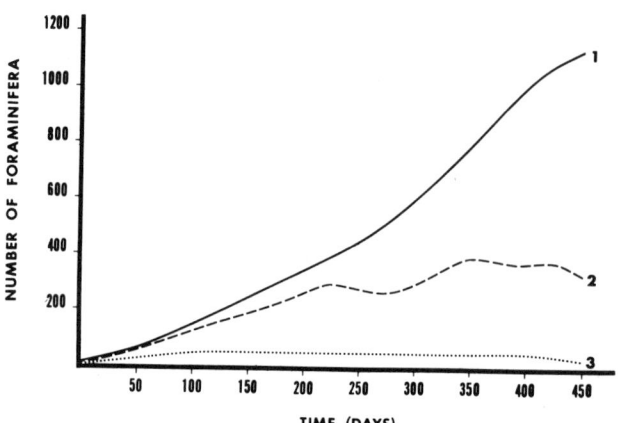

Fig. 2. Growth of <u>Rosalina leei</u> fed various food organisms.
1) <u>Nannochloris</u> sp, <u>Amphora</u> sp.
2) <u>Chlorococcum</u> sp.
3) <u>Nitzschia hungarica</u>, <u>Dunaliella parva</u>, <u>Dunaliella salina</u>.

Inductive methods take longer to do but they give more promise of favorable results. We have successfully used tracer feeding techniques (Lee et al. 1966) for the establishment of gnotobiotic cultures of foraminifera and nematodes. Briefly, potential food organisms are grown in ^{32}P labeled media and fed to the organisms we are attempting to culture. Experience with various radionuclides over the years has led us to believe that there is less experimental error with ^{32}P than when ^{3}H, ^{35}S, ^{55}Fe or ^{14}C are used as the label. Error is greatest with ^{14}C because of label loss during respiration. We vary the initial level of medium label according to the generation time of the bacterium or alga. Our goal is to feed organisms labeled with ^{32}P at & 0.5-10 dpm/food organism to potential predators. If the level of radionuclide label is too low, counting error becomes significant as one approaches background levels. When the level of label is much above 10 dpm/food organisms then radiation effects on predators can be observed. To obtain the desired level we use ~10 µCi ^{32}P/ml in the initial growth medium of most of the salt marsh algae we use. At harvest the cells are spun down and the medium decanted. From this point on, aseptic technique is not necessary. The cells are resuspended in fresh unlabeled medium and then spun down again. After 3 serial washes the labeled food organisms are inoculated into experimental vessels (Finger bowls, test tubes, Petri dishes, spot plates, etc.) with prospective predators. Aliquots of the food organism inoculum are enumerated with the aid of a hemocytometer and used to establish the level of radionuclide label per organism. Depending upon the predator, these tracer feeding experiments are harvested after 4-24 hours. The predators are washed free of their potential food in one of the ways described above and their radioactivity measured. The relative number of prey eaten is obtained by dividing the net cpm (- background) by the radioactivity per food organism determined above. For most species of foraminifera tested there is almost a direct relationship between the concentration of potential food organisms fed and those eaten, if they are eaten at all (Fig. 3; see Lee et al. 1966 for details).

Below a minimum concentration ($\sim 10^2$ - 10^3) feeding is erratic. We infer from the results of tracer feeding and ecological efficiency experiments (Lee and Muller 1973), that salt marsh foraminifera are "bloom feeders". That is that individual species have very specialized nutritional requirements and that their feeding behavior and trophic efficiency is programmed to optimize utilization of blooms of prey. We must simulate bloom conditions if we wish to maximize productivity of laboratory cultures.

Sometimes the results of tracer feeding experiments can be misleading. Rhabditis marina, a marine nematode, for example, ingests many more organisms than it is able to digest (Tietjen et al. 1970). In such cases the animals are harvested from tracer feeding experiments and incubated for 24 hours with non-labeled food before measuring their radioactivity. This is usually enough time for undigested labeled food to be eliminated from the digestive tracts.

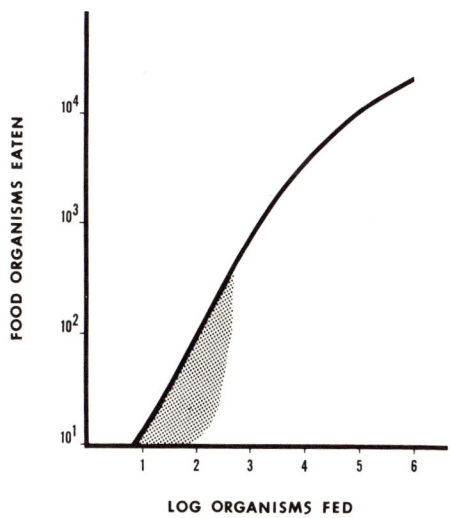

Fig. 3. Idealized tracer feeding curve.

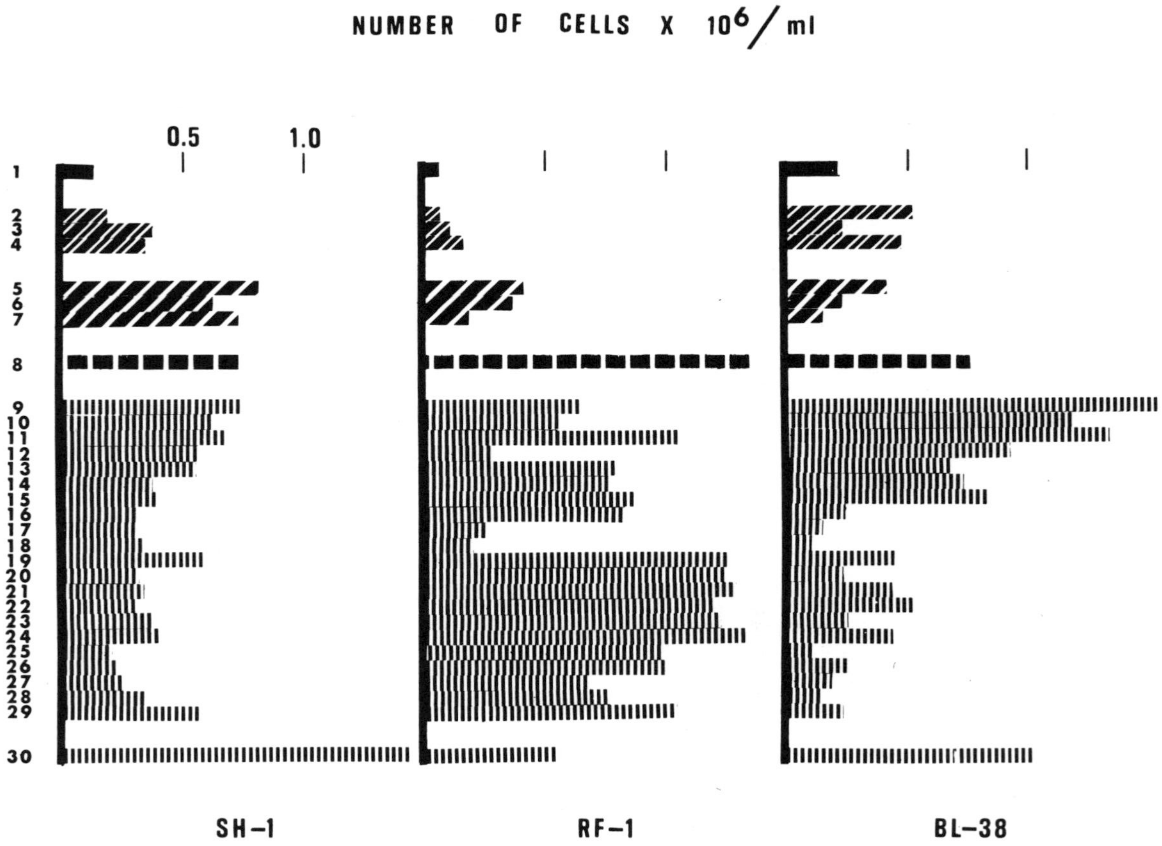

Fig. 4. Nutritional fingerprints of 3 species of salt marsh algae: BL-38 - Amphora sp; RF-1 - Nitzschia brevirostris; SH-1 - Dunaliella sp.

KEY TO FIGURE 4

Stimulation of growth by various vitamins (2-4), natural extracts (5-7), a selected mixture of carbohydrates (8), amino acids (9-29), and a selected mixture of amino acids (30).

1. No additions (control)
2. vitamin B_{12} — 5.0 μg %
3. thiamine · HCl — 20.0 mg %
4. vitamins 8a — 1 ml/100 ml

5. soil extract — 5.0 mg %
6. tryptone — 25.0 "
7. yeast extract — 25.0 "

8. Na acetate — 50.0 mg %
   Na lactate — 25.0 "
   glucose — 50.0 "
9. DL-alanine — 1.0 mM
10. DL-methionine — "
11. DL-glutamic acid — "
12. DL-arginine — "
13. DL-proline — "
14. DL-phenylalanine — "
15. DL-serine — "
16. glycine — "
17. L-hydroxyproline — "
18. DL-aspartic acid — "
19. DL-asparagine — "
20. DL-lysine — "
21. DL-leucine — "
22. DL-isoleucine — "
23. DL-cystine — "
24. DL-valine — "
25. DL-histidine·HCl — "
26. DL-threonine — "
27. DL-norleucine — "
28. DL-tryosine — "
29. DL-glutamic acid — 50.0 mg %
    DL-alanine — "
    glycine — "

30. 2 + 4 + 8 + 29 above

Another inductive method for screening potential food organisms is one we are using at the present time to study ciliates (Rubin and Lee, in progress). Agnotobiotic cultures of ciliates are inoculated into respirometry flasks and potential food into the flask side arms; a piece of filter paper saturated with KOH is placed in the center well. Control flasks with either the predator or prey are also set up. After the flasks reach equilibrium and respiration is measured in the experimental and control flasks, the potential food organisms in the experimental flask side arms are poured into the main compartment. Feeding rate can be calculated from the data by measuring the rate of respiration depression. This type of experiment is usually completed within 2 or 3 hours to avoid several potential sources of experimental error.

One inductive method has been very useful in the isolation of limited agnotobiotic cultures of nematodes. It takes advantage of the relative differences in the nutrient requirements of the large variety of bacteria and diatoms growing in the natural community. Nutritional fingerprints of axenic isolate of diatoms (Fig. 4) were utilized in the development of a series of 24 differential media which were employed to study seasonal changes in the nutritional profile of a natural salt marsh epiphytic community (Lee et al., ms in preparation). The same seasonal nutritional profile changes were also studied in a different way by incubating aliquots of natural communities in the field with ^{14}C labeled organic substrates (Lee et al. 1973). This information, coupled with data on the natural distribution of the foraminifera, nematodes, diatoms and bacteria at our field station (Lee et al. 1969; Matera and Lee 1972; Lee et al., ms in preparation; Kennedy, work in progress) gives us a basis for selecting a group of media from the differential series which are more likely to grow new isolates and their food organisms. We have been successful using both liquid and solidified media overlain with liquid. Inocula for these experimental cultures are usually aseptically collected in the field (Lee et al. 1970).

In the end, reductive approaches are necessary to establish axenic (pure culture containing only one organism) or limited-gnotobiotic cultures (monoxenic - with one food, dixenic - with 2 foods, synxenic - with known food organisms, usually used to imply a limited number of species). Successful agnotobiotic cultures are usually the point of departure. Antibiotics and aseptic washing techniques (Lee et al. 1970; Lee and Pierce 1963; Muller and Lee 1969; Tietjen et al. 1970) are the most useful routes to axenic or synxenic cultures. In a recent disc test screen of the antibiotic sensitivity of 20 bacteria isolated at random from agnotobiotic cultures of salt marsh microherbivores the following results with respect to effectiveness were obtained: Polymyxin B, 100 units, 85%; Neomycin 10 µg, 65%; Chloramphenicol 10 µg, 60%.

Penicillin 1600 units 50%; dihydrostreptomycin sulfate 775 µg, 45%; Erythromycin 5 µg 35%; triple sulfa 150 µg (combined) 30%; Novobiocin 10 µg, 5%; Bacitracin 5 units, 0%. Antibiotics can be used in the washes, in the incubation medium, or both. Bacteria can contain growth factors necessary for the organism being cultured (Muller and Lee 1969; Lee and Pierce 1963; etc.). It is not always easy to differentiate between antibiotic toxicity and death due to starvation. Growth factors can be so esoteric it can take years to identify them (Droop 1962; Droop and Doyle 1966; Droop and Pennock 1971).

Salt marsh microorganisms and micrometazoa seem to have wide optimum ranges and ranges of tolerance to physical factors. (Tietjen et al. 1970; Tietjen and Lee 1972; Pierce 1965; Muller 1972). Nonetheless there are clear boundaries to be observed in setting up cultures if high productivity is to be attained. At the outset it is very difficult to predict the optimum physical conditions for any organism we are attempting to culture. The outer boundries for reasonably good reproduction of the organisms we have studied lie between 12-28 C, 12-40 °/oo salinity and pH 5.5 - 9.5. We start at 25 C, pH 8.1, and 25 °/oo salinity which are close to the optima of many of the organisms we have in culture (e.g., Figs. 5 through 7). The actual optima must be empirically determined

Culture techniques, no matter how easy or complex, are after all only a means by which interesting and important environmental problems can be addressed. Studies of the individual organisms

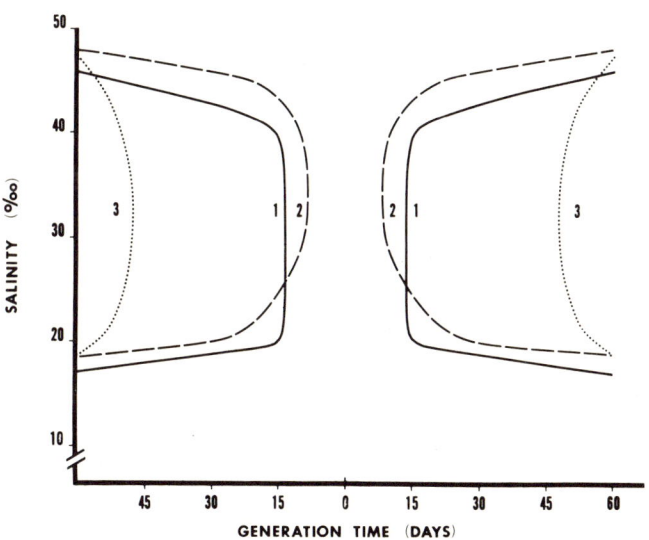

Fig. 5. The effect of salinity on the growth of 3 species of foraminifera.

can seldom, if ever, be extrapolated to yield meaningful information on the community or ecosystem level. With an eye toward natural communities we must isolate the ecologically significant species, and study their functional relationships together as a whole in model microcosms as a first step toward community understanding. The task which lies ahead seems formidable, but after all, we must start somewhere.

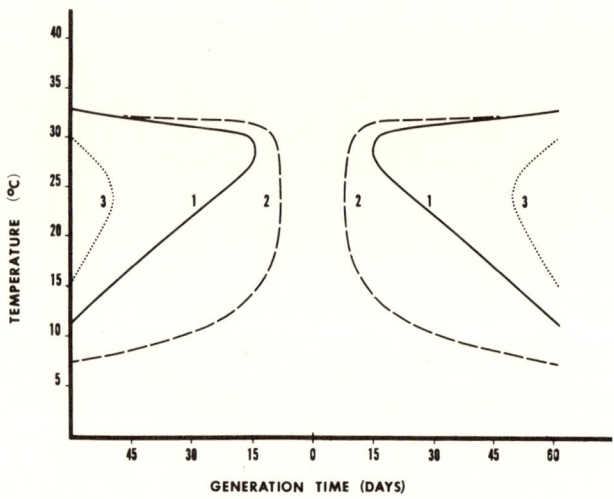

Fig. 6. The effect of temperature on the growth of 3 species of foraminifera.

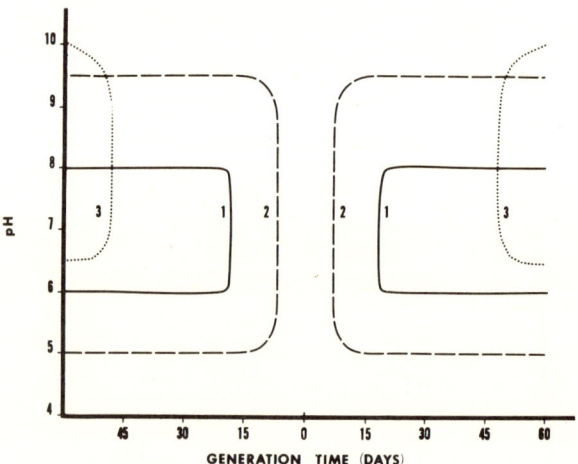

Fig. 7. The effect of pH on the growth of 3 species of foraminifera. Key: #1 - Spiroloculina hyalina; #2 - Allogromia laticollaris; #3 - Rosalina leei.

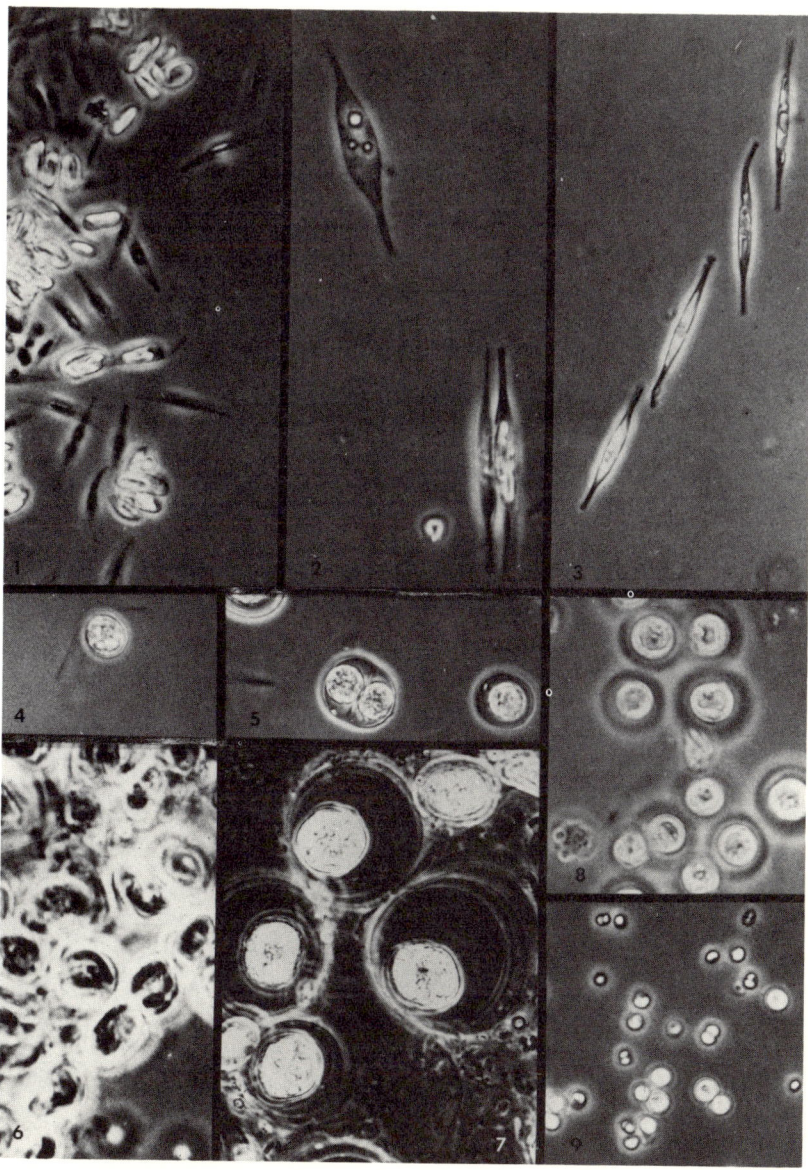

Fig. 8. Algae which are among the best food organisms for salt marsh microherbivores.

1. *Phaeodactylum tricornutum*
2. *Cylindrotheca closterium*
3. *Nitzschia acicularis*
4. *Dunaliella salina*; motile form
5. *Dunaliella salina*; vegetative form
6. *Amphora* sp
7. *Chlorococcum* sp
8. *Chlamydomonas* sp; vegetative form
9. *Nannochloris* sp

TABLE 2

Media for the Initial Isolation of Bacteria
and Algae From the Epiphytic Communities of
Enteromorpha

I. Aged Seawater Based Media

   Seawater plus

   a) Erdschreiber

   b) Seawater alone

   c) Marine nutrient agar

   d) KNO_3 10.0 mg %

       $NaH_2PO_4 \cdot H_2O$ 10.0 mg %

   e) Trypticase soy agar

II. Synthetic Based Media

Basal Medium

Components	Strength
NaCl	2.500%
$MgSO_4 \cdot 7H_2O$	0.900%
KCl	0.070%
Ca (as Cl^-)	0.030%
$NaSiO_3 \cdot 9H_2O$	0.007%
$NaHCO_3$	0.010%
$Na_2H_2PO_4 \cdot H_2O$	0.001%
NTA	0.007%
Tris	0.100%
P II metals	3 ml/100 ml

SERIES A

Base Medium

Additions:

1. none
2. $NaNO_3$ 10.0 mg%
3. NH_4Cl 10.0 mg%
4. $NaNO_2$ 10.0 mg%
5. NH_4NO_3 10.0 mg%

    $NaH_2PO_4 \cdot H_2O$ 5.0 mg%

SERIES B

Base Medium

$NaH_2PO_4 \cdot H_2O$ 5.0 mg%

Further additions:

1. Amino acid mixture 18 (A and B)

    18 A: alanine   1.0 mM
            glycine   1.0 mM
            glutamate   1.0 mM
            lysine   1.0 mM
            arginine   1.0 mM
    18 B: aspartate   1.0 mM
            leucine   1.0 mM
            histidine   1.0 mM
            methionine   1.0 mM

2. Peptone   0.1%
3. Urea   0.1%

SERIES C

Base Medium
$NaH_2PO_4 \cdot H_2O$ 5.0 mg%
NH_4NO_3 10.0 mg%

Further additions:

1. B_{12} 10.0 μg%

2. Vitamin mixture:
   Thiamine   100 μM
   Biotin   10 μM
   B_{12} 10 μM

3. Soil extract   2 ml/100 ml

4. Vitamins 8A   1 ml/100 ml

SERIES D

Base Medium
$NaH_2PO_4 \cdot H_2O$ 5.0 mg%
NH_4NO_3 10.0 mg%
B_{12} 0.1 mM
Biotin   1.0 mM
Thiamine   0.1 mM

Further additions:

1. Na acetate   3.0 mM
   Na lactate   3.0 mM
   Glucose   3.0 mM

2. Glutamate   3.0 mM

3. Asparagine   3.0 mM

4. Glycerol   3.0 mM

5. Mannitol   3.0 mM

6. Acrylic acid   3.0 mM

7. Autoclaved *Enteromorpha* + epiphytes   1.0 ml/100 ml

8. Acetone extract of *Enteromorpha*   0.1 ml/100 ml

ACKNOWLEDGEMENTS

University Institute of Oceanography of CUNY, Contribution No. 14

Supported by grants from the NSF and the US AEC (COO 3254-10)

REFERENCES

Bradshaw, J.S. 1968. Environmental parameters and marsh foraminifera. Limnol. Oceanog. 13:26-38.

Burkholder, P.R. 1956. Studies on the nutritive value of Spartina grass growing in the marsh areas of coastal Georgia. Bull. Torrey Bot. Club, 83:327-334.

Burkholder, P.R. and G.H. Bornsider. 1957. Decomposition of marsh grass by aerobic marine bacteria. Bull. Torrey Bot. Club 74: 366-383.

Droop, M.R. 1962. Organic micronutrients. In: Physiology and Biochemistry of Algae. R.A. Lewin, ed. Academic Press, New York. pp. 141-159.

Droop, M.R. and J. Doyle. 1966. Ubiquinone as a protozoan growth factor. Nature 212:1474-1475.

Droop, M.R. and J.F. Pennock. 1971 Terpenoid quinones and steroids in the nutrition of Oxyrrhis marina. Journ. Mar. Biol. Assoc. U.K. 51:455-470.

Jeffries, H.P. 1962. The atypical phosphate cycle of estuaries in relation to benthic metabolism. Symposium on the environmental chemistry of marine sediments. Univ. Rhode Island Grad. School Oceanogr. Occ. Pap. 1:58-68.

Jitts, H.R. 1959. The absorption of phosphate by estuarine bottom deposits. Australian Journ. Mar. Freshw. Res. 10:7-21.

Lee, J.J., M. McEnery, S. Pierce, H.D. Freudenthal, and W. Muller. 1966. Tracer experiments in feeding littoral foraminifera. Journ. Protozool. 13:659-670.

Lee, J.J., W.A. Muller, R.J. Stone, M. McEnery, and W. Zucker. 1969. Standing crop of foraminifera in sublittoral epiphytic communities of a Long Island Salt marsh. Jour. Mar. Biol. 4:44-61

Lee, J.J., J.H. Tietjen, R.J. Stone, W.A. Muller, J. Rullman and M. McEnery. 1970. The cultivation and physiological ecology of members of salt marsh epiphytic communities. Helgolander wiss. Meers. 20:136-156.

Lee, J.J., M. McEnery, and E. Kennedy. 1973. Educing the functional relationships among the diatom floral assemblages within sublittoral marsh epiphytic communities. In: Modern Methods in the Study of Microbial Ecology, Bull. Ecol. Res. Comm. Th. Rosswall, ed. (Stockholm) 17.

Lee, J.J., M. McEnery, E. Kennedy, and H. Rubin. A nutritional analysis of a sublittoral epiphytic diatom assemblage from a Long Island salt marsh. (in ms)

Lee, J.J. and W.A. Muller. 1973. Trophic dynamics and niches of salt marsh foraminifera. American Zool. 13: (in press)

Lee, J.J. and S. Pierce. 1963. Growth and physiology of foraminifera in the laboratory: Part 4 - Monoxenic culture of an allogromiid with notes on its morphology. Journ. Protozool. 10: 404-411.

Matera, N.J. and J.J. Lee. 1972. Environmental factors affecting the standing crop of foraminifera in sublittoral and psammonlittoral communities of a Long Island salt marsh. Journ. Mar. Biol. 14:89-103

Muller, W.A. 1972. Graphic representation of niche width and its application to salt marsh littoral foraminifera. Ph.D. Thesis City University of New York.

Muller, W.A. and J.J. Lee. 1969. Apparent indispensability of bacteria in foraminiferan nutrition. J. Protozool. 16:471-478.

Odum, E.P. and E. Smalley, 1959. Comparison of population energy flow of herbivorous and deposit-feeding invertebrates in a salt marsh ecosystem. Proc. Natl. Acad. Sci. 45: 617-622.

Phleger, F.G. 1960. Ecology and distribution of recent foraminifera. John Hopkins Press, Baltimore, Md.

Pierce, S. 1965. A comparative study of two members of the family Allogromiidae (Protozoa, Foraminifera). Ph.D. Thesis, New York University.

Pomeroy, L.R. 1959. Algal productivity in salt marshes of Georgia. Limnol. Oceanogr. 4:386-397.

Pomeroy, L.R. 1963. Experimental studies of the turnover of phosphate in marine environments. In: Radioecology, ed. V. Schults and A.W. Klements. Reinhold, N.Y. pp 163-166.

Pomeroy, L.R., L.R. Shenton, R.O. Jones, and H. Reimold. 1972. Nutrient flux in estuaries. In: Nutrients and Eutrophication, ed. G.E. Kines and J.E. Hobbie. Amer. Soc. Limnol. Oceanogr. Spec. Publ. Lawrence, Kansas.

Saks, N.M. and J.J. Lee. 1973. Radiosensitivity and light quality responses of salt marsh epiphytic algae. Nat. Symp. Radioecol. III, Proc. (in press).

Stowe, W.C., C. Kirby, S. Berkich, and J.G. Gosselink. 1971. Primary production in a small saline lake in Barataria Bay, La. Coastal Studies Bull. #6. Spec. Sea Grant Issue. Feb.

Teal, J.M. 1962. Energy flow in the salt marsh ecosystem of Ga. Ecol., 43:614-624.

Teal, J.M. and J. Kanwisher. 1966. Gas transport in the marsh grass, Spartina alterniflora. Jour. Expt. Bot. 17:355-361.

Teal, J.M. and J. Kanwisher. 1970. Total energy balance in salt marsh grasses. Ecol. 51:690-695.

Tietjen, J.H., J. Rullman, A. Greengart, and J. Trompeter. 1970. Gnotobiotic culture and physiological ecology of the marine nematode, Rhabditis marina Bastian. Limnol. Oceanogr. 15:535-543.

Tietjen, J.H. and J.J. Lee. 1972. Life cycles of marine nematodes. Oecologia 10:167-176.

Udell, H.F., J. Zarudsky, and T.E. Doheny. 1969. Productivity and nutrient values of plants growing in the salt marshes of the town of Hempstead, Long Island. Bull. Torrey Bot. Club, 96(1): 42-51.

ANTIBIOTICS IN CULTURES OF INVERTEBRATES

Anthony D'Agostino

St. John's University; Jamaica, New York and

New York Ocean Science Laboratory, Montauk, New York

Invertebrates occupy very strategic positions in aquatic food chains. The filter feeding herbivores are major links in the translation of nutrients from the first to the second trophic levels. They are predators to the primary producers and prey to all other carnivores.

Laboratory culture of invertebrates facilitates the study of the nutritional requirements and other more subtle interspecies relationships which influence growth, development, metamorphosis and fecundity of the organisms in nature.

Knowledge of the fundamental biological prerequisites of ecologically important species is essential in order to manage a deliberate maritime husbandry comparative in magnitude to that practiced with terrestrial species.

Invertebrates may be studied in crude (agnotobiotic) cultures. But to identify the source, role and target of dissolved organic substances of ecological significance, more often than not it becomes necessary to control the accompanying microflora and microfauna. Microbes may release essential nutrients, excrete or consume inhibiting substances, and otherwise influence the media.

Microorganisms in crude cultures of invertebrates can be inhibited or eliminated by the use of antibiotics combined with frequent washing of the experimental animals and adherence to aseptic techniques (glassware, media, etc.) Antibiotics cannot be used indiscriminately; in some instances they may be more toxic to the invertebrate than to the microbes.

CRUDE CULTURES

Costello and Henley (1971), Silvan (1966) and Needham et al. (1959) give methods for obtaining, handling and rearing invertebrates in crude cultures. A fairly comprehensive view of recent advances in culture techniques for invertebrates is available through the published proceedings of the International Helgoland Symposium on the Cultivation of Marine Organisms and Its Importance for Marine Biology (Kinne and Bulnheim, 1970).

In successful crude cultures of invertebrates, fortuitously favorable bacteria, flagellates and other protozoa dominate the system. Such cultures last as long as the biotic balance is not upset by changes of temperature, salinity, and concentration of inorganic and organic nutrients. Equilibria observed in natural habitats are seldom matched indoors. Often the most troublesome members of the original biota -- species of the bacteria, fungi and protozoa -- take over, eventually killing the invertebrates.

However small the inoculum of microorganisms originating from the water sample, the gut and body surfaces of invertebrates, under laboratory conditions it may give rise to bacterial populations far denser than those in nature. Freshly collected seawater may have 1 to 10^6/ml viable bacteria. Seawater may be antimicrobial, yet samples may carry $\sim 10^2$/ml bacterial cells (Jones, 1963). Wood (1965) tabulates bacterial counts from different seas; clearly, population densities may differ widely.

Few bacteria are freely suspended in natural waters. Zobell (1943) suggested that dissolved organic substances in seawater are too dilute to support large populations of microbes but that suspended matter (detritus and plankton) with its absorbed organic substances may provide optimal substrates for bacterial growth. In crude cultures bacteria thrive on surfaces of laboratory vessels (Zobell and Anderson, 1936). Particles which may adsorb organic substances form spontaneously in stored seawater (Sheldon et al., 1967). Formation of a wide variety of particulate matter can be hastened by bubbling air through seawater (Riley, 1963). In all instances the end result is increased bacterial action.

Zobell and Felthan (1934) found that the typical seawater community may contain representatives of all major groups of microorganisms. Gram-negative rods may predominate. Gram-negative and gram-positive cocci occur but are less frequent. Ammonifiers and glucose fermenters may prove most troublesome in static cultures; the former may overload the system with NH_4 and the latter may over-acidify. Acidification of the medium enhances mold growth. Excessive growth of bacteria and molds may be deleterious to non-bacteria feeders. Recently, Murchelano and Brown (1970) carried out an extensive seasonal survey of the bacterial heterotrophs occurring in the water column of an oyster bed in Long Island Sound. Samples were taken bi-weekly for one complete biological year. The bacterial densities ranged from 1.2×10^3/ml in the summer to 3.8×10^4/ml in the winter. Of the eight genera identified, 92.3% were <u>Achromobacter</u>, <u>Flavobacterium</u> and <u>Pseudomonas</u>. The pseudomonads dominated the summer flora. Bacteria exhibiting proteolytic, lipolytic and amylolytic activity were present independent of season.

Ultimately, the growth of microorganisms in crude cultures depends on nutrients in the water sample. These vary with locality and season. Seawater, even if depleted of all potentially nutritive substances before its use in crude cultures, can be counted on to accumulate more nutrients very rapidly from the metabolic activities of the phytoplankton (Fogg, 1966) and zooplankton; for example, although Corner and Newell (1967) disagree, Johannes and Webb (1965) estimated that zooplankton under experimental conditions excreted from 2.5 to 30.5 mg α-amino acid nitrogen per gram dry weight. Antibiotics are useful to prevent the growth of microorganisms under these conditions.

Methods for the identification and control of harmful microbial activities in aquaria intended for mass culture of organisms are outlined in <u>Fish and Invertebrate Culture</u> (Spotte, 1970).

ANTIBIOTICS IN CRUDE CULTURES

Inspection of the literature suggests that with few exceptions, antibiotics were used in a variety of experimental procedures without first gauging their limitations. Some typical applications of antibiotics for the control of microbes in crude cultures of invertebrates were summarized in Table I. Listed are: the primary organism in culture, the antibiotics and concentration used, the reason for their use and the appropriate bibliographical references.

Attempts at large scale cultures of oysters and clams for experimental and commercial purposes (Loosanoff and Davis, 1963) have led to speculations on the large scale use of antibiotics to curb sudden fungal and bacterial infections. Davis and Chanley (1956) significantly reduced the mortality of bivalve larvae with antibiotics. Streptomycin 100 µg/ml, Combistrep (a mixture of dihydrostreptomycin and streptomycin sulfate) 125-200 µg/ml, or Sulmet (sulfamerazine) 33 µg/ml and chlortetracycline 30 µg/ml, added to crude cultures, controlled dangerous bacteria. Some species of bacteria associated with oysters and clams harmed the shellfish, especially the larvae. Guillard (1959) isolated two strains from bacterial swarms on dying clams. These isolates singly, together,

Table I

Typical Application of Antibiotics in Crude Culture

Organism	Antibiotic	Concentration µg/mℓ	*Reason for Use A	B	C	Reference
Protozoa						
Tintinnopsis sp.	penicillin	50	✓			Gold, 1966
	streptomycin	50				
Hexamita inflata	penicillin	5×10^4 I.U.		✓		Khouw, et al., 1968
	dihydrostreptomycin	50				
	mycostatin	2×10^3 I.U.				
Cnidaria						
Hydra oligactis	Fumidil B^2	200			✓	Claybrook and Eakin, 1960
Chlorohydra viridissima						
Tubularia crocea	penicillin	100-125			✓	Fulton, 1959
	streptomycin	100-125				
Platyhelminthes						
Dugesia dorotocephala	Aureomycin	100			✓	Miller and Johnson, 1959
	chloramphenicol	100				
	tetracycline	100				
	streptomycin	50				
	neomycin	50				
	erythromycin	50				
	Albamycin	50				
	Vancomycin	5				
	M-16655	5				
	penicillin	1×10^2 I.U.				
	trypaflavin	5				
	filipin	5				

* A = use in crude culture
  B = to axenize
  C = use other than in culture

Table I (cont'd.)

Organism	Antibiotic	Concentration µg/mℓ	Reason for Use A	Reason for Use B	Reason for Use C	Reference
Rhynchocoela *Lineus vegetus Amphiporus formidabilis*	streptomycin	25			✓	Tuker, 1959
Rotifera *Lecane inermis*	penicillin streptomycin	1×10^3 I.U. 100		✓		Dougherty et al., 1960
Mollusca (Bivalvia) *Pinctata maxima*	neomycin streptomycin	33.5 I.U. 250	✓			Minaur, 1969
Mya arenaria	penicillin streptomycin	50 I.U. 100	✓			Stickney, 1964
Mytilus edulis	penicillin streptomycin	50 50	✓			Bayne, 1965
Ostrea edulis (adult)	penicillin streptomycin	50 I.U. 250	✓			Walne, 1963
Ostrea edulis (larvae)	penicillin streptomycin	50 I.U. 50	✓			Walne, 1958
Crassostrea virginica	cycloheximide	50			✓	Ray, 1965
"	Sulmet	33	✓			Calabrese and Davis, 1967

Table I (cont'd.)

Organism	Antibiotic	Concentration µg/mℓ	Reason for Use A	B	C	Reference
Mercenaria mercenaria	penicillin streptomycin chloramphenicol	100 100 50		✓		Hidu and Tubiash, 1963
"	penicillin streptomycin chloramphenicol	670 50 5		✓		Millar and Scott, 1967
Mollusca (Gastropoda) Crepidula fornicata Nassarius reticulatus	penicillin streptomycin	50 I.U. 50	✓			Pilkington and Fretter, 1970
Arthropoda (Crustacea) Decapoda Penaeus aztecus	penicillin dihydrostreptomycin	50 I.U. 50	✓			Cook and Murphy, 1966
Cirripedia Balanus eburneus	penicillin	$2-4 \times 10^4$ I.U.	✓	✓		Costlow and Bookhout, 1957
Copepoda Calanus finmarchicus	chloromycetin streptomycin	50 50			✓	Marshall and Orr, 1958
Temora stylifera Oncaea sp. Euterpina acutifrons	penicillin streptomycin	6.5 6.5		✓		Neunes and Pongolini, 1965
Invertebrates Many taxa	tetracycline streptomycin penicillin	40 55 500 I.U.			✓	Stephens, 1967

or mixed with ten other clones of nonpathogenic bacteria, killed larvae of Venus mercenaria. Penicillin 50 µg/ml and streptomycin sulfate 50 µg/ml inhibited the two pathogenic isolates.

Walne (1958) reported that seawater collected off Brittany fluctuated widely in bacterial count. During the first few days after the water was brought into the laboratory, the counts increased. The bacteria then gradually diminished; presumably they were exhausting the nutrients. Poor Ostrea edulis spatfalls were correlated with large bacterial counts. When penicillin (50 units/ml) and streptomycin (50 µg/ml) were added, bacterial counts declined almost at once and more larvae survived. Similarly, Costlow and Bookhout (1957) obtained better survival and development of Balanus eburneus when 400 units/ml of penicillin was added to the seawater. For the successful development of decopod brachyuran larvae, Costlow and Bookhout used 200 units/ml of penicillin.

Pelagic copepods are notoriously difficult to keep in the laboratory. Two simple expedients permitted culture of the copepods Temora stylifera, Oncaea sp. and Euterpina acutifrons (Neunes and Pongolini, 1965). To the seawater was added 3.7 mg% EDTA and antibiotics. Best survival was obtained with 6.5 µg/ml of penicillin alternated weekly with 6.5 µg/ml of streptomycin.

These experiences suggest that antibiotics may be indispensable. Nevertheless, it is possible to culture many invertebrates without the aid of antimicrobial agents. Many crustacea feed on flagellates and bacteria. Several non-pelagic copepods require the bacteria. In crude cultures, the harpacticoid Tigriopus japonicus grew on Platymonas maculata supplemented with unknown bacteria; the same Platymonas in bacteria-free culture was insufficient (Provasoli et al., 1959)

Bacteriostasis may be obtained also with a healthy population of algae. Many produce antibacterial substances (review: Sieburth, 1964).

Streptomycin and penicillin appear to have been employed most often, perhaps because of their widespread medical uses. Of the two, streptomycin is the better suited for crude cultures of marine invertebrates. It is a wide-spectrum antibiotic with relatively long half-life, and not readily deactivated at alkaline pH's. Penicillin is not a wide-spectrum antibiotic, has a very short half-life, and is quickly inactivated in alkaline media. By adding penicillin alone to raw cultures of invertebrates, a limited number of the microorganisms are inhibited. Penicillin-resistant forms left without competitors have their growth accelerated. This effect is well illustrated by the experiments of Neunes and Pongolini (1965): penicillin added to seawater inhibited microbial growth for only a very limited period. At zero time, water treated with penicillin (6.5 μg/ml) contained 7×10^3 cells/ml; untreated water had 4×10^4 cells/ml. By the 24th hour, the bacterial populations were numberically equal in treated and untreated cultures. On the 38th hour, the penicillin-treated water had more viable cells (1.6×10^6) than the untreated water (1×10^6). The antibiotic may have inhibited the growth of some species (gram-positive) and favored others (gram-negative), or it may have inhibited all forms (unlikely) until it was inactivated by unfavorable pH, temperature, and salinity. Other wide-spectrum antibiotics may prove more effective in long-term crude cultures. Some are considered in the following pages.

Selection of Antibiotics

Under most agnotobiotic culture conditions, no one antibiotic would be expected to inhibit all microbial growth. One may assume that: 1) heterogeneous populations of microorganisms associated with cultures of marine invertebrates are predominantly gram-negative while those of fresh water cultures include both gram-positive and gram-negative forms; 2) antibiotics with relatively long half-lives (20 or more days) are preferable to antibiotics whose decay proceeds rapidly. From this the antimicrobial agents capable of inhibiting the greatest number of microorganisms for the longest time may be tentatively chosen from their known antimicrobial spectra and half-lives. Accordingly, the properties of common antibiotics were summarized from the literature (Baron, 1950; Spector, 1957; Goldberg, 1959; Lorian, 1966). Included in Table II are: a) spectra, b) minimal and maximal levels inhibiting growth of bacteria, fungi, and protozoa; in Table III are: c) pH at which is expected maximal activity, d) stability of the antibiotics in aqueous solution, e) emergence of resistant strains and cross-resistances.

Antibiotics with long half-lives -- streptomycin, chloramphenicol, kanamycin, and neomycin, alone or in combination -- should repress most marine and fresh water bacteria. Polymixin and novobiocin appear able to inhibit some gram-negatives and gram-positives respectively. Since both have narrow spectra they should not be used singly if the entire microflora is to be inhibited. Dihydrostreptomycin is preferred to streptomycin because the latter is deactivated by many factors; anaerobiosis, reducing substances, and pH's greater than 7.8. Strains resistant to dihydrostreptomycin emerge quickly, therefore for total repression of microbial growth it should not be used alone. Polymixin would seem an ideal choice for crude cultures since it is highly effective against gram-negative bacteria. Neomycin, kanamycin, and chloramphenicol are also suitable but their concentration needs careful regulation since they are toxic to the invertebrate.

Table II

Biological Activity

Antibiotic	Spectra (1) G+	G⁻	Range (2) μg/ml Bacteria	Fungi	Protozoa
Chlortetracycline (3)	+	+	0.002-50		25-1000
Bacitracin	+	±	0.002-125		
Carbomycin	+	±	0.01-12		32-250
Chloramphenicol (3)	+	+	0.06-50		125-2000
Colistin	-	+	0.5-50	20	125
Candicidin				0.5-50	
Erythromycin	+	±	0.003-200		
Kanamycin	+	+	0.3-500		
Nystatin				1-13	250
Neomycin	+	+	0.2-100		43-3000
Novobiocin	+	±	0.02-200	10-1000	125
Oleandomycin	+	+	0.08-50		
Penicillin	+	±	1-5000		
Polymixin B	-	+	0.02-50	125-250	125
Streptomycin (3)	+	+	0.5-300		
Tetracycline (3)	+	+	0.05-7.5		62-250
Oxytetracycline (3)	+	+	0.002-50		31-250
Trichomycin				0.6-10	250

(1) Effectiveness of antibiotic against gram-positive (G+) and gram-negative (G⁻) bacteria; (-) no activity; (±) little activity, effective against a few representative species; (+) activity, effective against most representative species.

(2) Range in μg/ml of antibiotic which inhibits microbial growth.

(3) Known as broad-spectrum antibiotic because active against gram-negative, gram-positive and other microorganisms.

Blank indicated lack of data.

Table III

Antibiotic: Stability and Activity

Antibiotics	pH*	Stability in aqueous solutions			Develop** resist.	Cross-*** resist.
		Days	pH	°C		
1. Chlortetracyclin	6.0-6.6	14	2.5-3	25	slow	4,16,17
2. Bacitracin	6.0-6.6	14	5-7	35-37	slow	none
3. Carbomycin		11	5-7	25	yes	7,12
4. Chloramphenicol	7.4-8.0	30	6-8	30	slow	1,14,16,17
5. Colistin	7.0-8.0	16	7-7.8	20	slow	
6. Candicidin		7	7	4		
7. Erythromycin	7.4-8.0	1	7-8	25	rapid	3,12
8. Kanamycin	7.4-8.0	30	7.8	4	rapid	10
9. Nystatin						
10. Neomycin	7.4-8.0	30	2-9	25	slow	8,15
11. Novobiocin	6.0-6.6	60	7-10	24	rapid	none
12. Oleandomycin	6.0-6.6	1	5-7	25	rapid	3,7
13. Penicillin	6.0-6.6	3	6-7	25	yes	
14. Polymixin		365	6-7	37	slow	5
15. Streptomycin	7.4-8.0	90	3-7	25	rapid	8,10
16. Tetracycline	6.0-6.6	3	7	37	slow	1,14,16,17
17. Oxytetracycline	6.0-6.6	7	7	25	slow	1,16
18. Trichomycin					none	

Blanks indicate lack of data.

*pH range for maximal antimicrobial activity.

**Bacterial strains develop resistance to the antibiotic.

***Microorganisms which have become resistant to an antibiotic numbered in column one also acquire resistance to antibiotics whose numbers are listed in this column.

Highly effective antibiotics, with short half-lives, may be employed successfully if one is willing to add fresh preparations periodically to the cultures. Chlortetracycline and oxytetracycline (wide-spectrum), bacitracin, and carbomycin (limited spectrum) are deactivated very quickly in alkaline media. They are most active at pH 6.0 - 6.6. In seawater (pH 8.0 or above) their half-lives would be much shorter. Colistin would remain effective in seawater.

Antibiotics with very short half-lives -- oleandomycin and tetracycline (wide spectrum); erythromycin and penicillin (limited spectrum) -- although highly effective against sensitive microorganisms, are not likely to exert much control in long lasting crude cultures.

Natural populations of microorganisms invariably contain fungi, usually growing slowly when bacteria predominate. However, if the bacteria are inhibited, fungal growth is invariably enhanced. A fungicide would be included in antibiotic mixtures. Nystatin is routinely added to tissue cultures. Lechevalier (1960) reported that amphotericin B was most effective against filamentous forms; candicidin was especially inhibitory to yeast; nystatin was the least active against all forms tested. Minimal inhibitory doses were in µg/ml, 0.009-0.5; 0.001-0.7 and 1.6-5.7, respectively. These fungicides appear more active at near neutral pH's, perhaps due to the reduced vigor of the fungi at pH's greater than 5. Some yeast can be inhibited by 2.5-10 µg/ml of trichomycin. Most fungicides, however, are often more toxic to the invertebrate than to the fungi.

Antibiotics now in commerce were developed mainly to combat pathogens common in clinical work. Therefore, the choice of which antibiotic to add to inhibit complex communities of unknown microorganisms should be preceded by: 1) characterization of the antibiotic sensitivity of the microbial population at hand; 2) determination of the toxicity of the few most effective antibiotics to the invertebrates, alone or, 3) in combination; and finally, if the invertebrates are to be fed with phytoflagellates while undergoing prolonged treatment, 4) the concentrations of antibiotics must be such as not to inhibit the food organisms.

1) *Susceptibility of Microorganisms*. The agar sensitivity-disc method was designed for simultaneous testing of many antibiotics. An example of an actual determination follows: The susceptibilities of the microorganisms in samples of pond and bay water were tested in six different nutrient agar media prepared as follows: Nutrient agar (Difco) 2.3% was dissolved in: 1) distilled water, 2) filtered pond water, 3) 25%, 4) 50%, 5) 75% and 6) 100% filtered seawater. The media were dispensed into (125 x 25 mm

Pyrex) screw cap tubes, autoclaved for 20 minutes at 120°C/15 lb. pressure and stored at room temperature. Before use, the media were melted, cooled to 45°C, 10 ml aliquots were poured into petri dishes (Falcon 90 x 15 mm) and allowed to set for 1-2 hours. Gunhel et al. (1961) reported that the agar must be allowed to cool at room temperature to avoid killing thermosensitive marine bacteria. Natural water samples (10 ml, marine or fresh water) were added first to an equal volume of Dextrose broth (Difco) and allowed to stand four to six hours at room temperature (24-26°C). The cultures were shaken, and 1 ml aliquots were spread on the nutrient agar plates. One hour after surplus fluid was removed by tilting the petri dishes and withdrawing the excess with sterile Pasteur pipettes. Antibiotic sensitivity discs (Difco) were then pressed on the agar surface. The cultures were incubated for 24-48 hours at room temperature and the zones of inhibition were recorded as they developed.

The size of an inhibitory zone is a function of both antimicrobial activity and the diffusibility of the antibiotics in agar. Therefore, bacteria can be considered sensitive even if the zones observed are only a few mm in diameter. To avoid overlapping of zones of inhibition, medium-potency discs (5-10 µg) were used. These permitted testing of more discs per plate than usual. Media carrying various dilutions of seawater were prepared also with commercially available enrichment formulae such as: N.I.H. Agar B258, AC Medium B316, and Penassay Base Agar B263 (Difco). The former two are widely employed for sterility testing, the last one is recommended for estimating antibiotic potency.

If the effect of different salinities on the microorganisms and the antibiotics was not being tested, microbial sensitivity was assayed directly in marine media; Bacto-Marine Agar 2216 or Bacto MOF Medium, both from Difco. Other media such as STPs and St-3s (Tatewaki and Provasoli, 1964) may be more suitable for growing bacteria normally inhibited by high concentrations of organic matter.

The results of a 48-hour antimicrobial assay test are reported in Table IV. They show the effect of eight antibiotics on the bacterial flora of pond and bay water.

As expected, streptomycin, chloramphenicol, kanamycin, and neomycin inhibited most fresh water and bay water organisms plated respectively on fresh, brackish and seawater media. However, there were species in the fresh water cultures resistant to these four wide spectrum antibiotics (Table IV, PW, column 1 and 2). Streptomycin was obviously ineffective against bacteria from the same pond water when these were plated on seawater media (Table IV, PW, columns 4, 5, and 6). Since streptomycin was very effective a-

Table IV

Antimicrobial Assay 48 Hours

Antibiotics Sensitivity Discs		PW Pond water bacteria on nutrient agars						SW Sea water bacteria on nutrient agars					
		1	2	3	4	5	6 *	1	2	3	4	5	6
Chloramphenicol	10 μg	++/r	++/r	++/r	++/r	++/r	++/r	++	++	++/r	++/r	++/r	++/r
Streptomycin	5 μg	++/r	++/r	-	-	-	-	+/r	+/r	+	+	+	+
Kanamycin	10 μg	+/r	+/r	+/r	±/r	±/r	±/r	+	+	+	+	+/r	+/r
Penicillin	5 units	+/r	+/r	+/r	±/r	-	-	+/r	±/r	±/r	-	-	-
Neomycin	10 μg	+/r	+/r	+/r	+/r	+/r	+/r	±/r	±/r	±/r	+	+	+
Tetracycline	10 μg	+/r	+/r	+/r	±/r	-	-	++	++	+/r	+/r	-	±/r
Erythromycin	5 μg	-	-	-	±/r	±/r	±/r	+	+/r	+/r	+/r	+	+
Novobiocin	10 μg	-	-	-	±/r	±/r	±/r	++	+	+/r	-	-	-

(*) Numbers given in text for the different types of water employed to dissolve the nutrient agar.

(-) No inhibition

(±) Negligible inhibition

(+) Zone of inhibition 1-5 mm

(++) Zone of inhibition 5-10 mm

(/r) Scattered colonies within zone of inhibition; indicate presence of resistant strains.

gainst seawater bacteria growing on seawater agar media (Table IV, SW, columns 3, 4, 5, and 6), the composition of the bacterial population of fresh water must have changed in salty media.

These same four antibiotics seemed more effective when they were employed against bay water bacteria plated on the seawater media (Table IV, SW, columns 5 and 6). There were species of marine bacteria resistant to chloramphenicol and kanamycin but none to streptomycin and neomycin, at least within the first 48 hours of incubation.

Penicillin and tetracycline repressed fresh water and bay water organisms growing on fresh water media (Table IV, PW, columns 1 and 2), but had little, if any, inhibitory activities against marine bacteria when these were growing on full strength marine media (Table IV, SW, columns 5 and 6). Penicillin is mainly active against gram-positive bacteria; marine bacteria are mostly gram-negative. Tetracycline, a wide spectrum antibiotic, inhibited marine and fresh water bacteria only when these were plated on fresh water media. Poor activity in media containing seawater (Table IV, PW, SW, columns 5 and 6) may be due to the chelating properties of tetracyclines toward Mg^{++} and other heavy metals. Erythromycin was particularly effective against marine bacteria plated on full strength marine media (Table IV, SW columns 5 and 6) but not against fresh water organisms. This antibiotic has a very short half-life so that it may be of little use in long term agnotobiotic cultures. Novobiocin was unexpectedly ineffective.

2. _Tolerance of Invertebrates_. Antibiotics found effective against the microorganisms, as shown by the 48-hour antimicrobial assay, were tested for toxicity to the invertebrates. Whenever possible, at least two stages of the life cycle of the animals (nauplii and adults) were exposed to varying concentrations of antibiotics. For example, nauplii and adults of _Artemia salina_ were inoculated in tubes containing 10 ml of synthetic media and the antibiotics.

Organic substances were excluded to avoid excessive bacterial growth. The antibiotics were added from freshly prepared Seitz-filtered stock solutions. The tubes were incubated at room temperature in racks illuminated (160 ft. candles) continuously by white fluorescent lamps. The effect of the antibiotics could be detected by the abnormal behavior of the second naupliar antennae, or the arhythmic beating of the adult phyllopodia and subsequent death. The tolerances of _Artemia_ are reported in Table V.

To assess the long range effect of sublethal doses of antibiotics, algae were added to the culture vessel in order to keep the invertebrates fed _ad libitum_. Thus it became possible to de-

Table V

ARTEMIA Tolerance

Antibiotic μg/ml

Antibiotics	Nauplii		Adult	
	Safe*	Inh.**	Safe	Inh.
Penicillin***	150	200	150	200
Streptomycin	200	300	250	300
Polymixin	30	50	40	50
Chloramphenicol	25	50	25	50
Naramycin	10	15	10	20
Candicidin	2.5	3	2.5	5
Nystatin	10	15	10	30
Kanamycin	250	500	250	750
Neomycin	5	10	5	15
Nalidixin	50	100	50	150
Tetracycline	5	7.5	5	9.5
Trichomycin	25	50	25	75
Methenamine-mandalate	500	1000	500	1000

* Non-inhibitory maximal concentrations, no apparent effect on *Artemia* adults and nauplii after 37 days.

** Inhibition of growth was evident on the seventh day. Animals would recover if transferred to media <u>without</u> antibiotic.

*** Penicillin 1 mg = 1650 units; polymixin 1 mg = 7760 units.

termine concurrently the tolerance of the algae. These may bleach within a few hours in toxic concentrations of antibiotics.

The tolerance data reported in Table V can suggest only approximate limits, and is not to be considered universally applicable because the toxicity of the antibiotics to the invertebrate is modified by many factors: number of microorganisms in the culture, dissolved antibiotic-binding compounds, pH's and temperatures favoring inactivation. Penicillin was not toxic at 200 µg/mg to the Artemia, probably because it was inactivated before much was taken up. But it was toxic to Daphnia at 40 µg/ml (D'Agostino and Provasoli, 1970).

The dissimilar tolerances of these two crustaceans points out the need to determine the antibiotic profile of each invertebrate prior to the use of antibiotics in crude cultures.

3. Tolerance to Antibiotic Mixtures. Mixtures were prepared with antibiotics bacteriocidal to dissimilar groups of organisms but least toxic to the invertebrates. The toxicity of the mixtures to the invertebrates was tested (Table VI). Used as indicated in the legend, the mixtures were lethal to microorganisms. Mixture 2M was tolerated for limited periods only by the Artemia. But mixture 3M used at 0.1 ml/ml of media was seldom inhibitory. Both larvae and adults remained unaffected. The larvae fed ad libitum developed normally into adults.

4. Tolerances of Typical Phytoflagellates. The susceptibility of algae serving as food organisms to invertebrate cultures is given in Table VII and Provasoli et al. (1951). An extensive treatment of phytoflagellate sensitivities is in Krauss (1962).

ANTIBIOTICS FOR AXENIZATION

Agnotobiotic culture is only the first of a series of steps which eventually lead to the rearing of an invertebrate in axenic and chemically defined conditions. Axenic cultures permit more exacting studies of trophic and chemosensory exchanges occuring between species.

Selection of Antibiotics

Choice of an antibiotic or a mixture of antibiotics for the axenization of invertebrates is simpler than the corresponding choice of a mixture to be used continuously in crude cultures. Axenization is achieved by exposure of eggs, larvae or adults for brief periods to massive doses of antibiotics, coupled with serial

Table VI
ARTEMIA Tolerance to Antibiotic Mixes

µg/ml

	Mix 2M[1] µg/ml	Mix 3M[2] µg/ml
Penicillin [3]	1000	
Streptomycin	1000	1000
Polymixin [3]	375	37
Chloramphenicol	250	25
Candicidin	25	2.5
Methenamine-mandalate	1000	100
Tetracycline		
Neomycin		

(1) Designed for axenization of viviparous larvae. Used at 0.1 ml/ml of media for adult *Artemia*, 0.05 ml/ml for nauplii, both will die within 2-3 days. Serial washes must be completed within this time limit.

(2) Used at 0.1 ml/ml of media this mixture is tolerated indefinitely by *Artemia* nauplii and adults.

(3) Penicillin 1 mg - 1650 units; polymixin 1 mg - 7760 units.

Table VII
Concentration of Antibiotics Tolerated by Phytoflagellates*

Algae	Pen. Units/ml	Bac. Units/ml	DhS. Units/ml	Aur. µg/ml	Ter. µg/ml	Chl. µg/ml	Pol.** Units/ml
Cyanophyta***		10		6	0.5		30
Rhodophyta		20	80	10	10	1	10
Chrysophyta	500	30	20	5	20	50	40
Pyrrophyta	1000	40	10	0.5	10	5	0.5
Euglenophyta	1000	30	60	40	80	50	
Chlorophyta	1000	30	0.5	2.5	10	30	

Blanks indicate lack of data.

* Concentrations as given are equivalent to half the maximum inhibitory concentrations.

** Pen. = penicillin; Bac. = bacitracin; DhS. = dihydrostreptomycin; Aur. = Aureomycin (chlortetracycline); Ter. = Terramycin (Oxytetracycline); Chl. = chloramphenicol (Chloromycetin); Pol. = polymixin.

*** See abstract for list of test species (Provasoli et al., 1951).

rinses. Hence, the half-lives of the antibiotics and their possible long term effects on food organisms and the invertebrates are not critical factors. Preferably, once axenization is accomplished, antibiotics should not be carried in the culture media. This practice is frowned upon by some investigators and commended by others, depending upon the type of study under way (review: Mickelsen, 1962). If the washing method with germicides or antibiotics did not free the invertebrate of microorganisms, addition of antibiotics in the growth medium is useless and often detrimental. Antibiotics may slow the growth of the invertebrate by inhibiting the food organisms, or by acting as binding agents may deprive the organisms of needed growth factors. If bacteria survive through the washings, the invertebrate is not axenized, even though bacterial growth is temporarily suppressed.

A test of the activity of the antibiotics against the prevailing bacterial flora remains a prerequisite to obviate use of needless antimicrobial agents, some of which may antagonize one another. A test of the tolerance of the invertebrate to single and/or combinations of antibiotics is advisable to forestall the possibility of killing the organism that is being washed free of bacteria.

Few aquatic metazoans are now maintained bacteria-free. Of these, most were made axenic by washing larvae or eggs with germicides or antibiotics. Usually the life cycle and nature of the integument of the organisms dictated the method. Old larvae and adults are seldom suitable stages for axenization. The latter, however, after exposure to antibiotics, may serve as donors of relatively clean cells, tissues, eggs, and larvae.

Coelenterate tissue cultures were established by washing whole organisms with a solution containing per ml: penicillin 500 units, streptomycin 500 µg and nystatin 50 units (Phillips, 1961). The antibiotic mixture was not carried over into the growth media because it was toxic to isolated cells; streptomycin may have been too high.

To obtain planaria free of bacteria, Miller and Johnson (1959) washed head regenerates of Dugesia in media containing practically every antibiotic known. Asevin and Buchsbaum (1961) reported that exposure of adult worms to 2500 units/ml of penicillin + 2500 units/ml of streptomycin sufficed to establish planarian cell cultures. However, neomycin 100 µg/ml was incorporated in the maintenance media to inhibit surviving bacteria.

To have large numbers of bacteria-free nematode larvae for nutritional studies, Fatt and Dougherty (1963) treated fertilized adult Caenorhabditis elegans through four changes of media containing 1000 units/ml of penicillin and 100 µg/ml streptomycin.

The adults were rinsed free of antibiotics and were incubated in an appropriate substrate to allow deposition of larvae.

A simpler method, which has yielded good results, consists of incubation of eggs having thick protective membranes (most insects and some aquatic metazoa) in strong germicides such as Merthiolate 1:1000; $AgNO_3$ 1:1000; phenol 1:50; hypochlorite 0.5-1.2%; alcohol 90%; and formalin 3%. Immersion from 1 to 10 minutes in one of the above killed all accompanying microorganisms. Eggs of nematodes (Weinstein and Jones, 1959) withstood treatment with 0.25 to 1.25% Chlorox for 10 minutes. However, to insure sterility, 300 units/ml of penicillin and 300 units/ml of streptomycin were added to the hatching medium. Mictic eggs of rotifers (Nathan and Laderman, 1959) survived one minute immersion in 0.005-0.01% hypochlorite if followed by six serial washes in sterile media. The amictic eggs did not withstand this treatment. Gibor (1956) and Provasoli et al. (1959) washed wintering eggs of Artemia salina in Merthiolate (1:1000) for ten minutes and several washes in sterile seawater.

Invertebrate eggs, with soft protective membranes should not be expected to survive immersion in powerful oxidizers. It is possible to adjust the concentration of oxidant and duration of exposure so as not to damage the eggs, but this also lets bacteria survive. Chernin (1959) washed eggs of Australorbis glabratus with 0.0025% (v/v) Chlorox for ten minutes. He then rinsed them repeatedly with antibiotic solutions. Penicillin and/or streptomycin were retained in the hatching medium. Streptomycin 50-100 µg/ml inhibited growth of the larvae. Procedures which utilize first a bactericide then antibiotics in the washing, hatching and culture media result from inadequate use of antimicrobial agents.

Eggs can be washed free of bacteria merely with antibiotic mixtures. Nathan and Laderman (1959) successfully axenized the amictic eggs of rotifers through several baths containing per ml: penicillin 100 units; streptomycin 100 units; chlortetracycline 50 units, and cycloheximide (Actidione) 100 µg. Seventy-two hours exposure to the antibiotics was followed by extensive serial washing in sterile media. Eggs of Tigriopus japonicus could be hatched aseptically (Provasoli et al., 1959) simply by rinsing repeatedly the excised egg sacs with sterile seawater. However, 50% of the isolated nauplii were not bacteria-free. Addition of six antibiotics per ml: dihydrostreptomycin 100 µg; K penicillin G 100 units; polymixin B SO_4 30 units; tetracycline 5 µg; chloramphenicol 10 µg; and neomycin 60 µg, in the washing procedure only, allowed collection of 100% aseptic nauplii.

If eggs and egg sacs are not available, the next best choice seems to be newly hatched larvae. Axenic cultures of a parthenogenetic tetraploid strain of Artemia salina were established by subjecting gravid females to periodic washes through sterile media and antibiotics (D'Agostino and Provasoli, 1968). The females with eggs in their ovisacs were transferred to 30 ml of a salt solution to which had been added 0.1 ml/ml of antibiotic mix 2M (Table VI) and the food organism. They were transferred every hour during the working day to similarly prepared sterile media. This effectively prevented accumulation of fecal matter and bacteria. On the second or third day each nauplius, as it emerged from the ovisac, was washed through ten successive baths (30 minutes stay in each of 10 ml of the same salt solution but with only 0.05 ml/ml of antibiotic mix 3M. After the last wash the food organism Dunaliella tertiolecta was added to the media; 24 hours later the nauplii were washed free of antibiotics and were transferred to STP (Provasoli et al., 1959) to test for sterility. Ten out of 30 larvae washed by this method proved bacteria free. A similar technique was used to obtain bacteria free larvae of Balanus eburneus (D'Agostino and Sheridan, 1969).

Axenic culture are not necessary for most experimental work. They are essential, however, in order to critically evaluate the nutritional requirements of predators and the nutritive value of the prey.

Guidelines

The following guides for axenization of invertebrates can be summarized. For organisms which produce wintering eggs with hard coverings, the eggs can be immersed directly in strong germicides rinsed and hatched in sterile media. Antibiotics are rarely necessary. If the organisms produce quick-hatching soft eggs or soft eggs in delicate sacs, these require the use of antimicrobial agents other than germicides. Eggs or egg sacs can be incubated in strong antibiotic mixtures, washed repeatedly and allowed to hatch in media containing mild doses of antibiotic mixtures. The larvae in turn can be rinsed and kept in sterile media with no antibiotics. If the invertebrates are viviparous, i.e., the eggs hatch in the ovisac and the larvae are subsequently released, the gravid females can be maintained in antibiotics until the larvae are freed. The latter then can be treated with antibiotics, washed and incubated in sterile media. If suitable food organisms are available, the larvae can be grown to adulthood in mono- or dixenic conditions.

One may choose at random a few antibiotics and empirically go through a number of washing procedures until a few organisms are obtained bacteria-free. A more rational approach, although it requires some preparatory steps (assay of microbial susceptibility and tolerance of the invertebrates), may permit isolation of greater numbers of bacteria-free organisms -- perhaps at the very first attempt.

OTHER USES OF ANTIBIOTICS

Anitbiotics have been used for a variety of reasons unconcerned with the improvement of rearing methods. For example, in order to rid invertebrates of symbiotic and parasitic infections complicating histological preparations and short-term respiratory studies, Claybrook and Eakin (1960) exposed Hydra oligactis and Chlorohydra viridissima for five weeks to Fumidil B 200 µg/ml.

Large bacterial populations may develop rapidly during extraction of organic crudes from homogenates of animal tissues. Fulton (1959) found that substances supposedly released into water by adult hydranths of Tubularia crocea, and reputed to inhibit regeneration of cut stems of the hydroid, were recovered only from preparations infected with greater than 2.3×10^8 bacterial cells per ml. If bacteria were kept less than 1×10^5 cells/ml with 100-250 µg/ml of penicillin and streptomycin (1:1), inhibitory substances were absent. Regeneration of hydranths was unaffected by antibiotics even at a 4-fold higher concentration.

Amputated invertebrates regenerate faster if the tissues are not under stress, i.e., either competing for O_2 with bacteria or being attacked by pathogens. Sengel (1960) employed penicillin to minimize bacterial interference with the regeneration of Dugesia lugubris. Similarly, Tuker (1959) added antibiotics to regenerating parts of nemertean worms. Regenerates of Lineus vegetus, Cerebratulus californiensis, and Amphiporus formidabilis tolerated high streptomycin. In the absense of antibiotics Tuker claimed that the bacteria inactivated worm extracts being tested for their ability to affect regeneration of totipotent tissues.

Marshall and Orr (1958) studied the influence of antibiotics on the respiratory, feeding, and excretory processes of marine copepods. Chloramphenicol and streptomycin were the most satisfactory although high concentrations of chloramphenicol in some experiments seemed to inhibit feeding of Calanus. Other antibiotics tested were not reliable for periods over a few days.

Stephens (1967), in an extensive study of uptake of dissolved organic matter by invertebrates, used a mixture seemingly well tolerated by representative species of 35 genera of 11 phyla of marine invertebrates and 12 genera of 6 phyla of fresh water invertebrates; Tetracycline 40 µg/ml, streptomycin 55 µg/ml, and 500 units/ml of penicillin.

Marine organisms were not affected for the duration of the experiment (96 hours). With fresh water organisms he observed that the disappearance of labeled amino acids from the media was reduced when the antibiotic mixture was used. Presumably the discrepancy in uptake was due to the bacteria -- perhaps the invertebrate's metabolism was inhibited, too. Fresh water organisms may be less tolerant than their marine counterparts.

Extrapolation from findings with bacteria-free organisms may be regarded as highly speculative on the assumption that the natural habitats are seldom aseptic. Nevertheless, bacteria-free cultures are unavoidable since they afford the only approach to the study of the absolute requirements of organisms.

REFERENCES

Asevin, K.D. and R. Buchsbaum, 1961. Observation on Planarian cells cultivated in solid and liquid media. Jour. Exp. Zool., 146: 153-161.

Baron, A.L., 1950. Handbook of Antibiotics. Reinhold Publishing Corp., New York, N.Y.

Bayne, B.L., 1965. Growth and the delay of metamorphosis of the larvae of Mytilus edulis (L.). Ophelia, 2(1):1-47.

Calabrese, A. and H.C. Davis, 1967. Effects of "soft" detergents on embryos and larvae of the American oyster (Crassostrea virginica). Proc. Nat. Shellfisheries Assoc., 57:11-16.

Chernin, E., 1959. Cultivation of the snail Australorbis glabratus under axenic conditions. Ann. N.Y. Acad. Sci., 77:237-239.

Claybrook, D.L. and R.E. Eakin, 1960. Nutrition requirements for asexual reproduction of hydra. J. Exp. Zool., 145:179-190.

Cook, H.L. and M.A. Murphy, 1966. Rearing penaeid shrimp from eggs to post-larvae. Proc. Southeastern Assoc. Gam Fish. 19th Ann. Conf., pp. 283-288.

Corner, E.D.S. and B.S. Newell, 1967. On the nutrition and metabolism of zooplankton. IV. The form of nitrogen excreted by Calanus. Journ. Mar. Biol. Assoc. U.K., 47:113-120.

Costello, D.P. and C. Henley, 1971. Methods for Obtaining and Handling Marine Eggs and Embryos. Marine biological Laboratory, Woods Hole, Mass., pp 247.

Costlow, J.D. and C.G. Bookhout, 1957. Larval development of Balanus eburneus in the laboratory. Biol. Bull., 112:313-324.

D'Agostino, A. and P.O. Sheridan, 1969. Balanus eburneus, factors affecting metamorphosis and larval axenization. Am. Zool., 9:618 (abstract).

D'Agostino, A.S. and L. Provasoli, 1968. Effects of salinity and nutrients on mono- and dixenic cultures of two strains of Artemia salina. Biol. Bull., 134:1-14

D'Agostino, A.S. and L. Provasoli, 1970. Dixenic culture of Daphnia magna, Straus. Biol. Bull., 139:485-494.

Davis, H.C. and P.E. Chanley, 1956. Effects of some dissolved substances on bivalve larvae. Proc. Nat. Shellfisheries Assoc., 46:59-74.

Dougherty, E.C., B. Solberg, and D.J. Ferral, 1960. Synxenic and attempted axenic cultivation of rotifers. Anat. Rec., 137:350-351.

Fatt, H.V. and E.C. Dougherty, 1963. Genetic control of differential heat tolerance in two strains of the Nematode Caenorhabditis elegans. Science, 141:266-267.

Fogg, G.E., 1966. The extracellular products of algae. Oceanogr. Mar. Biol. Ann. Ref., 4:195-212.

Fulton, C., 1959. Re-examination of an inhibitor of regeneration in Tubularia. Biol. Bull., 116:232-238.

Gibor, A., 1956. Some ecological relationships between phyto- and zooplankton. Biol. Bull., 111:230-234.

Gold, K., 1966. The role of ciliates in marine ecology. I. Isolation and cultivation of a member of the Order Tintinnida. Am. Zool., 6(4):39.

Goldberg, H.S., 1959. Antibiotics, Their Chemistry and Non-medical Uses. Van Nostrand Co., Inc., New Jersey.

Guillard, R.R.L., 1959. Further evidence of the destruction of bivalve larvae by bacteria. Biol. Bull., 117:258-266.

Gunhel, W., G.E. Jones, and C.E. Zobell, 1961. Influence of volumes of nutrient agar medium on development of colonies of marine bacteria. Helgoländer wiss. Meers., 8:85-93.

Hidu, H. and H.S. Tubiash, 1963. A bacterial basis for the growth of antibiotic-treated bivalve larvae. Proc. Nat. Shellfisheries Assoc., 54:25-39.

Johannes, R.E. and K.L. Webb, 1965. Release of dissolved amino acids by marine zooplankton. Science, 150:76-77.

Jones, G.E., 1963. Suppression of bacterial growth by seawater. In: C.H. Oppenheimer (ed.), Sym. on Marine Microbiol., Thomas, Springfield, Illinois, pp 572-579.

Khouw, B.T., H.D. McCurdy, and R.E. Drinnan, 1968. The axenic cultivation of Hexamita inflata from Crassostrea virginica. Can. J. Microbiol., 14:184-185.

Kinne, O. and U.P. Bulnheim (eds.) 1970. Cultivation of marine organisms and its importance for marine biology. International Helgoland Symposium, Helgoländer wiss. Meers., 20(1-4):1-721.

Krauss, R.W., 1962. Inhibitors. In: R.A. Lewin (ed.), Physiology and Biochemistry of Algae. Academic Press, New York, pp 673-685.

Lechevalier, H., 1960. Comparison of the in vitro activity of four polyenic antifungal antibiotics. Antibiotic Ann. 1959-1960, pp 614-618. Antibiotica, Inc., New York.

Loosanoff, V.L. and H.C. Davis, 1963. Rearing of bivalve mollusks. Advan. Marine Biol. 1:1-136. Academic Press, Inc., London.

Lorian, V., 1966. Antibiotic and Chemotherapeutic Agents in Clinical and Laboratory Practice. C.C. Thomas Publ., Illinois, 337 p.

Marshall, S.M. and A.P. Orr, 1958. Some use of antibiotics in physiological experiments in seawater. J. Marine Res., 17:341-346.

Mickelsen, O., 1962. Nutrition-germfree animal research. Ann. Rev. Biochem., 31:515-548.

Millar, R.H. and J.M. Scott, 1967. Bacteria-free culture of oyster larvae. Nature, 216:1139-1140.

Miller, C.A. and W.H. Johnson, 1959. Preliminary studies on the axenic cultivation of a planarian (Dugesia). Ann. N.Y. Acad. Sci. 77: 87-92.
Minaur, J., 1969. Experiments on the artificial rearing of the larvae of Pinctada maxima (Jameson) (Lamellibranchia). Aust. J. Mar. Freshwater Res., 20(2): 175-187.
Murchelano, R.A. and C. Brown, 1970. Heterotrophic bacteria in Long Island Sound. Marine Biology, 7: 1-6.
Nathan, H.A. and A.D. Laderman, 1959. Rotifers as a biological tool. Ann. N.Y. Acad. Sci. 77: 96-101.
Needham, J.G., P.S. Galtsoff, F.L. Lutz, and P.S. Welch, 1959. Culture Methods for Invertebrate Animals. Dover Publications, Inc., New York.
Neunes, H.W. and G.F. Pongolini, 1965. Breeding a pelagic copepod Euterpina acutifrons (Dana) in the laboratory. Nature, 208: 571-573.
Phillips, J.H., 1961. Isolation and maintenance in tissue culture of coelenterate cell lines. Biology of Hydra. Univ. of Miami Press, Florida, pp. 254-264.
Pilkington, M.C. and V. Fretter, 1970. Some factors affecting the growth of prosobranch veligers. Helgoländer wiss. Meers., 20: 576-593.
Provasoli, L., I.J. Pintner, and L. Packer, 1951. Use of antibiotics in obtaining monoalgal bacteria cultures. Proc. Am. Soc. Protozoa, 2: 6.
Provasoli, L., K. Shiraishi, and J.R. Lance, 1959. Nutritional idiosyncrasies of Artemia and Tigriopus in monoxenic cultures. Ann. N.Y. Acad. Sc., 77: 250-261.
Ray, S.M., 1965. Cycloheximide inhibition of Dermocystidium marinum in laboratory stocks of oysters. Proc. Nat. Shellfisheries Assoc., 56: 31-36.
Riley, G.A., 1963. Organic aggregates in seawater and the dynamics of their formation and utilization. Limnol. Oceanog., 8: 372-381.
Sengel, C., 1960. Culture in vitro de blastêmes de regeneration de Planaire. Jour. Embryol. Exp. Morphol., 8: 468-476.
Sheldon, R.W., T.P.T. Evelyn, and T.R. Parson, 1967. On the occurence and formation of small particles in seawater. Limnol. Oceanogr., 12: 367-375.
Sieburth, J. McN., 1964. Antibacterial substances produced by marine algae. Develop. Indu. Microbiol., Am. Inst. Biol. Sci., 5: 124-134.
Silvan, J., 1966. Raising laboratory animals. Natural History Press, N.Y.
Spector, W.S. (ed.) 1957. Handbook of Toxicology, Vol. II, Saunders Co. Phila., Penna., 264 p.
Spotte, S.H., 1970. Fish and Invertebrate Culture. Wiley-Interscience, Div. of John Wiley and Sons, Inc., N.Y., 145 p.
Stephens, G.C., 1967. Dissolved organic material as a nutritional source for marine and estuarine invertebrates. In: G.H. Lauff (ed.) Estuaries, Vol. 83: 367-373. AAAS Washington, D.C.

Stickney A.P., 1964. Salinity, temperature, and food requirements of soft-shell clam larvae in laboratory culture. Ecology, 45(2): 283-291

Tatewaki, M., and L. Provasoli, 1964. Vitamin requirements of three species of Antithamnion. Botanic Marina, 6:193-203

Tuker, M., 1959. Inhibitory control of regeneration in nemertean worms. J. Morphol., 105:569-600.

Walne, P.R., 1958. The importance of bacteria in laboratory experiments on rearing the larvae of Ostrea edulis. J. Mar. Biol. Assoc., U.K., 37:415-426.

Walne, P.R., 1963. Observations on the food value of seven species of algae to the larvae of Ostrea edulis. I. Feeding experiments. J. Mar. Biol. Assoc., U.K., 43:767-784.

Weinstein, P.P. and M.F. Jones, 1959. Development in vitro of some parasitic nematodes of vertebrates. Ann. N.Y. Acad. Sci., 77:137-162.

Wood, E.J.F., 1965. The Significance of Marine Microbiology. Reinhold Publishing Corp., N.Y.

Zobell, C.E., 1943. The effect of solid surfaces upon bacterial activity. J. Bacteriol., 46:39-56.

Zobell, C.E. and D.Q. Anderson, 1936. Observations on the multiplication of bacteria in different volume of stored water. Biol. Bull., 71:324-342.

Zobell, C.E. and C.B. Feltham, 1934. Preliminary studies on the distribution and characteristics of marine bacteria. Bull. Scripps. Inst. Oceanogr. Univ. Calif. Tech. ser., 3:279-296.

PART II

A REVIEW OF COELENTERATE LABORATORY CULTURE

A. Harry Brenowitz

Adelphi University Institute of Marine Science

Garden City, New York

The laboratory maintenance of some Coelenterates was established in the latter part of the 19th century. Ashworth and Annandale published in 1904, "Observations on some aged specimens of Sagartia troglodytes (later identified as Cereus pedunculatus) and on the duration of life in the Coelenterates." They made observations on specimens of the Anthozoan which were at least fifty years old. These specimens were collected on the coast of Iran before 1862. Sixteen of the original specimens were living when Ashworth and Annandale made their observations. They were kept in a bell jar about 13 inches in diameter and 9 inches in depth. The specimens had been under constant observation from 1862 to the time of the recorded observations. The specimens were later transferred to the Edinburgh Zoo and lived until 1942.

The significance of being able to maintain colonial hydroids and other Coelenterates in laboratory culture is the continual supply of specimens for study and experimentation. Thus, with an ample supply of specimens and the ability to control environmental conditions, researchers have been able to study systematics, growth, developmental biology, physiology, histochemistry, fine structure, nutrition, sexual and asexual reproduction, aging, encystment, and to some extent ecological relationships.

The studies of Browne (1897, 1907) focused on methods for growing the colonial hydroids, Syncoryne exima and Bougainvillia sp. Browne developed two systems for maintaining laboratory cultures; the bell jar with plunger and the continuous tube. The disadvantages of the two systems were that they provided only adequate conditions for a limited number of colonies and they required a relatively large amount of laboratory space. Neither disadvantage hindered Browne's work, and in spite of the fact that his cultures remained alive for a relatively short time, he was able to complete a series of observations on the growth rate of the organisms he cultured. Actually Browne determined the basic requirements for culturing the organisms; a suitable medium, a nutritionally adequate food supply, and a means for preventing the depletion of required substances or the accumulation of deleterious substances.

Since the work of Browne many researchers have developed methods for culturing and maintaining in the laboratory a variety of Coelenterates. Davis (1971) listed thirty-seven colonial hydroids maintained in laboratory culture. A number of Coelenterates in addition to the colonial hydroids have also been successfully cultured (Table 1).

Descriptions of several selected culture methods follow: Fulton (1960) grew colonies of Cordylophora lacustris attached to microscope slides slanted in beakers containing 100 ml of culture solution. He used a defined aqueous solution (CCS5) containing 0.05 M NaCl, 0.0001 M $KHCO_3$, 0.005 M $CaCl_2$, and 0.005 M $MgCl_2$ made up in demineralized water. Sodium, potassium, chloride, and calcium ions were absolute requirements, the absence of magnesium ions results in reduced growth rates, and the bicarbonate ions are required to serve as a buffer for the solution. Artemia was used as the food source. They were hatched on a daily basis, collected each day and washed. The colonies were fed to repletion for about an hour. The culture solution was changed one hour after feeding and again six to eight hours later. The colonies were maintained at $22°C$. A second culture medium used by Fulton was CVD. This medium consisted of the same chemical composition in distilled water plus 1.5×10^{-4} M disodium ethylenediamine tetraacetate to sequester the heavy metals. The two solutions, CCS5 and CVD, were used interchangeably.

Crowell (1957) working with Campanularia used running seawater over the microscope slides to which the organisms were attached. The water was cooled to $19°C$. The animals were fed twice a day for a five to ten minute period.

West and Renshaw (1970) utilized a closed circulatory system for culturing Clytia attenuata. A colony attached to a Macrocystis blade was tied to a glass slide and placed in a 800 liter capacity closed circulatory system. The return tube to the resin lined reservoir contained a glass-wool and sand filter. The number of rearing jars varied as subcultures were made and water level in the jars was maintained by a constant level siphon. A population of copepods (Tigriopus sp) was placed in the rearing jars to keep the colonies and water clear of waste products. The hydranths did not prey upon the copepods. The slides containing colonies were placed

TABLE 1

Some Coelenterates Successfully Maintained in Laboratory Culture (Davis 1971 modified).

Species	Reference
Acaulis ilonae	Brinkmann-Voss 1966
Amphinema dinema	Rees & Russell 1937
Amphinema rugosum	Rees & Russell 1937
Aurelia aurita	Spangenberg 1965
Bougainvillia carolinensis	Brock & Strehler 1963
Bougainvillia muscus	Browne 1907
Campanularia flexuosa	Crowell 1953
Chrysaora quinquecirrha	Cargo & Schultz 1967
Clava multicornis	Kinne & Paffenhofer 1965
Cordylophora lacustris	Fulton 1960
Craspedacusta sowerbii	McClary 1959
Hydra sp.	Loomis 1953, 1954
Hydra sp. (mass culture)	Loomis & Lenhoff 1956
Hydractinia echinata	Toth 1965
Merga galleri	Brinkmann 1962
Obelia sp.	Palincsar & Palincsar 1960
Podocoryne carnea	Braverman 1962
Protohydra leuckarti	Muus 1966
Sarsia tubulosa	Kukinuma 1966
Stauridiosarsia japonica	Nagao 1962
Staurocoryne filiformis	Rees 1936
Syncoryne exima	Browne 1907
Tubularia crocea	Mackie 1966

in suspensions of Artemia nauplii for 45 to 60 minutes once a day. West and Renshaw used a colony of Tubularia crocea as a water indicator. T. crocea sheds its hydranths whenever the water becomes fouled. This occurred before Clytina attenuata was affected, and the problem was eliminated by changing the filter sand.

Medusae of Clytia attenuata were kept in one gallon jars maintained at either room temperature ($20°-24°C$) or in a cold water bath maintained at approximately $18°C$. The medusae were fed a constant supply of Artemia nauplii to which was added about 20 ml of hard boiled egg yolk suspension. The suspension was prepared by placing one gram of yolk in 200 to 300 ml of seawater and shaking vigorously. The utilization of the yolk suspension resulted in a significant increase in the survival rate of the newly liberated medusae. The number of medusae that reached maturity was increased by 900 percent.

Kinne and Paffenhofer (1965) maintained Clava multicornis in culture for experiments on hydranth structure and digestion rate as a function of temperature and salinity. The basic culture medium was aged seawater of 32 °/oo S, filtered through a Seitz-Filter K5, stored at 20°C in glass jars, and kept in complete darkness.

Polyps were attached to glass microscope slides, and 5 or 6 slides were placed almost vertically in a glass dish containing about 500 ml of the culture medium. Four-day old Artemia nauplii were used as the food source. Each day 500 to 600 nauplii were placed in the glass dish containing the polyps growing on separate slides. The culture dishes were darkened during feeding. After one hour of feeding, the remaining Artemia nauplii and other debris were carefully removed with pipettes and the culture medium was renewed.

Cultures of Staurocladia portmanni sp. n. were maintained by Brinkman (1964). The hydroids were difficult to maintain. Single polyps have to be fed very carefully, one Artemia every second or third day. If fed on a daily basis, the hydranths begin to disintegrate. Newly liberated medusae were successfully raised in a large culture jar with the alga, Udothea sp. and Schreiber's solution[1] in seawater. For the first three weeks the medusae do not require any additional food. After the three week period Artemia nauplii were used.

Scyphistomae of Chrysaora quinquecirrha, the sea nettle, collected on the south shore of Long Island and from the Chesapeake Bay (provided by Dave Cargo of the Chesapeake Biological Laboratory and Dale Calder of the Virginia Institute of Marine Science) were maintained at the Adelphi University Institute of Marine Science Laboratories for about eighteen months.

The polyps were attached to oyster or clam shells and glass slides which were kept in small culture dishes. The mudium used was seawater collected from Great South Bay, filtered and stored in the dark before use. Salinity was maintained at approximately 20 °/oo. The polyps were fed freshly hatched Artemia nauplii, frozen Artemia or zooplankton collected from Great South Bay. Feeding was to repletion every other day. After feeding the remaining food organisms and debris were carefully removed with a pipette. The

---

[1]Schreiber's solution: solution A, 8.0 g $NaNO_3$ dissolved in 2000 cc aqua bidestillata; solution B, 1.6 g Na_2HPO_4 in 2000 cc aqua bidestillata; solution C, 200 g soil in 2000 cc seawater boiled for 2 hours and then filtered. The amounts used were 45 cc solution A, 45 cc solution B, and 45 cc solution C mixed in 1700 cc boiled and filtered seawater.

water was changed twice a week, and aerated for five minutes on the days no change was made. The cultures were kept at approximately 20°C in an incubator.

SUMMARY

For each species of Coelenterate to be cultured, factors to be most closely examined are nutritional requirements, the media, and the prevention of stagnation.

Crowell (1953) using Campanularia flexuosa, Hauenschild and Kanellis (1953) using Hydractinia echinata, and Loomis (1953) using Hydra littoralis were the first to demonstrate that when Artemia salina, the brine shrimp, nauplii were used as a food source, the organisms maintained a healthy physiological condition and reproduced. Other foods used have included various annelids, harpacticoid copepods, available zooplankton, fish and clam bits, frozen brine shrimp, ctenophore suspensions, and prepared aquarium food.

Medusae usually have requirements in addition to the foods mentioned above. Brinkmann-Voss (1966) found that mature medusae of Acaulis ilonae grew better and produced larger gonads when fed the Chaetognath Sagitta and a variety of copepods. West and Renshaw (1970) increased survival and maturing of Clytia attenuata medusae by 900 percent with the addition of hard boiled egg yolk suspension.

The growth of many hydroids is influenced by the quantity as well as the quality of the food. Cordylophora lacustris, Podocoryne carnea, and Bougainvillia carolinensis will produce maximum rates of budding when the cultures are fed to repletion once each day. Contrary to these results, Staurocladia portmanni disintegrated within a few days when fed to repletion.

Media utilized by various workers have included natural seawater, filtered seawater, Instant Ocean, CCS5, CVD, and Schreiber's solution.

Stagnation can be prevented by a variety of methods: large and small volume closed circulating systems; open circulating systems using running seawater; maintenance of cultures on shakers; and changing water frequently in small culture vessels.

REFERENCES

Ashworth, J.H. and N. Annandale, 1904. Observations on some aged specimens of Sagartia troglodytes and on the duration of life in coelenterates. Proc. Royal Soc. Edin., 25: 295.

Braverman, M.H., 1962. Podocoryne carnea culture methods and carbon dioxide induced sexuality. Exptl. Cell Res. 27(2): 301-306.

Brinkmann, A., 1962. The life cycle of Merga galleri sp. n. Publicazion della Stazione zoologica di Napoli 33: 1-9.

___ 1964. Observations on the biology and development of Staurocladia portmanni sp. n. (Anthomedusae, Eleutheridae). Canadian Journ. Zoology 42: 693-705.

Brinkmann-Voss, A., 1966. The morphology of Acaulis ilonae sp. n. (Anthomedusae/Athecata, Fam. Acaulidae). Canadian Journ. Zoology 44: 291-301.

Brock, M.A. and B.L. Strehler, 1963. Studies on the comparative physiology of aging. IV. Age and mortality of some marine cnidaria in the laboratory. Journ. Gerontology 18: 23-28.

Browne, E.T., 1897. On keeping medusae alive in an aquarium. Journ. Mar. Biol. Assoc., U.K. 5: 176-180.

___ 1907. A new method for growing hydroids in small aquaria by means of a continuous current tube. Journ. Mar. Biol. Assoc. U.K. 8: 37-43.

Cargo, D. and L.P. Schultz, 1966. Notes on the biology of the sea nettle, Chrysaora quinquecirrha in Chesapeake Bay. Chesapeake Sci. 7: 95-100.

___ 1967. Further observations on the biology of the sea nettle and jellyfishes in Chesapeake Bay. Chesapeake Sci. 8: 209-220.

Costlow, J.D. ed., 1970. Marine Biology V. Gordon and Breach Sci. Publishers, New York.

Crowell, S., 1953. The regression-replacement cycle of hydranths of Obelia and Campanularia. Physiological Zoology 26: 319-327.

___ 1957. Differential responses of growth zones to nutritive level, age, and temperature in the colonial hydroid Campanularia. Journ. Exptl. Zoology 134: 63-90.

Davis, L.V., 1971. Growth and development of colonial hydroids. In: Experimental Coelenterate Biology, H.M. Lenhoff, L. Muscatine and L.V. Davis, eds. pp 16-36. University of Hawaii Press, Honolulu.

Fulton, C., 1960. Culture of a colonial hydroid under controlled conditions. Science 132: 473-474.

___ 1961. The development of Cordylophora. In: The biology of Hydra and of some other coelenterates, H.M. Lenhoff and W.F. Loomis, eds., pp 287-296. University of Miami Press, Coral Gables.

Hauenschild, C. and A. Kanellis, 1953. Experimentelle Untersuchungen an Kulturan von Hydractinia echinata Flemm., zur Frage der Sexualität und Stockdifferenzierung. Zoologische Jahrbücher. Abteilung 3, Allegmeine Zoologie und Physiologie 64: 1-13.

Kinne, O. and G.A. Paffenhofer, 1965. Hydranth structure and digestion rate as a function of temperature and salinity in Clava multicornis (Cnidaria, Hydrozoa). Helgoländer Wissenschaftliche Meeresuntersudungen 12: 329-341.

Kukinuma, Y., 1966. Life cycle of a hydrozoan, Sarsia tubulosa (Sars) Bulletin. Marine Biological Station of Asamushi, Japan 12: 207-210.

Lenhoff, H.M., L. Muscatine and L.V. Davis, eds., 1971. Experimental Coelenterate Biology. University of Hawaii Press, Honolulu.

Loomis, W.F., 1953. The cultivation of Hydra under controlled conditions. Science 117: 565-566.

_____ 1954. Environmental factors controlling growth in Hydra. Journ. Exptl. Zoology 126: 223-234.

Loomis, W.F. and H.M. Lenhoff, 1956. Growth and differentiation of Hydra in mass cultures. Journ. Exptl. Zool. 132: 555-568.

Mackie, G.O., 1966. Growth of the hydroid Tubularia in culture. In: The Cnidaria and their evolution, W.J. Rees, ed., pp 397-410. Academic Press, New York.

McClary, A., 1959. The effect of temperature on growth and reproduction in Craspedacusta sowerbii. Ecology 40: 158-162.

Muus, K., 1966. Notes on the biology of Protohydra leuckarti (Hydroidea, Protohydridae). Ophelia 3: 141-150.

Nagao, Z., 1962. The polyp and medusa of the hydrozoan Stauridiosarsia japonica n. sp., from Akkeshi, Hokkaido. Annotationes zoologicae japonenses 35: 176-181.

Palincsar, E.E. and J.S. Palincsar, 1960. The effect of 8-azaguanine and chloramphenicol on the regression-replacement cycle of hydranths. Biological Bulletin, Woods Hole 119: 329-330.

Rees, W.J., 1936. On a new species of hydroid, Staurocoryne filiformis, with a revision of the genus Staurocoryne Rotch, 1872. Journ. Mar. Biol. Assoc., U.K. 21: 134-142.

_____ ed., 1966. The Cnidaria and their evolution. Academic Press, New York.

Rees, W.J. and F.S. Russell, 1937. On rearing the hydroids of certain medusae, with an account of the methods used. Journ. Mar. Biol. Assoc., U.K. 22: 61-82.

Spangenberg, D.B. 1965. Cultivation of the life stages of Aurelia aurita. Trans. American Microscopical Soc. 83: 448-455.

Toth, S.E., 1965. Cultivation of marine hydroids. Bios. 36: 63-65.

West, D.L. and R.W. Renshaw, 1970. The life cycle of Clytia attenuata (Calyptoblastea: Campanulariidae). Mar. Biol. (Berlin) 7: 332-339.

COMMENTS ON THE LABORATORY CULTURE OF SCYPHOZOA

David G. Cargo

Chesapeake Biological Laboratory, Natural Resources

Institute, University of Maryland, Solomons, Maryland

The culture of a number of major invertebrate types of animals has been accorded increasing attention in recent years. However, the culture of scyphozoans has received very little attention until recent years. The group, as a whole, is of little commercial value. Other invertebrates, especially pelycepods and crustaceans, have received most of this type of study. Work with this group before 1920 dealt mostly with geographic distribution and taxonomic aspects. Some of the earlier workers in this field were Delap (1906a and b), Mayer (1910), Chuin (1930), Thiel, M.E. (1935), Littleford (1939), Berrill (1949a), and H.J. Thiel (1966). Recently, some research in North America and in Europe has been examining means of culturing certain members of this group because developing such techniques has become increasingly important. Aside from a purely academic interest, certain species are injurious (Chironex), damaging pests in certain areas (Chrysoara, Cyanea), a food item (Rhopilema), or of interest to medical research for various avenues of investigation. This treatment will attempt to summarize the present state of knowledge on such developments.

The scyphozoa are divided into five orders -- Stauromedusae, Cubomedusae, Coronatae, Semaeostomeae, and Rhizostomeae. The ontogensis of the first three is not well known although research is currently progressing in Australia on the life cycle of the deadly cubomedusan, Chironex (Barnes, 1966). Much culture work up to now has been motivated by a need to understand certain life cycle phenomena. Most of the available published information deals with the Semaeostomeae, to which the more familiar jellyfish belong. The Rhizostomes have been examined in the western Pacific and more recently in the western Atlantic. These remarks will be confined to these last two orders.

The manner in which different species can be induced to provide viable eggs, sperm, and larvae varies. There is usually no trouble in obtaining sperm. The males appear to mature at a smaller size, and possess an abundance of viable sperm. The females are not always as simple to manipulate. I shall first consider Chrysaora quinquecirrha, the sea nettle in the Chesapeake Bay, because I am most familiar with it and because there is now available a considerable quantity of experience and information. I will then use this species for comparative purposes in discussing other species. Kramp (1961) and Mayer (1910), and Russell (1970) give excellent coverage of the occurrence of scyphozoans throughout the world.

CHRYSAORA QUINQUECIRRHA

Mature females of Chrysaora quinquecirrha are seen in the mid Chesapeake Bay area about July 1. We have not yet observed hermaphrodites as reported by others for this genus (Chuin 1930). Temperatures above $25^{\circ}C$ and salinities of 10-20 $^o/oo$ are common at this time. Limiting parameters appear to be approximately $10^{\circ}C$ and $35^{\circ}C$ and 5 and 35 $^o/oo$. Sexual maturity is indicated by a dusky grey or brown color of the gonads, gradually darkening as the eggs develop. Most of our success has occurred when the mature female medusae are maintained in aerated containers and portions of the gonads from mature males are inserted into their gastric cavities. Running water aquaria are not usually successful because the medusae are swept to the over-flow drains. This problem can be partially avoided by employing under gravel filters. When conditions are suitable, many planulae may be removed from the aquarium or from medusae within 48 hours.

The embryos and planulae are generally incubated outside the gonads on the oral arms or even set free into the surrounding water. The planulae, about 190 μ long, are free swimming, and set directly into miniature scyphistomae or polyps. They do not produce larval cysts. They proceed rapidly with the development of tentacles up to a full complement of sixteen within 10-14 days. They feed on a variety of small organisms: protozoans, rotifers, larvae of many types, small crustaceans, planulae of their own and other species and can be cultured indefinitely on Artemia nauplii. Artificial sea salts are satisfactory for the long term maintenance of scyphistoma. The addition of wild planktonic organisms to the diet may be important since polyps held for long periods on an exclusive diet of Artemia have an abnormal appearance. They are gross and fleshy and quite pink in contrast to polyps from the wild.

Under natural conditions, the polyps which have set in the summer, produce large numbers of reddish-brown podocysts at their base during the fall. They do this also in culture, particularly if an abundance of food and lowering of temperature to about $20^{\circ}C$ is provided. A continued lowering will result in a cessation of feeding at $8-10^{\circ}C$ and encystment of the polyp frequently occurs at $2-4^{\circ}C$.

Strobilation customarily occurs in late spring when the water temperature reaches $22^{\circ}C$. Polyps held in cultures at temperatures above $22^{\circ}C$ frequently cease strobilating. We can usually induce strobilation by lowering the temperature to $15^{\circ}C$ for a period of 4-6 weeks. A diapause relationship is indicated here. Upon returning the scyphistomae slowly to above $22^{\circ}C$ over a period of a week, strobilation ensues. Strobilation can be completed within 2-4 days although more time is usually required. If encystment has occurred, excystment can be expected at $15-18^{\circ}C$. Podocysts will also excyst at this temperature.

Ephyrae and medusae are difficult to sustain in culture. Ephyrae feed upon protozoa and other very small organisms, and can be maintained on newly hatched Artemia nauplii or finely strained ctenophore. (Compton, Calder, pers. comm.) The addition of salt water infusoria is suggested by this acceptable diet. Calder successfully used Erlenmyer flasks in a wrist action shaker to provide a gentle agitation. At Solomons, we have not attempted long-term maintenance except with aeration. Success has been mixed.

During the warmer parts of the year, many scyphistomae alternate strobilation with periods of stolon and podocyst production. We have seen a single polyp strobilate as many as four times in one summer. It is also possible for newly set polyps to strobilate within several weeks of setting but we cannot believe that such an occurrence is customary in the Bay.

Chrysaora polyps exhibit what N.J. Berrill (1949a) calls "an almost unbelievable protean plasticity." Regeneration is rampant, particularly for the tentacles. They can be fully reformed after loss in three to five days. The stalk is also capable of rapid repair and growth. Excised tentacles round into planulae-like structures and swim for extended periods of time. Chuin (1930) and Herouard (1913) state the tentacles of C. hysoscella can set and bud new polyps in this fashion. Several attempts to repeat this with C. quinquecirrha have failed but I'm not convinced that it can't happen.

In culture, these polyps are sessile, moving only very slightly by means of stolons and producing podocysts. If dislodged mechanically, they can easily reattach within a short time. They have never been seen to swim as Haven (pers. comm.) has observed in Aurelia. Their feeding is mostly accidental although I have seen some apparent recognition of the passage of small organisms such as barnacle nauplii. Capture and feeding are quite rapid, the contacted tentacle quickly recoiling into the mouth which gapes open to receive the food organism.

Polyps of Chrysaora can easily be stained with an aqueous solution of neutral red. The dye is taken up quickly and penetrates the animal completely. It does not appear to affect their activities except to make them more negatively sensitive to light. Medusae and ephyrae can be anesthetized easily with propylene phenoxitol or calcium sulfate. Polyps, however, are unaffected. They may be relaxed by heat to $32^{\circ}C$ or slowed down and retracted by reducing the temperature to $5^{\circ}C$.

Medusae in aerated containers pulsate almost continually. The speed of pulsation is inversely related to the size of the animal. Occasionally, they may cease activity, allow the oral arms to spread and drift slowly to the bottom, perhaps in a feeding behavior. It is difficult to maintain growth of these medusae in culture. The medusae will take a variety of organisms. Polychaetes, ctenophores, small fish and certain crustacea are suggested. Despite our best efforts, at summer temperatures most medusae regress slowly, becoming smaller with the passage of time and eventually disintegrating. Four medusae which were hand fed polychaetes (Nereis) several times daily, showed an average growth from 8 mm to 52 mm in 45 days.

We have little information about disease in Chrysaora. Cultures sometimes become fouled and bacteria occasionally infest containers, even in running water aquaria. Polyps usually respond to unsatisfactory conditions by encysting. Medusae are sometimes seen in the wild in a moribund condition, marked by an absense of tentacles or oral arms. We have not been able to ascertain the cause for this, a more or less seasonal phenomenon. At the present time, we have an ambitious program under way aimed at answering this question. No other parasites or predators are likely to occur in laboratory cultures.

AURELIA AURITA

Spangenberg (1965) has provided us with a recent and quite complete account of Aurelia. Doing her work in the Gulf of Mexico, she has been able to establish required cultural techniques so the Aurelia can be maintained anywhere and for extended periods of time.

Aurelia medusae are more easily stripped of their embryos than Chrysaora. The fully developed planulae, 250-350 μ in length, are incubated and retained on the oral arms. The planulae may be easily secured by shaking the mature specimen in a container of natural (NSW) or artificial seawater (ASW) at a salinity close to that in which the female has existed. Merely placing the female in a bowl of water will sometimes result in the release of many planulae. These larvae are vigorous and active. They are commonly available from mature females in the Chesapeake area from August through December.

The larvae usually set within two days on shells or other suitable substrate. The polyps, characterized by a circular mouth (Haven, pers. comm.), are small but feed avidly on newly hatched Artemia. Spangenberg fed her polyps twice weekly, changing the water after each feeding. Most cultures can be kept at 21-27°C. Her ASW formula contained the following C.P. chemicals:

Sodium Chloride	23.47 gm/l
Magnesium Chloride	4.98 "
Sodium Sulfate	3.91 "
Calcium Chloride	1.10 "
Potassium Chloride	.664 "
Sodium Bicarbonate	.192 "
Potassium Bromide	.096 "
Boric Acid	.026 "

These chemicals were first dissolved separately in de-ionized water and then mixed. The initial pH is 7.8 and the specific gravity is 1.025 (33.7 °/oo). Standard commercial salts compounded for ASW are also satisfactory. Strobilation in Aurelia is easily induced at modest temperatures. Spangenberg reports strobilation occurring more or less continuously at 21-24°C but mass strobilation occurred only sporadically. Both natural and artificial seawater suffice, and raising the temperature to 28°C or 29°C generally causes an immediate initiation of strobilation. She reported in 1967 that cultures held in ASW for extended periods were reluctant to strobilate. By lowering the temperature to 19°C for 30 days, she was able to cause a large percentage of the reluctant polyps to strobilate upon return to the higher temperature. This phenomenon is similar to that observed for Chrysaora.

Mrs. Spangenberg in 1967 also provided us with a valuable insight on how the element, iodine, influences strobilation. Either free iodine or iodine bound in a soluble compound seems to be satisfactory. She also pointed out that it may act in a hormonal way since thyroxin ($C_{15}H_{11}I_4NO_4$) can also induce strobilation.

Environmental factors such as light or pH are apparently of little importance to strobilation.

Within reasonable limits, Aurelia polyps, like those of Chrysaora, can be held in crowded conditions if adequate food and water volume is maintained. The maintenance of ephyrae and medusae is more difficult. In her experiments with ephyrae and medusae, Spangenberg successfully maintained 10 to 50 ephyrae in 3 1/2 gallons of water. These were fed 10 cc of concentrated living Artemia daily. She noted a difference in survival of ephyrae from polyps maintained in ASW as opposed to NSW. In ASW, only 10 percent of the ephyrae reached maturity whereas 50-75 percent matured from polyps maintained in NSW!

After reaching one inch in diameter, young Aurelia were transferred to larger containers, maintaining a one gallon/medusae ratio. Growth was slow. After 42 to 105 days, her stage "9" animals had reached only 30-50 mm in diameter but were sexually mature. Since much larger adults are commonly seen, it appears that it is as difficult to satisfy the nutritional requirements of the Aurelia medusae as it is for Chrysaora (Cargo & Schultz 1966, 1967). Calder found Aurelia medusae will also take strained ctenophores and show some growth.

Maintenance of the polyp stage of Aurelia follows the same technique already described for Chrysaora. They have nearly the same temperature requirements, slightly higher salinity preferences and can utilize the same food. Aurelia polyps show one habit which makes their culture more difficult. They cannot be held for any length of time in running water tanks since they frequently detach from the substrate and swim away using their cilia (Haven, pers. comm.). Neither Cyanea or Chrysaora have displayed this habit.

Diseases, per se, have not been described for Aurelia in cultures. The polyps are exceptionally long-lived, many existing in cultures for nearly three years (Spangenberg 1965). The medusae generally deteriorate after spawning. The gastric filaments are extruded at the same time that the sex products are released. This senescence possibly occurs similarly in the other scyphozoa since the gastric filaments are known to release digestive enzymes capable of destroying any larvae released into the gastric cavity (Chuin 1929, Smith 1936, Bodansky and Rose 1922).

The toxin of Aurelia medusae is mild and generally not noticed when handled. The introduction of nematocysts into the eyes or other body openings should be avoided.

CYANEA CAPILLATA

This species is ubiquitous and nearly as well known as Aurelia. Many of the features of its life cycle and culture are similar to those already discussed. It differs in several ways, particularly with respect to the Mid-Atlantic area.

This species is a cold-weather form in the Chesapeake area. Mature female medusae carrying embryos are found from February to May. They are more abundant in years of higher than normal salinity. The eggs are fertilized within the gonads, which are bright yellow or mustard-colored central organs. The eggs can be removed from the medusae or the medusae can be allowed to shed them into the receiving container. The embryos and resulting planulae are either yellow or white; we do not yet know why. Planulae are frequently found on the oral arms of the medusae.

The planulae, about 250 μ long, swim for several days and usually set on the bottom of the container as shiny, convex, larval cysts. Chrysaora and Aurelia have never displayed larval cyst stages. They are translucent to pearly, circular, and about 1 mm in diameter. Polyps may then bud from these larval cysts in several days or the cysts may remain quiescent for six months or more. Calder in Virginia has found that most planulae have set as larval cysts at 9°C. They soon bud polyps which begin forming stolons and yellow-green podocysts at 15°C. At 25°C these polyps begin encysting under the impact of high summer temperatures.

Salinities for cultures of this species should be maintained higher than for Aurelia and Chrysaora. Polyps have not been found in natural waters where the salinities reach much below 20 °/oo. Calder states that their optimum is 20 °/oo and that the temperature range should be maintained between 5°C and 17°C. Budding from larval and podocysts can be expected at salinities up to 35 °/oo.

Encysted polyps, podocysts, and larval cysts begin encysting in September and October as the temperature falls, below about 20°C. At 15°C, they initiate strobilation. They are unusually small in size at this time in comparison to Chrysaora prior to strobilation, and utilize much of their body materials in the manufacture of monodisk and polydisk strobila (Berrill 1949b). The strobila closely resemble those of Cassiopea and Rhopilema, causing some doubt, at least to me, about their rightful taxonomic classification. After strobilation, the polyps can be handled as with the other species, providing space, food such as Artemia and other organisms of a similar size.

The ephyrae are robust and can be cultured on small plankton organisms or Artemia or strained ctenophore material. They even ingest and seem to derive some benefit from diatomaceous scrapings from shells. We have raised them to 18 mm in diameter. At this size, they are easily identifiable.

Cyanea medusae behave unlike Aurelia and Chrysaora in that they frequently pulsate to the surface, reverse themselves and drift slowly to the bottom to rest there in an inverted position.

This is another reason to suspect their close relationship to the Rhizostomes since Cassiopea spends much of the time inverted so that symbiotic zoozanthellae are exposed to the light.

Cyanea is probably the easiest of the semaeostomes to cultivate, mostly because the low water temperatures make it a simple matter to provide life's requirements. Its toxin at the smaller sizes is relatively innocuous to most people and it can generally be handled with impunity.

RHOPILEMA VERRILLI

This Rhizostome is never very abundant but ranges along the Atlantic coast from Vinyard Sound to the Gulf of Mexico. It is an occasional visitor to Chesapeake Bay in late summer and fall. Until recently only the medusa was known. Investigators in Maryland and Virginia have succeeded in describing various other stages and developing cultural techniques.

The medusae are robust and large and the female medusae frequently carry fertilized embryos and planulae within the gonadal tissue. They do not appear to carry them on the oral arms as do other species. The planulae are 165-310 μ long, white and quite active, (Calder, pers. comm.). They metamorphose into polyps after 7-10 days at 12°C. Budding of polyps can also apparently occur directly from the gonadal tissue without passing through a motile planulae stage. The polyps, small and white, survive and feed actively down to 5°C. They can be fed Artemia successfully. Salinities of 18-20 °/oo are satisfactory.

Polyps held at 10 and 12°C grew slowly in cultures at Virginia Institute of Marine Science. When raised to room temperature of about 20°C, growth was more rapid and the polyps appeared healthier. They can be fed Artemia or other suitably sized organisms. When held under unsatisfactory conditions the polyps can make small golden podocysts, 280-350 μ in diameter.

We are able to induce strobilation in early spring by allowing the temperatures to rise slowly from 9°C to 20°C. We raised the salinity from 13.5 °/oo to 19.5 °/oo over a period of six days by adding 1 gram of *"Rila Marine Mix" artificial sea salts daily to 500 ml of tank water (Cargo, 1971). Strobila of the monodisk type were present within seven days. Strobila resembled closely published descriptions for Cassiopea in Mayer (1910) with some minor differences.

Calder (pers. comm.) has recently described the ephyrae of R. verrilli. We were not able to culture them in 1966 but feel certain that methods similar to those just described for related species will prove satisfactory. R. verrilli is more common to the south of Chesapeake Bay, so that warmer temperatures seem indicated.

ACKNOWLEDGEMENTS

Many people have contributed to the information contained in this report. I have leaned heavily on unpublished information in many cases. Particular thanks go to Dr. Dale Calder, Mr. Reinaldo Morales and Mr. Dexter Haven at the Virginia Institute of Marine Sciences and to Dr. Leonard P. Schultz, Miss Joanne Allwein, Miss Rosalie Vogel and Mr. Michael J. Reber at the Chesapeake Biological Laboratory for allowing me access to their unpublished reports.

*Rila Products, Teaneck, N.J.

REFERENCES

Barnes, J.H. 1966. Studies on three venomous cubomedusae, Symp. Zool. Soc. London #16, 1966. In: The Cnidaria and their evolution. Academic Press, London and New York. pp 307-332.

Berrill, N.J. 1949a. Developmental analysis of scyphomedusae. Biol. Ref. 24:393-410.

_____ 1949b. Form and growth in the development of a scyphomedusae. Biol. Bull. 96(3):283-292.

Bodansky, M., and W.C. Rose. 1922. Comparative studies of digestion. I. The digestive enzymes of coelenterates. Amer. J. Phys. 67:547-550.

Cargo, D.G. 1971. The sessile stage of a scyphozoan identified as Rhopilema verrilli. Tul. Stud. Zool. 17(2):31-34.

Cargo, D.G. and L.P. Schultz. 1966. Notes on the biology of the sea nettle, Chrysaora quinquecirrha in Chesapeake Bay. Ches. Sci. 7(2):95-100.

_____ 1967. Further observations on the biology of the sea nettle and jellyfishes in Chesapeake Bay. Ches. Sci. 8(4):209-220.

Chuin, T.T. 1929. Sur les vacuoles digestive chez le scyphistome. C.R. Soc. Biol. 102:825-827.

_____ 1930. Le cycle evolutif du scyphistome de Chrysaora. Trav. Sta. Biol. Roscoff 8:1-179.

Compton, Ann M. 1966. Mnemiopsis leidyi as a food for the cultivation of Chrysaora quinquecirrha under controlled conditions. Unpub. Ms. Ref. No. 66-49. Ches. Biol. Lab. Solomons, Md. 11 pp

Delap, Mary J. 1906a. Notes on the rearing in an aquarium, of Aurelia aurita L and Pelagia perla (Slabber). Rep. Sea Inland Fish. Ire. 1905, 1:22-26.

_____ 1906b. Notes on the rearing in an aquarium, of Cyanea lamarcki Peron & Leseur. Rep. Sea Inland Fish. Ire. 1902-1903, 2:20-22.

Herouard, E. 1913. Relations entre la depression et la formation de pseudoplanula tentaculaires chez le scyphistome. C.R. Acad. Sci. 156:1093-1095.

Kramp, P.L. 1961. Synopsis of the Medusae of the World. Jour. Mar. Biol. Assoc. of the U.K. 40:469 pp.

Littleford, R.A. 1939. The Life Cycle of Dactylometra quinquecirrha L. Agassiz in the Chesapeake Bay Bio. Bull. 77:368-381.

Mayer, A.G. 1910. Medusae of the World. Vol. III. The Scyphomedusae. Pub. Carnegie Inst. Wash. No. 109:499-735.

Russell, F.S. 1970. The medusae of the British Isles. II Pelagic scyphozoa. Cambridge Univ. Press. 284 pp.

Smith, H.G. 1936. Contribution to the anatomy and physiology of Cassiopea frondosa. Papers from Tort. Lab. 31:19-52.

Spangenberg, Dorothy D. 1965. Cultivation of the life stages of Aurelia aurita under controlled conditions. Journ. Exp. Zool. 159(3):303-318.

___ 1967. Iodine induction of metamorphis in Aurelia. Jour. Exp. Zool. 165(3):441-449.

Thiel, H.J. 1966. The Evolution of Scyphozoa; A Review. In: The Cnidaria and their evolution. Symp. Zool. Soc. Lond. No. 16 pp 77-117.

Thiel, M.E. 1935. Uber die Wirkung des Nesselgiftes der Quallen auf den Menschen. Ergenbnisse und Fortschritte der Zoologie VIII: pp 1-35.

Contribution No. 530, Natural Resources Institute, University of Maryland; Reference No. 73-2.

BRYOZOA

Marie B. Abbott

Gray Museum, Marine Biological Laboratory

Woods Hole, Massachusetts

The plan for this chapter is to present, for the generalist, an overview of those aspects of the biology of bryozoans that are related directly to their procurement for and maintenance in the laboratory and to draw attention to the pertinent literature, particularly that of the 15 years since the Hyman (1959) volume. The data and opinions presented are derived from the author's field and laboratory research and experience with the bryozoa of the northeastern United States coastal waters, from word of mouth information from colleagues, and from the literature.

THE PHYLUM BRYOZOA

The Bryozoa are very small, exclusively colonial, sessile epibenthic coelomate animals characterized by the possession of a lophophore, a ring of ciliated oral tentacles which performs the combined functions of filter-feeding, respiration and excretion, and of an inelastic supportive outer body wall in which a membranous or gelatinous matrix is commonly reinforced by chitin or calcium carbonate or both.

Each individual (zooid) in a colony consists of three elements: 1) the soft, retractile, and usually transient polypide, including the mouth and tentacles of the lophophore itself, a U-shaped digestive tract with the anus opening outside the lophophore ring, a nerve ganglion lying between the latter and the anus, and several sets of muscles to operate the lophophore; 2) the more permanent cystid or zooid body walls, from which successive generations of polypides and also the gametes arise; and 3) the indefinitely durable and usually calcareous exoskeleton.

Individual zooids rarely attain a size larger than 1.0 mm, but fully developed colonies are eminently visible and can comprise a significant element in the marine epifauna in both abundance and diversity of species if not in biomass. Although they occur at all latitudes and at depths from low intertidal to 8300 m, their maximum diversification and density is reached in the photic zone of shelf seas. The two abiotic requirements essential to bryozoa are food-carrying water currents and substrate for larval settlement and colony growth. Wherever these occur together, bryozoan populations will sooner or later be recruited from passing larvae. The group is predominently a marine one; only a very few of the approximately 4000 living species have successfully invaded brackish or fresh water.

To distinguish them from the pseudocoelomate Entoprocta with which they were formerly grouped, the Bryozoa have sometimes been called Ectoprocta in recent years, following Hyman (1959, p. 277). This minor nomenclatural controversy appears to have been fully resolved, however, by Mayr, who designated (1968, p. 213) Zoobotryon Ehrenberg, 1831, as the type genus of the class and phylum Bryozoa. Both the U.S. National Museum of Natural History and the British Museum (Natural History) use the term Bryozoa, the latter institution having recently abandoned the name Polyzoa long used by British and Commonwealth zoologists.

Most bryozoologists today accept the division of the phylum into three classes, as first proposed by Borg (1926):

Phylactolaemata:	restricted to fresh water.
Stenolaemata:	exclusively marine; represented in modern seas by only one surviving order (Cyclostomata).
Gymnolaemata:	marine, but some species tolerate or prefer brackish water; the overwhelmingly dominant group.
Order Ctenostomata:	Zooid walls membranous or gelatinous.
Order Cheilostomata:	Zooid walls more or less calcified.

With the exception of a very few free-living warm-water groups, colonies of marine bryozoa assume one of three general habits. They may be encrusting, forming sheet-like layers on rock, shell or macroalgae; stolonate, appearing as creeping threads or in an open mesh pattern, most commonly on hydroids, macroalgae or other bryozoa; or erect, bushy or tuft-like, attached to either hard or soft substrate only at the proximal end of the colony. Some of the latter may at first sight be confused with thecate hydroids.

An exhaustive and profusely illustrated compilation of bryozoan morphology, embryology, histology and physiology may be found in Hyman (1959), although Brien's (1960) treatment, especially of sexual reproduction and growth, is perhaps even more comprehensive. A lucidly presented account of the present status of knowledge of the group, intended for the general reader, is that of Ryland (1970b).

The question might easily be asked -- why culture or maintain marine bryozoa? The biomass of the polypide is clearly too small, both of itself and (in all but a few ctenostome species) in proportion to the inedible exoskeleton, to be a potential food source for man or even to constitute a significant element in a productivity chain.

The principal incentive for maintaining bryozoa under laboratory conditions is for observation and experimentation in research on some aspect of their basic biology. Although the phylum is relatively large, these animals are not well understood or widely studied, probably because of the difficulties inherent in their small size.

While bryozoa are of little direct economic importance, they can be a nuisance in commercially valuable shellfish beds (see Closed Seawater Systems below). They are also an important component in fouling communities, especially in relatively warm waters (Redfield, 1952; Mawatari, 1951a, 1951b, 1952, 1953; Skerman, 1958, 1959; Nicolic, 1959; Kawahara and Iizima, 1960; Nair, 1962; Starostin, 1963; Morales and Arias, 1965a, 1965b; Personne, 1965; Ryland, 1965a, 1967a: 353-357, 1970a; Jebram, 1969) and are widely distributed on ships' hulls (Allen, 1953) and on seed oysters (Powell, 1970). Further "counterculture" studies of factors and conditions that inhibit larval settlement and colony growth would be valuable.

REPRODUCTION IN BRYOZOA

Reproduction in the Bryozoa is of two types. Each colony originates with an ancestrula (commonly single, but twinned or multiple (Maturo, 1973) in some species), a specialized zooid which in marine forms develops from a metamorphosed sexually-produced larva. From this ancestrula the colony grows by vegetative budding -- lateral, terminal, or frontal, depending on the species and on the age (astogenetic growth phase) of the colony -- into a theoretically infinite series of successive generations of intercommunicating zooids.

Asexual Growth and Colony Maintenance

<u>In Open Seawater Systems</u>: Marine bryozoa are on the whole not easy to maintain in open seawater systems over long periods of time. Newly collected colonies of all but a few species will usually exhibit signs of degeneration within one to three days, permitting only short-term observations or experimentation. However, many ctenostome species and a few cheilostomes -- notably the bugulids and species with planktotrophic larvae and/or tolerance for reduced salinities -- may survive in good condition for longer periods (see Tables I and II). Most of these species have either membranous or chitinous exoskeletons, or are but lightly calcified. They are, therefore, transparent to some degree and hence comparatively easy to observe and manipulate. The first modern experimental studies on marine bryozoa (Marcus, 1926) were based on the ctenostome <u>Farella repens</u> and the cheilostome <u>Electra pilosa</u>. Indeed, most of our knowledge of the basic biology of the bryozoa is based on this latter group of species, many of which are amphiatlantic or cosmopolitan.

For the convenience of those wishing to secure living, growing marine bryozoan colonies for experimental, observational, or teaching purposes, the most common shallow sublittoral species of northeastern United States waters are listed in Table I, together with information on the habit of the colony, substrate preference (whether for hard rock, shell, or wood or for soft macroalgae, eelgrass, hydroids or other bryozoa) and abundance in each of three major geographic zones. Data for the latter are derived from the writer's field experience and from the literature (Stickney and Stringer, 1957; Watts, 1957; Powell and Crowell, 1967; Maturo, 1968; Powell, 1968; and Wass, 1972).

Criteria for inclusion of a species in Table I are 1) it must be readily identifiable by the non-specialist, using such well-illustrated sources as Maturo (1957), Maturo and Schopf (1968), Osburn (1912, 1933, 1944), Rogick (1964), Rogick and Croasdale (1949), and Ryland (1960a, 1965a); 2) it must be accessible from shore and require no specialized collecting equipment, i.e. colonies can be expected at no more than 4 m depth and commonly at less than 2 m; and 3) it must be sufficiently abundant to be available in adequate quantity and to be found readily throughout the year (except perhaps locally between mid-December and early March) in at least one of these geographic zones:

> Zone 1 -- coast of Maine, New Hampshire and Massachusetts as far south as Manomet Point near Plymouth: shore predominantly rocky; macroalgae abundant; bottom water temperatures sub-arctic to boreal, equalized by strong tides at least 3 m in range.

Zone 2 -- coast of Massachusetts from Manomet southward (including Cape Cod), shores of Long Island Sound: rocky shore (rock outcrops in Connecticut, terminal moraine boulders elsewhere) alternating with sandy coves and spits, salt marsh or mud flats; macroalgae abundant on rock, eelgrass in sand areas; bottom water temperatures modified boreal, except in colder pockets of Cape Cod Bay (Young et al., 1971: ii) and Block Island Sound (Abbott, 1973: 42); tide range 1-2 m.

Zone 3 -- mid-Atlantic coast from the south shore of Long Island to the mouth of Chesapeake Bay: shore predominantly sand, silt or mud, with extensive but irregular areas of shell; macroalgae common only on rock jetties and wharf pilings; large bodies of euryhaline waters (estuaries); bottom temperatures in summer may exceed $26^\circ C$; tide range averages 1 m.

The bryozoa may be collected by hand, by wading or snorkeling at low tide, from exposed bedrock, boulders, jetties, and mooring ropes and from the lower stipes and holdfasts of eelgrass and macroalgae, notably _Fucus_, _Ascophyllum_, _Laminaria_, _Chondrus_ and _Phycodrys_. Buoys and wharf pilings can be scraped with a pile scraper or oyster knife, either from shore or from a small boat. Where the only available substrate is loose shell and gravel, the surest way to procure colonies is by a small hand-operated dredge or grab. It should be noted that species which require or prefer the soft substrates will be at their most abundant when these substrates are most developed in the warmest months -- July to September. Similarly, after an unusually cold winter species near the critical northern limit of their range may be scarce, very late in appearing or be absent entirely from very shallow water (Allee, 1919).

The writer has evolved a number of procedures which have been found, pragmatically, to slow down the rate of deterioration and increase survival time of bryozoan colonies in open system laboratory conditions:

Transport colonies from the field to the laboratory in damp algae, in plastic bags or small closed buckets with only adherent seawater, keeping containers in as cool and dark a place as possible. The polypides and lophophore will be retracted behind closed opercula and therefore less affected by decreased oxygen tension than they would be if protruded into stagnant seawater.

Remove known or possible predators: pycnogonids (Fry, 1965; Wyer and King, 1973), polyplacophora (Barnawell, 1940), amphipods (Kaufmann, 1971: 20-24), nudibranchs (Miller, 1961, 1965), chitons,

Table I. COMMON SHALLOW SUBLITTORAL BRYOZOA - NORTHEASTERN U. S.

	Colony Habit[a]	Substrate Preference[b]		Geographic Range and Abundance[c]		
		Hard	Soft	Zone 1	Zone 2	Zone 3
STENOLAEMATA						
Crisia eburnea (L., 1758)	E	+	+++	C	A	C
GYMNOLAEMATA:						
Ctenostomata						
*Bowerbankia gracilis Leidy, 1855	S	+	+++	C	A	C
*Flustrellida hispida (Fabricius, 1780)	Enc.		++++	A	C	-
Cheilostomata						
Aetea anguina (L., 1758)	S	++	++	C	C	C
*Electra pilosa (L., 1767)[d]	Enc.	+	+++	C	A	R
Callopora aurita (Hincks, 1877)	Enc.	+++	+	R	C	R
Amphiblestrum flemingii (Busk, 1854)	Enc.	+++	+	R	C	-
Tegella unicornis (Fleming, 1828)	Enc.	+++	+	A	R	-
*Bugula simplex Hincks, 1886	E	+++	+	R	C	-
*Bugula turrita (Desor, 1848)	E	+	+++	C	A	C
Dendrobeania murrayana (Johnston, 1847)	E	++++		C	R	-
Cribrilina punctata (Hassall, 1841)	Enc.	+++	+	A	C	R
Hippothoa hyalina (L., 1767)	Enc.	+	+++	A	C	R

Table I (cont'd)

Hippoporina americana (Verrill, 1875)	Enc.	+++	+	R	A	R
Hippoporina porosa (Verrill, 1879)	Enc.	++++		R	C	R
*Schizoporella unicornis (s.l.) (Johnston, 1847)	Enc.	+++	+	R	A	C
Parasmittina nitida (Verrill, 1875)	Enc.	+++	+	R	C	C
Microporella ciliata (Pallas, 1766)	Enc.	+++	+	R	C	R
Cellepora americana Osburn, 1912	Enc.		++++	R	A	-
*Cryptosula pallasiana (Moll, 1803)	Enc.	+++	+	R	A	C

* = survive relatively well in both open and closed seawater systems.

a/ E = erect, bushy; S = stolonate; Enc. = encrusting.

b/ ++++ = exclusively on indicated substrate.

+++/+ or +/+++ = predominantly on +++ substrate, but also occurring on alternate.

++/++ = occurring more or less equally on both.

c/ A = abundant; C = common; R = rare.   See text for further explanation.

d/ Larvae planktotrophic

Table II. BRYOZOA OF THE OYSTER COMMUNITY

	Mode of Sexual Reproduction	Lower Salinity Limit (°/oo)	Geographic Range and Abundance[a]		
			Zone 1	Zone 2	Zone 3
CTENOSTOMATA					
Alcyonidium polyoum (Hassall, 1841)	Lecithotrophic larva (brooding internal)	20	C	C	R
Victorella pavida Kent, 1870	"	3[b]	-	-	A
Bowerbankia gracilis Leidy, 1855	"	10	(see Table I)		
CHEILOSTOMATA					
Membranipora tenuis Desor, 1848	Planktotrophic larva (cyphonautes)	6	R	C	A
"Electra crustulenta"[c]	"	6	R	C	A
Electra monostachys (Busk, 1854)[d]	"	27	A	C	R
Callopora aurita (Hincks, 1877)	Lecithotrophic larva (brooding external, in ovicell)	24	(see Table I)		
Cribrilina punctata (Hassall, 1842)	"	24	"		
Schizoporella unicornis (s.l.) (Johnston, 1847)	"	18	"		
Microporella ciliata (L., 1758)	"	20	"		
Cryptosula pallasiana (Moll, 1803)	Lecithotrophic larva (brooding internal)	24	"		

[a] see Table I, footnote c.
[b] optimum 10-12 °/oo.
[c] also known under "Membranipora lacroixii" and "Conopeum reticulum".
[d] = E. hastigsae Marcus, 1938.

juvenile carnivorous gastropods, large benthic forams, ophiuroids, and echinoids (Gordon, 1972).

Maximize aeration by directing inflow over rapids created by slant tray filled with pebble and cobble, or by use of a mechanical aerating pump.

Keep the rate of seawater flow at the maximum the table will admit without overflowing and the proportion of tank volume to number of colonies as high as possible. Bryozoa are voracious feeders and colonies maintained in open systems are entirely dependent for food on the plankton, especially nannoplankton, brought in through intake pipes.

Inhibit bacterial growth and eutrophication by keeping tanks covered.

Reduce silting by lining the intake pipe with a loosely fitting role of ordinary fiberglass window screening. The film of bacteria which rapidly forms in the meshes in turn traps silt particles. The screen can be removed periodically and scrubbed. (The writer is indebted to F.P. Bowles for this valuable suggestion.)

In Closed Seawater Systems: The writer has not attempted to maintain bryozoa in closed seawater systems. However, Cook (1963) reports having kept lunulitiform bryozoa alive for three months; Dudley (1970) and Bullivant (1968) conducted feeding experiments in closed systems, and Schneider (1955, 1957a, 1957b, 1959), Kaissling (1963), and Schneider and Kaissling (1964) reared Bugula avicularia on the dinoflagellate Oxyrrhis for their studies on cuticle structure and oriented growth of calcite crystals.

By far the most elaborate and successful systems have been those designed by Jebram (1968, 1973), following the formula for artificial seawater and the methods of Schneider and Kaissling (cited above) and of Hauenschild (1962). Jebram has successfully raised to maturity the ctenostomes Farella repens and Bowerbankia gracilis, and the cheilostomes Electra crustulenta, E. pilosa, E. monostachys, Conopeum reticulum and Bugula stolonifera, all species belonging to the group that do well in open systems and in poly-or mesohaline waters (see Table II). Jebram has developed a method of asexually multiplying encrusting bryozoa by inducing them to grow on coverslips temporarily held onto slides by slit polyethylene tubing, then lifting off the coverslips (1968: 123-124). From a biologic viewpoint, the most interesting aspect of Jebram's work is his apparent ability to induce bizarre structures or major changes in morphological proportions in zooids by altering laboratory conditions.

Bryozoa have remarkable regenerative powers. Even when a colony, whether it be in an open or closed seawater system, is apparently dead, it may be only dormant, i.e. all the polypides may be dead but if the cystids of even a few zooids remain alive they can initiate regeneration of that colony after weeks or months of dormancy (P.L. Cook, pers. comm.).

Sexual Reproduction

After the bryozoan colony has reached a certain size through asexual budding -- and this size appears to be a function of both the species and the ambient temperature -- the gonads in at least some of its component zooids will mature. The great majority of species are hermaphroditic (Hippothoa hyalina is the best known exception), the testis usually developing earlier than the more distal ovary. The mechanisms of fertilization are not yet well understood, except in a few species (Silen, 1966; Bullivant, 1967); both self-fertilization and fertilization by gametes from another zooid in the same colony are undoubtedly very common.

Bryozoan larvae are highly modified free-swimming trochophores, of two basic types. The planktotrophic, so-called "Cyphonautes" larva develops from a fertilized non-brooded egg shed directly into the sea, where it remains in the meroplankton, feeding and growing actively, for a relatively long time (up to two months) before settling and metamorphosis. According to Lagaaij and Cook (1973:490) some 20 species, including many ctenostomes and some of the allegedly more primitive cheilostomes (membraniporids and electrinids) possess this type of larva. The lecithotrophic larvae are brooded, internally in a few species but usually in a special external brood chamber or ovicell, in which they are nourished by their self-contained yolk until they are expelled into the sea where they can survive only for a brief interval (1-24 hrs.) before either settling or dying (Wisely, 1958). All cyclostomes, most cheilostomes and some ctenostomes, produce this type of larva. Woollacott and Zimmer (1972a, 1972b) suggest that nutrients derived from the maternal zooid may supplement, or possibly substitute for, the yolk supply in some lecithotrophic larvae.

Planktotrophic Larvae: The cyphonautes larva is a laterally compressed sail-shaped plankter, triangular in side view, housed in a more or less transparent bivalved shell closed by an adductor muscle (Figure 1). The most detailed description of this type of larva is still that of Kupelwieser (1905) on Electra pilosa.

The larva possesses a fully functional ciliated gut consisting of a pharynx separated by an esophageal valve or "mouth" from the midgut (where food particles are spun in a rotating cord),

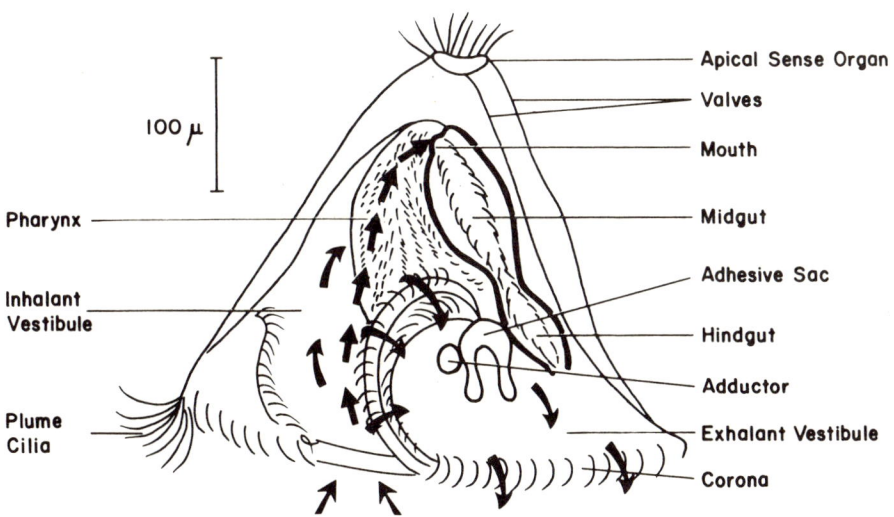

Fig. 1. Schematic generalized diagram, planktotrophic cyphonautes bryozoan larva. The longer arrows are food currents, the shorter water currents. See text for further explanation. (After Atkins, 1955b and Ryland 1970b).

which is in turn separated from the hindgut by a "pyloric" constriction. As a continuous water current passes from an anterior inhalant chamber into a posterior exhalant chamber it crosses a pair of ciliated ridges which intercept phytoplankton and direct it to the pharynx; fecal material is expelled from the hindgut into the exhalant chamber (Atkins, 1955b).

The larva swims by means of a basal ring of cilia (the corona), its apical organ directed forward and the ciliary plume forward and downward. The adhesive organ in the exhalant chamber and a cement-secreting pyriform organ (not shown in Fig. 1) both grow proportionately much more rapidly than the larva itself, eventually greatly reducing the size of these chambers. It may be that it is this progressive restriction of water currents, as well as increase in specific gravity and consequent difficulty in flotation, that precipitates settling, usually when the cyphonautes is between 200μ and 400μ in length, depending on the species.

Cyphonautes larvae are easily recognized in concentrates from plankton tows and are available from at least mid-April to late November off northeastern United States shores, all year in warmer waters. Publications helpful in identification of planktotrophic larvae are Lohman (1911), Atkins (1955a), Cook (1960, 1962, 1964), Ryland (1964, 1965b), Cook and Hayward (1966), and Mawatari and Itô (1972). In the laboratory the stage of development of the larva can be judged by the size of the adhesive sac and the pyriform organ relative to the larva itself. Larvae approaching metamorphosis can be placed in petri dishes for immediate observation or, if early growth stages are desired, in aquaria containing slides for their attachment.

Lecithotrophic Larvae: Lecithotrophic larvae (Figure 2), usually brooded one at a time except in cyclostomes where polyembryony obtains, vary considerably in external appearance but share certain common characteristics -- a rudimentary or non-functioning gut, and an apical sense organ, adhesive sac, and plume cilia similar to those of the cyphonautes. Calvet (1900: pl. 11) illustrates some common lecithotrophic larvae. The coronal cilia may be concentrated in a median band or distributed in elongate meridional bands.

Some lecithotrophic larvae, notably those of the bugulids, possess pigmented "eye spots" (Figure 2a) which may act as photoreceptors (Woollacott and Zimmer, 1972c). Live embryos exhibit a characteristic color, derived from carotenoid pigments, which Ryland (1958) considers species specific.

Since most of the common shore species are initially photopositive (Thorson, 1964; Table 1, p. 193) the simplest way to procure lecithotrophic larvae in quantity is to isolate a large parent colony in a finger bowl of seawater and keep it in the dark until larvae are desired. Within 15 minutes to 3-4 hours after exposure to light larvae will be released and may be pipetted from the surface (Grave, 1930: 361; Lynch, 1960: 523; Ryland, 1959: 616, 1960b: 786). The actual process of emergence of the larva from the ovicell has been described for only very few species (Cook, 1968; Mawatari, 1952; Wisely, 1958).

In order to know when and where to collect fertile parent colonies, knowledge of the breeding season is obviously necessary. With the exception of Flustrellida hispida, which breeds in the spring, fertile colonies of the species in Table I may be expected during the warmer months of the year in all three geographic zones. Eggleston (1972) discusses the pattern of reproduction in many of the common north Atlantic bryozoa. Other data on breeding season are scattered throughout the literature, the most useful sources being Barrois (1877), Marcus (1926), McDougall (1943), Fuller (1946), Rogick and Croasdale (1949), Maturo (1959), Gautier (1962), Ryland (1963), Gordon (1970), and Abbott (1973).

Settling and Metamorphosis: The behavior of both types of bryozoan larvae, reviewed recently by Ryland (1967a: 360-362) and Ryland and Stebbing (1971: 105-108), largely determines its site of attachment and metamorphosis. Perhaps the most important of the ex-

BRYOZOA 167

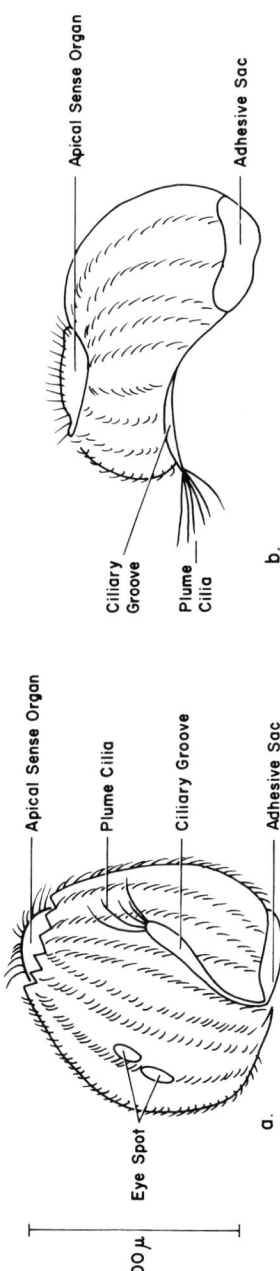

Figure 2. Schematic generalized diagrams, lecithotrophic bryozoan larvae. a) Bugula. b) Hippothoa. See text for further explanation.

ternal factors known to influence settlement is the nature and availability of the required substrate. The experiments of Ryland (1959, 1962b) demonstrate that some species prefer certain algae as a substrate, and the observations of Stebbing (1973) show that bryozoa must compete with other epiphytes for substrate space. The influence of texture of the substrate (Crisp and Ryland, 1960) and its orientation with regard to light (Maturo, 1959; McDougall, 1943; and Ryland, 1960b, 1962a), appear somewhat unpredictable and may possibly be species specific. Lynch's 15 years of experimental research on the effects of temperature and chemical ions are on the rate of metamorphosis summarized in his last papers (1960, 1961).

The process of metamorphosis itself, similar for both types of larvae, is one of enormous changes. After exploring the potential substrate with its ciliary plume, the larva first attaches itself with a secretion from the pyriform gland, then the adhesive sac suddenly everts. The internal anatomy of the erstwhile larva is completely altered into that of the primal zooid(s) or ancestrula. Waters' plates (1924, 1925a, 1925b, 1926) depict a wide range of such ancestrulae. Figure 3 of this chapter is a photograph of a just metamorphosed cyphonautes of Electra pilosa, with the valves not yet fallen away. Recent studies on metamorphosis and early colony development are those of Nielsen (1970), Gordon (1971), Woollacott and Zimmer (1971), Eitan (1972), and Soule and Soule (1972).

BRYOZOA AND THE OYSTER COMMUNITY

The association of bryozoa and the oyster is a close and -- to the oyster -- not necessarily always a completely felicitous one. A heavy bryozoan overgrowth does not ordinarily trouble an adult oyster, but occasionally some species which grow very rapidly at warm temperatures (Membranipora tenuis, Electra "crustulenta", Schizoporella unicornis) (Osburn, 1944: 33) may grow around the lip of the oyster and keep it from closing. It is also possible but unlikely that a very heavy bryozoan overgrowth may interfere with the nutrition of the oyster by intercepting a disproportionate share of phytoplankton.

The relationship of the bryozoan with oyster larvae and spat is somewhat more ambiguous. Bryozoan colonies preempt space and hard shell substrate otherwise available to spat; they can rapidly overgrow the spat and keep it from opening, as has occurred in Brittany (G. Lutaud, pers. comm.). It has been alleged that bryozoa capture and eat oyster spat. This appears highly improb-

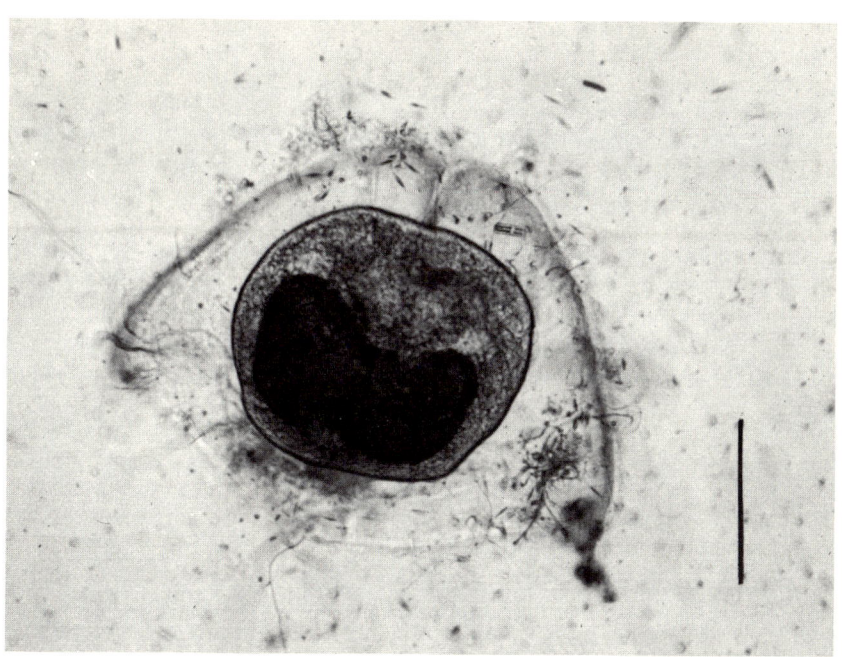

Figure 3. Ancestrula of <u>Electra pilosa</u>. The just-metamorphosed cyphonautes larva has not yet shed its bivalve shell. Photograph by J.L. Simon. Slide catalogue No. 878. 701b, Marine Biological Laboratory (Systematics-Ecology Program), Woods Hole.

able. The maximum size of particle that can be ingested by the largest known larva, a cyphonautes (<u>Membranipora membranacea</u>) is 30µ (Atkins, 1955b: 465); the minimum size of the larva of <u>Crassostrea virginica</u> and Ostrea edulis is 40-50µ at the beginning of the trochophore stage (Galtsoff, 1964: 355), the late trochophore is 50-75µ, the early veliger is 200µ, the fully developed veliger at settling, 300µ in diameter (Ibid: 358-363).

Certain bryozoan species appear consistently in the oyster community (Osburn, 1932, 1944; Hutchins, 1945; Galtsoff, 1964), although the component populations will vary in number and proportional representation of species with the particular geographic locality. The bryozoan species of the oyster community of the northeastern United States are listed in Table II, together with their mode of sexual reproduction, salinity tolerances and geographic range. Only three species (<u>Victorella pavida</u>, <u>Membranipora</u> tenuis and Electra "crustulenta" (which is not the true "crustulenta" of the European literature and which includes <u>Conopeum reticulum</u> and <u>Membranipora lacroixii</u> of American writers) can pose real problems in the oyster community in that they tolerate salinities as low (ca. 10 °/oo) as those supported by the oyster spat. Other members of the bryozoan community, with the possible exception of Schizoporella unicornis in Zones 2 and 3, are too stenohaline to constitute a serious threat.

THE OUTLOOK

There is reason to believe that research on both the basic and applied biology of the bryozoa stands at a threshold of major expansion. Several review works and manuals are in preparation or in press. A comprehensive revision of the Bassler treatise (1953), in at least five volumes (R.S. Boardman, Smithsonian Institution, ed.) will appear over the next several years, while the volume edited by Woollacott and Zimmer (publication anticipated in 1975) will stress the biology of modern bryozoa. Ryland and his colleagues (pers. comm.) have begun a three-year quantitative survey of the British bryozoa. Abbott is coordinating the Bryozoan volume for the National Marine Fisheries Service series, The Marine Flora and Fauna of Northeastern United States. Increasing numbers of workers of diverse skills in biology, paleontology, chemistry, and population genetics have been applying techniques as the SEM, computer, SCUBA, electrophoresis and epon-embedded sectioning (Nye, Dean, and Hinds, 1972) toward the development of a truly interdisciplinary approach to the biologic problems presented by this successful lophophorate phylum.

ACKNOWLEDGEMENTS

The writer is grateful to J.L. Simon, University of South Florida for permission to reproduce the photograph constituting Figure 3 and to A.G. Humes, Marine Biological Laboratory for assistance in the preparation of the other figures. The Rogick Collection on bryozoa at the library of the Marine Biological Laboratory, and the library itself, has been an invaluable resource.

REFERENCES

Abbott, M.B., 1973. Seasonal diversity and density of bryozoan populations of Block Island Sound (New York, U.S.A.), In: "Proceedings, International Bryozoology Association Conference for 1971" (G.P. Larwood, ed.) Art. 4, pp 37-51, Academic Press, London.

Allee, W.C., 1919. Note on animal distribution following a hard winter, Biol. Bull. 36: 96-104.

Allen, F.E., 1953. Distribution of marine invertebrates by ships, Austral. J. Mar. & Freshw. Res. 4: 307-316.

Atkins, D., 1955a. The cyphonautes larvae of the Plymouth area and the metamorphosis of Membranipora membranacea (L.), J. Mar. Biol. Assoc. U.K. 34: 441-449.

___ 1955b. The ciliary feeding mechanism of the cyphonautes larva, J. Mar. Biol. Assoc. U.K. 34: 451-466.

Barnawell, E.B., 1960. The carnivorous habit among the Polyplacophora, Veliger 2: 85-88.

Barrois, J., 1877. "Recherches sur l'embryologie des Bryozoaires", Imprimerie et libraire de Six-Horemans, Paris. 305 pp.

Bassler, R.S., 1953. "Treatise on Invertebrate Paleontology, Part G, Bryozoa", Univ. Kansas Press, Lawrence. 253 pp.

Borg, F., 1926. Studies on recent cyclostomatous bryozoa, Zool. Bidrag Uppsala 10: 181-507.

Brien, P., 1960. Classe des Bryozoaires, In: "Traité de Zoologie" (P. Grassé, ed.) 5(2): 1053-1335 (Gymnolaemata: 1167-1335).

Bullivant, J.S., 1967. Release of sperm by bryozoa, Ophelia 4: 139-142.

___ 1968. The rate of feeding of the bryozoan Zoobotryon verticillatum, New Zealand J. Mar. Freshw. Res. 2: 111-134.

Calvet, L., 1900. Contribution a l'histoire naturelle des bryozoaires ectoproctes marine, Trav. Inst. Zool. Univ. Montpellier (Ser. 2), Mém. 8: 11-488.

Cook, P.L., 1960. The development of Electra crustulenta (Pallas) (Polyzoa, Ectoprocta), Essex Naturalist 30: 258-266.

___ 1962. The early larval development of Membranipora seurati (Canu) and Electra crustulenta (Pallas), Polyzoa, Cah. Biol. Mar. 3: 57-60.

___ 1963. Observations on live lunulitiform zoaria of polyzoa, Cah. Biol. Mar. 4: 407-413.

___ 1964. The development of Electra monostachys (Busk) and Conopeum reticulum (L.), Polyzoa, Anasca, Cah. Biol. Mar. 5: 391-397.

___ 1968. Observations on living bryozoa, In: "Proceedings, First International Conference on Bryozoa" (E. Annoscia, ed.) Atti. Soc. Ital. Sci. Nat. e Museo Civ. di Storia Nat. Milano 108: 155-160.

Cook, P.L. and Hayward, P.J., 1966. The development of Conopeum seurati (Canu) and some other species of Membraniporine polyzoa, Cah. Biol. Mar. 7: 437-443.

Crisp, D.J. and J.S. Ryland, 1960. Influence of filming and of surface texture on the settlement of marine organisms, Nature 185: 119 only.

Dudley, J.W., 1970. Differential utilization of phytoplankton food resources by marine ectoprocts, Biol. Bull. 139: 420 (abst.)

Eggleston, D., 1972. Patterns of reproduction in the marine Ectoprocta of the Isle of Man, J. Nat. Hist. 6: 31-38.

Eitan, G., 1972. Types of metamorphosis and early astogeny in Hippopodina feegensis (Busk) (Bryozoa-Ascophora), J. Exp'l. Mar. Biol. & Ecol. 8: 27-30.

Fry, W.G., 1965. The feeding mechanism and preferred food of three species of Pycnogonida, Bull. Brit. Mus. Nat. Hist. (Zool.) 12: 195-223.

Fuller, J.L., 1946. Season of attachment and growth of sedentary marine organisms at Lamoine, Maine, Ecol. 27: 150-158.

Galtsoff, P.S., 1964. The American Oyster: Crassostrea virginica Gmelin, Fishery Bull. Fish & Wildlife Serv. 64: 1-480.

Gautier, Y.V., 1962. "Recherches ecologiques sur les bryozoaires chilostomes en mediterranée occidentale", Centre Regional de Documentation Pedagogique, Aix-en-Provence. 434 pp.

Gordon, D.P., 1970. Reproductive ecology of some northern New Zealand bryozoa, Cah. Biol. Mar. 11: 307-323.

____ 1972. Biological relationships of an intertidal bryozoan population, J. Nat. Hist. 6: 503-514.

Grave, B.H., 1930. The natural history of Bugula flabellata at Woods Hole, Mass., including the behavior and attachment of the larvae, J. Morph. & Physiol. 49: 355-383.

Hauenschild, C., 1962. Die Zucht mariner Wirbelloser im Laboratorium (Methode und Anwendung), Kieler Meeresforsch. 18: 28-37.

Hutchins, L.W., 1945. An annotated check list of the salt-water bryozoa of Long Island Sound, Trans. Conn. Acad. Arts & Sci. 36: 533-551.

Hyman, L.H., 1959. "The Invertebrates: Smaller Coelomate Groups", McGraw-Hill, New York. 783 pp. (Bryozoa, as Phylum Ectoprocta), pp. 275-515.

Jebram, D., 1968. A cultivation method for saltwater bryozoa and an example for experimental biology, In: "Proceedings, First International Conference on Bryozoa" (E. Annoscia, ed.) Atti. Soc. Ital. Sci. Nat. e Museo Civ. di Storia Nat. Milano 108: 119-128.

____ 1969. Bryozoen als Holzschädlinge im Brackwasser, Kieler Meeresforsch. 25: 224-231.

____ 1973. The importance of differential growth directions in the Phylaetolaemata and Gymnolaemata for reconstructing the phylogeny of the Bryozoa, In: "Proceedings, International Bryozoology Association Conference for 1971" (G.P. Larwood, ed.) Art. 54: 565-576, Academic Press, London.

Kaufmann, K.W., 1971. The form and functions of the avicularia of Bugula (Phylum Ectoprocta), Postilla Yale Univ. 151: 1-26.

Kaissling, K.E., 1963. Die phototropische Reaktion der Zooide von Bugula avicularia (L.), Zeitsch. Vergl. Physiol. 46: 541-594.

Kawahara, T. and H. Iizima, 1960. On the composition of marine fouling communities at various depths in Ago Bay, Rept. Fac. Fish. Perfect. Univ. Mie 3: 582-594.

Kupelwieser, H., 1905. Untersuchungen über den feineren Bau und die Metamorphose des Cyphonautes, Zoologica (Stuttgart). 19: 1-50.

Lagaaij, R. and P.L. Cook, 1973. Some Tertiary to Recent Bryozoa, In: "Atlas of Paleobiogeography" (A. Hallam, ed.) pp. 489-498, Elsevier, Amsterdam.

Lohmann, H., 1911. Die Cyphonautes der nordischen Meere, Nord. Plank. 9: 31-40.

Lynch, W.F., 1960. Problems of the mechanism involved in the metamorphosis of Bugula and Amaroecium larvae, Proc. Iowa Acad. Sci. 67: 522-531.

―――― 1961. Extrinsic factors influencing metamorphosis in bryozoan and ascidian larvae, Amer. Zoologist 1: 59-66.

Marcus, E., 1926. Beobactungen und Versuche an lebenden Meeresbryozoen, Zool. Jahb. (Abt. f. Syst., Okol., & Georgr.) 52: 1-102.

Maturo, F.J.S., Jr., 1957. A study of the bryozoa of Beaufort, North Carolina, and vicinity, J. Elisha Mitchell Sci. Soc. 73: 11-68.

―――― 1959. Seasonal distribution and settling rates of estuarine bryozoa, Ecol. 40: 116-127.

―――― 1968. The distributional pattern of the bryozoa of the east coast of the United States exclusive of New England, In: "Proceedings, First International Conference on Bryozoa" (E. Annoscia, ed.) Atti. Soc. Ital. Sci. Nat. e museo Civ. di Storia Nat. Milano 108: 261-284.

―――― 1973. Offspring variation from known maternal stocks of Parasmittina nitida (Verrill), In: "Proceedings, International Bryozoology Association Conference for 1971" (G.P. Larwood, ed.) Art. 55, pp 577-584, Academic Press, London.

Maturo, F.J.S., Jr. and T.J.M. Schopf, 1968. Ectoproct and entoproct type material: a re-examination of species from New England and Bermuda named by A.E. Verrill, J.W. Dawson, and E. Desor, Postilla, 120: 1-95.

Mawatari, S., 1951a. Natural history of a common fouling bryozoan, Bugula neritina (L.), Misc. Rep. Inst. Nat. Res. (Tokyo) 19-21: 47-54.

―――― 1951b. On Tricellaria occidentalis (Trask), one of the fouling bryozoans in Japan, Misc. Rep. Inst. Nat. Res. (Tokyo) 22: 9-16.

―――― 1952. On Watersipora cucullata I, II, Misc. Rep. Inst. Nat. Res. (Tokyo) 25: 14-27.

―――― 1953. On Electra angulata Levinsen, one of the fouling bryozoans in Japan, Misc. Rep. Inst. Nat. Res. (Tokyo) 32: 5-10.

Mawatari, S.F. and Ito, T., 1972. The morphology of Cyphonautes larva of Membranipora serrilamella Osburn from Hokkaido, J. Fac. Sci. Hokkaido Univ. (Ser. 4, Zool.) 18: 400-405.

Mayr, E., 1968. Bryozoa versus Ectoprocta, System. Zool. 17: 213-216.

McDougall, K.D., 1943. Sessile marine invertebrates at Beaufort, North Carolina, Ecol. Monog. 13: 321-374.

Miller, M.C., 1961. Distribution and food of the nudibranchiate mollusca of the south of the Isle of Man, J. Animal Ecol. 30: 95-116.

―――― 1965. Grazing carnivores -- some sea slugs feeding on sedentary invertebrates, Poirieria 3: 1-11.

Morales, E. and E. Arias, 1965a. Los organismos adherentes del puerto de Barcelona y ensayos efectuados con pinturas antiincrustantes, Réun Product. Pesq., Invest. Pesq. (Barcelona) 5: 150-157.

___1965b. Ecologiá del puerto de Barcelona y desarrollo de adherencias organicas sobre placas sumergidas, Invest. Pesq. 28: 49-79. (English summary).

Nair, N.B., 1962. Ecology of marine fouling and wood-boring organisms of western Norway, Sarsia 8: 1-88.

Nikolic, M., 1959. Polyzoan colonies in fouling on floating vessels in the northern Adriatic (Rovinj), Proc. Gen. Fish. Council for the Mediterranean 5: 231-234.

Nielsen, C., 1970. On metamorphosis and ancestrula formation in cyclostomatous bryozoans, Ophelia 7: 217-256.

Nye, O.B., D.A. Dean and R.W. Hinds, 1972. Improved thin section techniques for fossil and recent organisms, J. Palen. 46: 271-275.

Osburn, R.C., 1912. The Bryozoa of the Woods Hole region, Bull. Bur. Fisheries 30: 205-266.

___1932. Bryozoa from Chesapeake Bay, Ohio J. Sci. 32: 441-446.

___1933. The Bryozoa of the Mt. Desert region, In: "Biological Survey of the Mt. Desert Island Region" (W. Proctor, ed.) pp 291-385, Wistar Institute, Philadelphia.

___1944. A survey of the bryozoa of Chesapeake Bay, Dept. Res. & Ed., Bd. of Nat. Res. Pub. 63: 3-58.

Persoone, G., 1965. The importance of fouling in the harbor of Ostend in 1964, Helgoländer Wiss. Meer. 12: 444-448.

Powell, N.A., 1968. Studies on bryozoa (polyzoa) of the Bay of Fundy region, Part II: Bryozoa from fifty fathoms, Bay of Fundy, Cah. Biol. Mar. 9: 247-250.

___1970. Schizoporella unicornis - An alien bryozoan introduced into the Strait of Georgia, J. Fish. Res. Bd. Canada 27: 1847-1853.

Powell, N.A. and Crowell, G.D., 1967. Studies on bryozoa (polyzoa) of the Bay of Fundy region. Part I: Bryozoa from the intertidal zone of Minas Basin and Bay of Fundy, Cah. Biol. 8: 331-347.

Redfield, A.C. et al., eds., 1952. "Marine Fouling and its Prevention" Woods Hole Oceanographic Institution Publication 580, 388 pp.

Rogick, M.D., 1964. Phylum Ectoprocta, In: "Keys to Marine Invertebrates of the Woods Hole region (R.I. Smith, ed.) pp 167-180, Systematics-Ecology Program, Mar. Biol. Lab., Woods Hole.

Rogick, M.D. and H. Croasdale, 1949. Studies on marine bryozoa, Part III: Woods Hole region bryozoa associated with algae, Biol. Bull. 96: 32-69.

Ryland, J.S., 1958. Embryo color as a diagnostic character in polyzoa, Ann. & Mag. Nat. Hist. (Ser. 13) 1: 552-556.

___1959. Experiments on the selection of algal substrates by polyzoan larvae, J. Exp. Biol. 36: 613-631.

___1960a, The British species of Bugula (Polyzoa), Proc. Zool. Soc. London 134: 65-105.

___1960b. Experiments on the influence of light on the behavior of polyzoan larvae, J. Exp. Biol. 37: 783-800.

___1962a. The effects of temperature on the photic responses of polyzoan larvae, Sarsia 6: 41-48.
___1962b. The association between polyzoa and algal substrate, J. Anim. Ecol. 31: 331-338.
___1963. Systematic and biological studies on polyzoa (Bryozoa) from western Norway, Sarsia 14: 1-59.
___1964. Identification of some Cyphonautes larva (Polyzoa), J. Mar. Biol. Assoc. U.K. 44: 645-654.
___1965a. Polyzoa, In: "Catalogue of Marine Fouling Organisms" 2, pp 3-82, O.E.C.D., Paris.
___1965b. Polyzoa (Bryozoa), Order Cheilostomata. Cyphonautes larvae, In: "Fiches d'identification du zooplancton" (J.H. Fraser and B.J. Muus, eds.) No. 107, pp 1-5, Conseil Permanent International pour l'Exploration de la Mer, Copenhagen.
___1967. Polyzoa, Ann. Rev. Oceanog. & Mar. Biol. 5: 343-369.
___1970a. Bryozoa (polyzoa) and marine fouling, In: "Marine Borers, Fungi, and Fouling Organisms of Wood" (E.B.G. Jones and S.K. Eltringham, eds.), pp 137-154. O.E.C.D., Paris.
___1970b. "Bryozoans" Hutchinson University Library, London. 175 pp. (Paperback)
Ryland, J.S. and A.R.D. Stebbing, 1971. Settlement and oriented growth in epiphytic and epizoic bryozoans, In: "Fourth European Marine Biology Symposium" (D.J. Crisp, ed.) pp 105-123. Cambridge University Press.
Schneider, D., 1955. Phototropisches Wachstum von Bugula, Die Naturwissenschafter 42: 48-49.
___1957a. Analysis of cuticle microstructure of marine bryozoa, Werkzeitsch. 5: 60-63.
___1957b. Orientiertes Wachstum von Calcit-Kristallen in der Cuticula mariner Bryozoen, Verh. Deutsch. Zool. Ges. Graz (for 1957): pp 250-255.
___1959. Die Aufbau der Bugula Tierstöcke und seine Beeinflussung durch Aussenfaktoren, Biol. Zbl. 78: 250-283.
Schneider, D. and K.E. Kaissling, 1964. Wachstum und Phototropismus bei Moostieren, Naturwissensch. 51: 127-234.
Silen, L., 1966. On the fertilization problem in the gymnolaematous Bryozoa, Ophelia 3: 113-140.
Skerman, T.M., 1958. Marine fouling at the port of Lyttelton, New Zealand J. Sci. 1: 224-256.
___1959. Marine fouling at the port of Auckland, New Zealand J. Sci. 2: 57-94.
Soule, D.F. and J.D. Soule, 1972. Ancestrulae and body wall morphogenesis of some Hawaiian and eastern Pacific Smittinidae (Bryozoa), Trans. Amer. Micros. Soc. 91: 251-260.
Starostin, I.V. (ed.), 1963. Marine fouling and wood borers, Trudy Inst. Okeanol. 70: 1-279. (22 papers, in Russian; English table of contents (278-279) and short summaries).
Stebbing, A.R.D., 1973. Competition for space between the epiphytes of Fucus serratus L., J. Mar. Biol. Assoc. U.K. 53: 247-259.

Stickney, A.P. and L.D. Stringer, 1957. A study of the invertebrate bottom fauna of Greenwich Bay, Rhode Island, Ecol. 38: 111-122.

Thorson, G., 1964. Light as an ecological factor in the dispersal and settlement of marine bottom invertebrates, Ophelia 1: 167-208.

Wass, M.L. (ed.) 1972. A check list of the biota of lower Chesapeake Bay, Spec. Sci. Rep. Virg. Inst. Mar. Sci. 65: 1-290.

Waters, A.W., 1924. The ancestrula of "Membranipora pilosa" L. and of other cheilostomatous Bryozoa. Part I., Ann. & Mag. Nat. Hist. (Ser.9) 14: 594-612.

___ 1925a. Ancestrulae of cheilostomatous Bryozoa. Part II., Ann. & Mag. Nat. Hist. (Ser. 9) 15: 341-352.

___ 1925b. Ancestrulae of cheilostomatous Bryozoa. Part III., Ann. & Mag. Nat. His. (Ser. 9) 16: 529-545.

___ 1926. Ancestrula and frontal of cheilostomatous Bryozoa. Part IV. Ann. & Mag. Nat. His. (Ser. 9) 17: 425-439.

Watts, E., 1957. A survey of the Bryozoa in the southwest portion of Delaware Bay, with special reference to those species occurring on the blue crab, Callinectes sapidus, Univ. Del. Mar. Lab. Ref. 57: 1-19.

Wisely, B., 1958. The settling and some experimental reactions of the bryozoan larva, Watersipora cucullata (Busk), Austral. J. Mar. & Freshw. Res. 9: 362-371.

Woollacott, R.M. and R.L. Zimmer, 1971. Attachment and metamorphosis of the cheilo-ctenostome bryozoan, Bugula neritina, J. Morph. 134: 351-382.

___ 1972a. Origin and structure of the brood chamber in Bugula neritina (Bryozoa), Mar. Biol. 16: 165-170.

___ 1972b. A simplified placental-like brood chamber in Bugula neritina (Bryozoa), Proc. Elec. Micros. Soc. Amer. 13: 130-131.

___ 1972c. Fine structure of a potential photoreceptoral organ in the larva of Bugula neritina (Bryozoa), Zeitsch. f. Zellenforsch. & Mikroskop. Anat. 123: 458-469.

___ (eds.), "The Biology of Bryozoans", Academic Press, New York (in preparation).

Wyer, D. and P.E. King, 1973. Relationships between some British littoral and sublittoral Bryozoans and Pycnogonids, In: "Proceedings, Int. Bryozoology Assoc. Conf. for 1971" (G.P. Larwood ed.) Art. 18, pp. 119-208, Academic Press, London.

Young, D.K., K.D. Hobson, J.S. O'Connor, A.D. Michael, and M.A. Mills, 1971. "Quantitative analysis of the Cape Cod Bay Ecosystem", Systematics-Ecology Program, Marine Biological Laboratory, Woods Hole. pp. 71.

METHODS OF CULTURING POLYCHAETES

David Dean and Michael Mazurkiewicz

Ira C. Darling Center for Research, Teaching and Service; University of Maine, Walpole, Maine

The purpose of this paper is neither to review the literature on polychaete reproduction nor to describe the diverse reproductive habits and developmental stages that occur among polychaetes. Rather, this is an attempt to summarize the "do's, don't's, and probably's" pertaining to the successful culture of this fascinating group of marine invertebrates. The Polychaeta, one of the most prevalent components of almost any marine benthic community, has recently been selected by the Environmental Protection Agency of the United States as one of the groups of organisms to be used in bioassays (both adult and developmental stages) of man's transgressions upon the marine environment. A synopsis of polychaete culture is therefore timely.

Much of what we have to say also applies to the culture of other marine invertebrates. A few species of polychaetes are extremely hardy and do not require particular pains to culture; the majority, however, are not in this category. While good common sense, care in the handling of adults, eggs, and embryos, and the cleanliness of tools and dishes are the golden rule for success in the culture of any marine invertebrate, we will describe the various steps and techniques which have worked best for polychaete culturists. Although some techniques can be used for all life history stages, we will focus first on those which are pertinent to adults and then on to those which are particularly applicable to eggs, embryos, and larvae.

ADULTS

Field Collection

The first step in working with adult polychaetes is to obtain specimens in good healthy condition. Infaunal species should be removed from the sediment unbroken, if possible. Separation from the sediment may be done either in the field or back in the laboratory. A gentle stream of salt water applied to the sediment sample facilitates the removal of specimens intact. It is preferable, although not always practical, to place each specimen in its own container. If the specimens are not isolated 1) premature mass spawning may take place if they are ripe, 2) high mortality may occur if, as in the case of Glycera dibranchiata, the contents of the ruptured worm in the spawning condition are released into a container of other worms, and 3) considerable damage may occur due to aggressive behavior among individuals, e.g. as in Nereis succinea.

In transporting specimens to the laboratory, temperature extremes should be avoided, especially high temperatures. If the worms are carried to the laboratory in sediment, the latter serves as a temperature buffer; regardless, insulated coolers should be used. Polychaetes in general do best in moist substrates exposed to air.

Laboratory Procedures

Culture Systems and Equipment: There are two basic types of culture systems: closed systems and once-through, open systems. There are several grades of closed systems that range from standing water to water kept in motion via aerators, stirrers, or shakers to recirculating water systems with filters. Each has been used successfully, but each also has its own advantages and disadvantages. Frequent changes of water are necessary, especially if food is added. Filtration systems diminish the frequency of water changes. The type of non-toxic culture vessel used is largely a matter of personal preference and availability. Finger bowls and petri dishes are perhaps the most commonly employed. Before re-using culture vessels they should be thoroughly cleaned. For most adults, washing the vessels with a bio-degradable detergent, followed by several rinses and air drying is sufficient. Some workers routinely follow the detergent treatment with an acid rinse, a rinse in tap water, two rinses in distilled water, and air drying. If a vessel has ever been used with preservatives, one can never be sure that the last traces of the preservative have disappeared. Therefore, as a standard operating procedure in our laboratory, we mark each vessel and every item of equipment that has been in contact with formaldehyde and never use these items for culturing live organisms. The seawater used in closed systems is usually filtered to insure against the introduction of foreign eggs and larvae. While the type of filtering varies among investigators, we routinely use Millipore "prefilters".

The open seawater system is the most commonly used method of maintaining large numbers of adult polychaetes, as long as the over-flow opening is properly screened. The big draw-back to this is the possibility of introducing foreign species, unless a filter system is installed. Filters in an open system can be a major maintenance problem. The use of positive over-flows from tanks or vessels is well worth the added cost of a through-hull connection. The alternative, siphon overflows, invariably fail sooner or later as a result of air-locks forming in the siphons.

Adequate temperature control is one of the most important considerations in the culture of polychaetes in the laboratory. For closed system cultures, either water baths or controlled-temperature cabinets may be used. Dean (1969) describes a large, versatile, inexpensive water bath system for the culture of polychaetes. Several scientific supply houses add a few controls to a standard household refrigerator and market the unit as a controlled-temperature cabinet at double the cost of the household refrigerator. We have found the following to be far superior but less expensive units. Wholesalers of soft-drinks in six packs sell industrial-quality single and double door merchandizing units to stores for display and refrigeration of soft-drinks. We obtain these units from the soft-drink wholesaler and install a variable thermostat (Type D5, Model #9001, B071 from United Electric Controls Company, Watertown, Mass.) (Fig. 1). These modified units permit temperatures to be controlled between $0°C$ to $20°C$ with an accuracy of about $\pm 0.5°C$.

Handling Adults: The ideal way to examine adult polychaetes under magnification is to relax the worms in a manner so that they can be fully recovered for continued culture studies. Several relaxants have been used with success, including a solution of 3.5 to 7% magnesium chloride or magnesium sulphate and propylene phenoxytol (the latter from Goldsmidt Chemical Corp., 153 Waverly Place, New York, New York, 10014). Propylene phenoxytol is excellent but requires experimentation by the user since sensitivity to the relaxant varies from species to species. We suggest preparing a 1.5%, by volume, solution of propylene phenoxytol in distilled water as a stock solution. Refrigeration of the stock solution enhances shelf-life. If the stock solution is made with salt water, the shelf-life is significantly shortened. Some workers dilute the stock 1-10 with seawater and place a worm directly

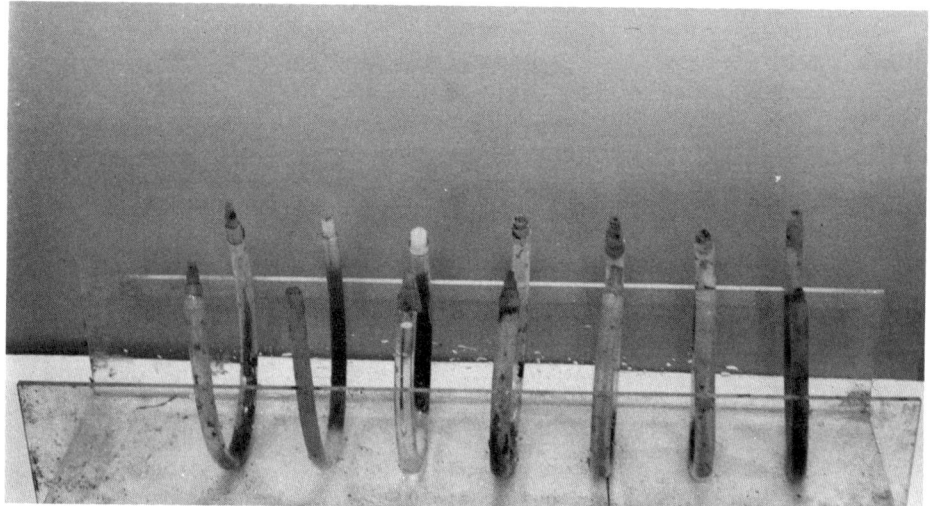

Fig. 1. Single door controlled temperature cabinet. The added thermostat is on the lower right.
Fig. 2. Sandworms, Nereis virens, being maintained in individual plastic tubes.
Fig. 3. Tubes (= artificial burrows) housing individual worms positioned upright in a plastic trough. The unit is ordinarily immersed in a seawater flow-through system.

in this solution for observation. We prefer to add the stock solution drop-by-drop to the vessel containing the adult polychaete. If the concentration is not allowed to become too strong, the adult will recover completely within a few minutes when placed in fresh seawater.

Frequently, the purpose of relaxing worms is to examine them for the presence of gametes which can ordinarily be seen through the body wall. However, punctures of the body wall with a capillary pipette may be necessary to obtain gametes for closer examination. We are currently exploring the technique of taking samples of coelomic contents from the individual worms, resuspending their coelomic fluid in filtered seawater, and obtaining a size-frequency distribution for coelomic components using a fine-particle counter (Coulter Counter).

Effects of Physical Factors: Laboratory maintenance of polychaetes requires consideration of several physical parameters. While some species of polychaetes, such as Ctenodrilus, Fabricia, and the archiannelid, Dinophilus, do well in standing cultures without a substrate, usually some form of substratum enhances maintenance. When Polydora spp. or Nereis spp. are maintained in a glass dish without sediment, there is a tendency for a mucoid constriction or ring to form around a portion of the body followed by severance of the body at that point. A thin layer of detritus and/or fine sediment works well for Capitella capitata, Laeonereis culveri, Polydora ligni and Aglaophamus sp. Thicker layers of sediment work well for Streblospio benedicti, Nereis spp., Glycera, and Potamilla, to name a few.

Several artificial substrates may also be used successfully. Absorbent cotton can be used if the amount of cotton is sufficient to extend from the bottom of the culture dish to the water surface so that the upper-most layer of cotton is exposed to the atmosphere. Nereids and glycerids do well under these conditions. Recently, the mineral, cryolite, has been reported as a useful substratum for maintaining infaunal organisms while still permitting observation within the substratum (Josephson and Flessa, 1972). Artificial burrows of plastic or glass tubing of appropriate diameter have also been used with success to maintain nereids, arenicolids, some spionids, and glycerids (Figs. 2 and 3). For the larger nereids and glycerids, partial plugging of the tube ends insures that the worms will remain within yet permits water flow through the tubes. Plugs are removed to add food. Handling individuals in this manner facilitates studies on food uptake, conversion efficiency, and growth, as shown by Goerke (1971).

Other physical factors of concern in maintaining adult polychaetes include temperature (mentioned previously), salinity, dissolved oxygen, and water quality, e.g., pollutants in the water such as toxic metal ions and the buildup of metabolic waste products.

Food: At temperatures of 5°C or less, many species of polychaetes can be kept for extended periods of time without the addition of food. Generally, however, there is a reduction of body size with time. As an example, a specimen of Nereis virens was kept in a refrigerator in our laboratory for over a year without the addition of food and appeared healthy but somewhat smaller at the end of the experiment. Various natural and artificial foods have been used with differing degrees of success, depending upon the species being cultured. Natural foods include detritus, cultured phytoplankters, fresh or dry macroscopic algae, and living or dead animals (Muller, 1962; Goerke, 1971). Artificial sources include liver powder, powdered alfalfa, pieces of frozen spinach (Akesson, 1967), and commercially available fish foods (Guerin, 1971). If sexually mature polychaetes have been collected from the field, it may not be necessary to feed them while being maintained in the laboratory preparatory to spawning.

The addition of food, especially non-living food, fosters bacterial buildup and a deterioration of water quality, particularly in static systems. Under these conditions the survival and subsequent development of spawned eggs, or released larvae, may be affected adversely.

Stimulating Adults to Spawn: Temperature manipulation is the most common method of stimulating polychaetes to spawn. Obviously, one can expect the best success in the laboratory spawning of polychaetes if the worms are collected just prior to the time they would ordinarily spawn in nature. To prevent premature spawning in the laboratory, ripe worms should be held at a slightly lower temperature than that of the water from which they were taken. Some species can be stimulated to spawn at times other than their normal season of spawning, for example, many polydorids, some nereids, Hydroides dianthus (Turner and Hanks, 1960) and Capitella capitata. In New England, during the period November through August, unripe adult Nereis succinea can be collected and held in the laboratory at 20-25°C. Within one to six weeks, heteronereid stages will develop and spawning can be induced. Adult Nereis succinea may also be maintained at 5°C and inhibited from maturing. As the demand arises, such individuals may be transferred to temperatures of 20-25°C to be conditioned for spawning. As with many marine invertebrates elevating the temperature to that which approximates the normal spawning temperature usually induces spawning. In some instances, the temperature elevation must be gradual (e.g., with Scolecolepides viridis, (George, 1966), or it may be more rapid (e.g. with Polydora ligni and Nereis succinea). Some species require no temperature manipulation to spawn. Hydroides hexagonus spawns almost immediately after being removed from its tube, while Lepidonotus squamatus and Podarke obscura spawn naturally at predictable times during night-time hours following their collection (Costello, et. al., 1957).

The presence of gametes and/or "pheromones" from the opposite sex may stimulate spawning. In the case of Nereis succinea, placement of the male heteronereid into water taken from a dish occupied by a mature female induces the male to release sperm. The male may also be induced to spawn by the presence of eggs. The addition of a small quantity of this sperm suspension to a dish containing a mature female heteronereid will, in turn, induce the latter to spawn. Since this species will also spawn when mature male and female heteronereids are placed in the same dish, resulting in polyspermy, the former technique should be followed.

Light can play a major role in both gametogenesis and the spawning of several polychaetes; Platynereis dumerilii is an example. Not only is the sexual development of this species affected by photoperiods, but its heteronereids appear, swarm, and reproduce periodically at a certain moon phase (Hauenschild, 1969). This rhythmicity can be duplicated in the laboratory, if the worms are kept under natural light conditions. Furthermore, the phase of the swarming rhythm can be shifted at will by using a lamp (either incandescent or fluorescent) to simulate moonlight (Hauenschild, loc. cit.).

Mechanical stimulation, will in some species, be enough to trigger the release of gametes. For example, a ripe male Laeonereis culveri will release sperm when gently prodded and forced to swim. Also, the larviparous species, Streblospio benedicti, will release larvae when handled roughly.

Fate of Adults After Spawning: The fate of adults after spawning is known for some species. Polydora ligni, for example, is capable of producing at least three successive broods of progeny. Streblospio benedicti has produced as many as six successive larval broods in our laboratory. As a generalization, those polychaetes which swarm during the reproductive process die after spawning. For instance, in the family Nereidae those species which spawn in a swarming heteronereid stage are short-lived, with death occurring one to four days after swarming. Greater longevity after spawning occurs amongst nereids with more specialized modes of reproduction, including atokal reproduction, incubation of developing eggs, and hermaphroditism.

EGGS, EMBRYOS, AND LARVAE

Procurement

Pelagic eggs, embryos, and larvae may be isolated from plankton samples. This is a tedious task and usually yields only small numbers of individuals in various stages of development. It is an approach that is taken only as a last resort if other, less tedious methods of procurement fail or if one wishes only to determine the identity of an unknown larva by rearing it to settlement and metamorphosis. Numerous closely-aged plankton-isolates can be obtained, however, if the species sought has a greatly abbreviated spawning season and if its pelagic life history stages are very abundant in the zooplankton. Such a species is the spionid, Scolecolepides viridis which reproduces in Maine waters during a two to three-week period from late February to early March with a spawning peak during the first week. Hence, the planktonic development of S. viridis can be followed readily by periodic sampling of the plankton (George, 1966). In Maine's Penobscot River estuary, we have observed S. viridis to be the dominant zooplankter during its spawning season. Numerous early developmental stages can be readily isolated from plankton samples. Pre-trochophore stages of S. viridis are especially easy to isolate by taking advantage of the photopositive responses of zooplankters associated with them, viz., by attracting the latter to light, thereby separating them from pre-trochophore stages of S. viridis which are not attracted to the light. The latter can then be removed from the sample by pipette. Large numbers of embryos at about equivalent stages in development can also be acquired by passing a plankton net through swarming polychaetes, such as epitokal glycerids and nereids. In this manner we procured embryos of the bloodworm Glycera dibranchiata. These approaches, of course, presuppose a sound knowledge of the time of spawning of desired species. Unfortunately, such knowledge is lacking for most polychaete species.

Another laborious method of procurement involves isolating demersal eggs or benthic larvae from meiobenthic samples. One is to collect a sample of the upper two centimeters of sediment for microscopic examination in the laboratory. The other is to dig a trench in a tidal flat, allowing the water to drain from the tidal flat into the trench, and then to pass the sediment-laden water through a fine plankton net for subsequent examination (Newell, 1949; Dales, 1950).

Large numbers of developmental stages can be obtained with greater facility from the natural habitat by collecting egg masses from the surface of the sediment or from tubes of brooding sedentary species. For example, gelatinous egg masses of Phyllodoce mucosa occur abundantly on tidal flats in Maine during April and May and afford ample material for culture. Eggs and embryos of Polydora spp. and Spirorbis spp. can be obtained readily upon opening parent tubes during appropriate seasons of reproduction.

By far the most ideal method of procuring early developmental stages is by inducing adults to spawn naturally in the laboratory using techniques previously discussed. Often, however, such attempts fail and alternative approaches must be sought. One approach is to rapidly collect swarming heteronereids prior to their release of sexual products. Individual heteronereids are isolated and brought to the laboratory where spawning is accomplished by reciprocal stimulation of the sexes. Another approach frequently used, with varying degrees of success, is to obtain gametes by rupturing the body walls of sexually ripe worms (compare the results of Dales, 1950, and Gilpin-Brown, 1959, with those of Bass and Brafield, 1972). Eggs released in this manner must be thoroughly washed free of coelomic contaminants prior to fertilization to enhance chances of success and to assure normal development. Washing is accomplished by either passing eggs onto fine-meshed sieves or by periodically suspending the eggs, allowing them to settle, and decanting the seawater. The latter technique is generally preferred since eggs are fragile and readily damaged when passed onto a sieve.

Standard precautionary measures taken when affecting the fertilization of freely-shed eggs include 1) the minimal addition of sperm suspension to a dish containing eggs (e.g., 2-3 drops of concentrated sperm-suspension to a large fingerbowl containing eggs) and 2) washing eggs several times within a few minutes after being mixed with sperm. Adherence to these measures alleviates problems associated with the presence of excess sperm, such as fouling of the culture water resulting from decomposition of dead sperm and the occurrence of polyspermy which leads to abnormal development (Costello et al., 1957).

With species that deposit egg capsules in tubes (e.g., Polydora spp., Fabricia sabella, Manayunkia aestuarina), mature females may be kept individually in culture dishes to be examined periodically for embryos and larvae. If the larvae are pelagic, they should be removed rapidly for culturing (e.g., Polydora spp. and Streblospio benedicti). If nonpelagic, the larvae may be maintained in culture with the parent if so desired. In our experience we have found that it is best to leave egg capsules in the tube of the brooding parent rather than to isolate them. When isolated, egg capsules are soon contaminated by bacteria or fungi and eggs rarely complete development. With the addition of antibiotics, however, eggs in isolated capsules can successfully develop (Blake, 1969).

Laboratory Rearing

<u>Culture Systems and Equipement</u>: A variety of culture systems and items of equipment can be used to rear developmental stages in a laboratory. Some of the vessels that we have used with success for closed-system, standing cultures are: finger bowls, plastic petri dishes, stender dishes, Boveri dishes, and funnels. Boveri dishes are especially useful for culturing small numbers of pelagic larvae. The curved walls permit microscopic observation of the entire water column while the flat bottom affords stability to the dish. Unfortunately, we have not found a supplier of these dishes in the U.S.A., but they can be obtained from Johann Gg. Bachofer, 7411 Degerschlacht ober Reutlingein, Germany. In view of their ideal utility for the culture of minute invertebrates, including larvae, it is curious why they aren't readily available in this country.

Small funnels (5 cm diameter) are very useful for culturing nonpelagic larvae that normally dwell in fine sediments. Prior to use, the stems are removed from the funnels and the bases are closed by fusing the glass with a torch. The funnel is filled with filtered seawater and sediment is added to form a layer about one to two centimeters thick. Larvae are then added and the funnel is covered with a watch glass. Since a relatively small amount of sediment can be concentrated to an appropriate thickness in the bottom of the funnel, the time required to search for individuals is greatly lessened. We have found funnels especially useful for rearing the larval sandworms, <u>Laeonereis culveri</u> and <u>Nereis virens</u>, but have also employed these vessels for housing individual gravid females of <u>Streblospio benedicti</u>.

Vessels used in closed-system, standing cultures are cleaned in the same manner as mentioned earlier in our discussion on adults. However, it has been our experience that vessels need to be even cleaner for early developmental stages than for adults. Similarly, water should be renewed more frequently in larval cultures, preferably every one or two days, using filtered seawater. Changing water in culture vessels can be extremely tedious and time consuming. In small vessels containing only a few larvae, the latter may be laboriously pipetted to a new dish containing fresh seawater. Also, a large-bore syringe fitted with finely meshed nylon netting over the open end can be used to withdraw old water from a culture dish leaving the larvae behind to be immersed in a fresh supply of water. This technique, unfortunately, has the disadvantage of leaving larvae in an old dish with much organic debris and increasingly flourishing microbial populations. Larger mass cultures are best handled by gently pouring the culture onto a nylon sieve partially immersed in water and backwashing larvae into a vessel with fresh seawater. Sieves require periodic cleaning to remove organic debris. This is easily done by immersing a sieve over-night in hydrogen peroxide (commercial grade cut 50% with distilled water), followed by a rinse in distilled water.

Closed-culture systems with water in motion require less frequent changes of water than do standing cultures and have the advantage that larvae do not tend to stick to the vessel as readily. Various stirrers, shakers, and aerators have been used to keep larvae in suspension and culture-water mildly in motion. Dr. Joseph Simon (pers. comm.) found that an air compressor near his culture bench imparted sufficient vibration to prevent polychaete larvae from adhering to the bottom of standing cultures. British workers claim to have reared polychaete larvae more successfully in plunger jars than in standing cultures, attributing their success to suspension of larvae in moving water (Wilson, 1932, 1968b).

Closed seawater systems in which the water is re-circulated through a filter have been used primarily to culture species with a nonpelagic development, e.g., Nereis arenaceodenta and Capitella capitata (Reish and Richards, 1966). To culture pelagic larvae in a closed system with an external filter, the re-circulating intake must be fitted with a screen having a large surface area to prevent larvae from being sucked against the screen. Such screens, however, become readily clogged and must be cleaned frequently. With hardy, short-lived pelagic larvae, the filter may be removed or by-passed until the larvae have settled and metamorphosed (Richards, 1967). Internal ground-bed filters have not been used extensively by polychaete culturists. Any of these closed, re-circulating systems must have a source of clean, oil-free air. We have used several makes and models of air pumps and have found that the "Silent Giant" model (available from many aquarium dealers) is excellent.

As with standing cultures, several types of open systems have been used. The simplest of these is a microchamber consisting of a tube with netting secured on both ends. Developmental stages are housed inside the tube which is suspended in the sea or placed on a flowing seawater table in the laboratory (Sveshnikov, 1955; Reish and Barnard, 1960). In using these chambers in the laboratory's flowing seawater system, we have found that the netting soon becomes fouled with sediment and undesirable biota, which restricts water flow through the chambers. Larval mortalities within these chambers have usually been higher than with other systems. The cirratulid, Ctenodrilus serratus, has been found to invade these chambers. Whether Ctenodrilus or associated biota contributed to larval mortality is not known. Grêve (1968) describes a more elaborate open system that employs the Plankton-kreisel, a round glass vessel with provisions for water replacement and ground-bed filter (Fig. 4). Using this device, Grêve

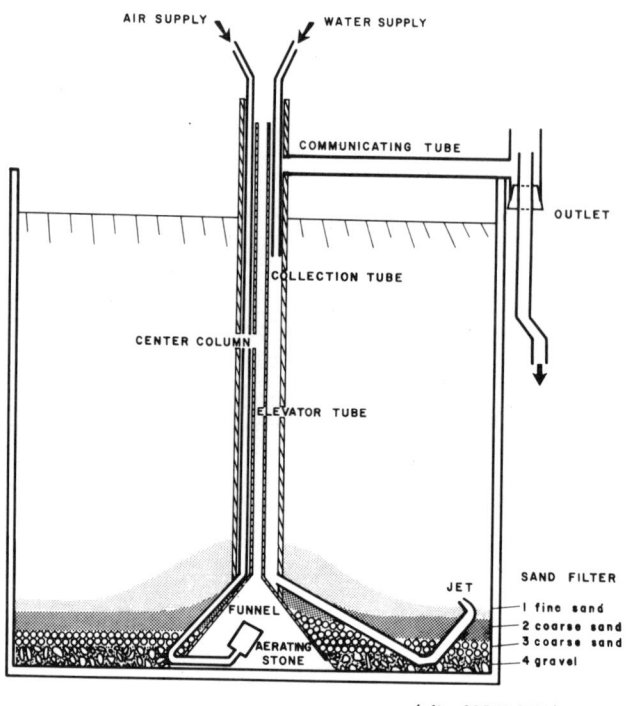

Fig. 4. Planktonkreisel, side view. The angle of the jet is critical for proper circulation within the cylindrical tank. The jet should be 55° from the tangent and elevated at an angle of 15° from the bottom.

successfully maintained larval <u>Spio filicornis</u> and a variety of larval crustaceans.

Suspended particulate matter in coastal seawater presents a serious problem to any open system used for culturing polychaete larvae. Screens fine enough to prevent the introduction of foreign species and the loss of cultured individuals rapidly become clogged with suspended particulate matter. The latter serves as an excellent culture medium for bacteria. Pre-system settling tanks remove a part of the suspended load, but the water may pick up metabolic by-products if allowed to stand too long. Filters used to remove particulate matter soon clog and can foul the water quickly. The researcher planning to use an open system with filtration for rearing polychaetes should be prepared for an excessive amount of constant maintenance.

Early developmental stages may be reared using temperature-control units similar to those used for adult cultures, as previously described.

Handling Early Developmental Stages for Observation

The highly mobile and delicate larval stages present a problem in handling and observing under high magnifications. Relaxants have been used with success (see earlier discussion) especially propylene phenoxytol in concentrations far more dilute than those recommended for adults. The appropriate concentration for the life stage and species being studied can best be determined by trial and error. If proper care is taken, larvae revive upon return to seawater.

Another good method of restricting larval movement for microscopic examination is the use of Saran Wrap coverslips (Dean and Hatfield, 1963). This involves placing the larva in a drop of water on a small strip of Saran Wrap and reflecting one end of the strip over the drop so that the larva remains within a Saran Wrap-sandwich. This sandwich can then be placed either side up on a regular glass slide for observation even under oil-immersion. A more conventional technique is the use of a hanging drop preparation, i.e., confining the larva to a drop of seawater on a coverslip which is then inverted over a depression slide. If a ring slide is used, enough water can be placed within the ring to just cover the bottom and the coverslip bearing the larva within a drop of water, can be inverted over the top of the ring. The top edge of the ring may also be sealed to form a microculture chamber in which a larva can be maintained and observed for two or more days.

We have found the use of Polaroid film to be a valuable aid in recording larval development and in the subsequent preparation of line drawings for publication. Type 47 film (ASA speed 3000) adequately stops the action of most larvae confined as described above. Only a few seconds are needed to determine whether a useful pictorial record has been obtained. A photograph, when placed on a dissecting microscope provided with a camera lucida, permits accurate reproduction of actual body form and characterizing larval features. Polaroid photographs of stage micrometers for each magnification used, provide a quick and accurate means of measuring larvae, eggs and embryos.

Effects of Physical Factors

As with other marine invertebrates, temperature and salinity markedly affect survival, growth, and development of polychaete

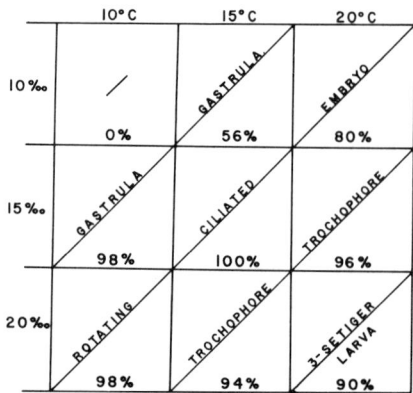

Fig. 5. Survival and development of <u>Nereis</u> <u>succinea</u> 2-cell embryos 48 hours after being placed in different temperature-salinity regimes. Diagonal lines pass through temperature-salinity combinations in which equivalent developmental stages were observed. % = % survival.

embryos and larvae. This is well demonstrated by the results of an experiment concerning the combined effect of temperature and salinity on survival and development of <u>Nereis</u> <u>succinea</u> embryos (Fig. 5). In this experiment, a brood of eggs was fertilized at $20^{\circ}C$ - 20 $^{\circ}/oo$, and upon undergoing the first cleavage, 50 embryos were placed into each of nine temperature-salinity combinations. They were examined 48 hours later for observations on survival and development. None of the embryos survived the lowest temperature-salinity combination of $10^{\circ}C$ - 10 $^{\circ}/oo$, and only 56% survived at $10^{\circ}C$ - 15 $^{\circ}/oo$. Survival was high (80-100%) in each of the other temperature-salinity combinations. Development during the run of the experiment was markedly influenced by both temperature and salinity. Rate of development was thus progressively increased at successively higher levels of salinity (10, 15, 20 $^{\circ}/oo$) and likewise, at successively higher temperature levels (10, 15, $20^{\circ}C$).

Indications that light has not received sufficient consideration as a factor influencing polychaete larval development is shown by the following preliminary experiment with <u>Polydora</u> <u>ligni</u> larvae. Equal numbers of larvae from the same brood were maintained under identical conditions, except that one group was placed in a normal day-night cycle, while the other was placed in total darkness. Figure 6 shows the stages of development found in a random sample from each group at the end of five and 13 days. A larger average size was obtained by larvae maintained under total darkness. The slower growth under the light-dark conditions may have been caused by the photo-positive response of the larvae. During

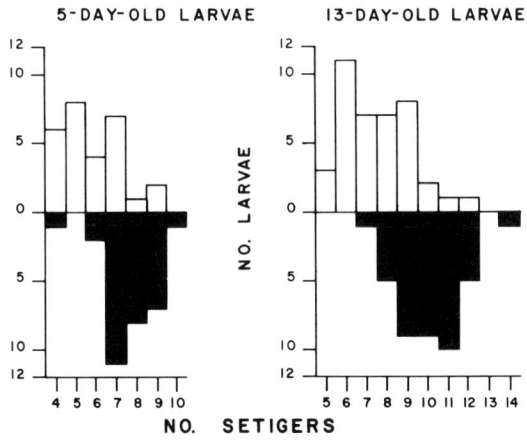

Fig. 6. Growth of *Polydora ligni* larvae under light-dark vs. total dark conditions.

daylight hours, larvae exposed to the light tended to congregate on the side of the vessel nearer the light source. Those kept in total darkness, however, appeared to be more randomly distributed throughout the water column, as observed when the vessel was briefly removed from the dark for examination. Since these were standing cultures, perhaps the addition of a stirring mechanism would have enhanced growth by keeping the larvae more randomly distributed when maintained under light-dark conditions.

Generally, dissolved oxygen concentrations occur at acceptable levels in culture systems. However, critically low D.O. levels may develop if 1) excess sperm is not washed off freely-spawned ova, 2) eggs, embryos, or larvae are over-crowded, and 3) larvae are overfed. Under these circumstances, high mortalities and abnormal development may be anticipated.

Poor water quality provides a real, although often a seemingly intangible, source of difficulty in the rearing of embryos and larvae. For example, D.P. Wilson (1951) reared *Ophelia bicornis* larvae in waters from two different sources and was consistently more successful with one of the sources. Since both water sources were about equivalent in salinity and were used under identical laboratory conditions, Wilson ascribed his results to differences in water quality between the sources. Poor water quality is usually attributable to pollutants and/or excessive concentrations of harmful metabolites. Frequent renewal of culture water, transfers to clean vessels, mild stirring, agitation or aeration, and filtration all aid in keeping metabolic waste products at minimal levels.

Microbial metabolites are particularly harmful and, therefore, microbes should be controlled by methods suggested later. Pollutants of greatest concern to culturists/researchers include heavy metals, pesticides, phenolic compounds, detergents, and lightweight hydrocarbons. The culturist should be concerned not only with the occurrence of pollutants in natural waters but also with their accidental introduction into laboratory cultures. For instance, the use of preservatives, especially formalin, should be discouraged in the vicinity of laboratory cultures. An observation by Wilson (1968a) concerning this matter is especially poignant. Wilson encountered high mortalities among sabellid larvae reared in a new culture room at his laboratory. This he attributed to cultures of the food organism, Isochrysis, which were made unhealthy by absorption into their culture medium of fumes from a formaldehyde-based adhesive used in construction of the laboratory furniture. Upon removal of the furniture, there were no further difficulties in rearing the sabellid larvae.

Larval Food

Pelagic lecithotrophic larvae (e.g., Capitella capitata and Spirorbis borealis) are relatively easy to rear to settlement since they do not require food and usually have a short pelagic existence. Nonpelagic larvae are also reared with comparative ease since they are benthic and may be maintained by methods analagous to those used with adults. These include the provision of an appropriate substrate and a food source such as 1) natural detritus and benthic diatoms, 2) dried powdered alfalfa or macroalgae, 3) settled concentrations of cultured algae (Phaeodactylum and Dunaliella are used in our laboratory).

The rearing of pelagic planktotrophic larvae presents certain difficulties concerning food sources. Recent advances in the culture of microalgae have greatly enhanced the rearing of many invertebrate larvae including polychaetes. Although such food sources as liver powder and dried, powdered alfalfa or algae, have been used with limited success, cultures of microalgae such as Dunaliella, Phaeodactylum and Isochrysis are much preferred food sources. Good quantitative data on food vs. polychaete-larval concentrations are not available. In our laboratory we have reared Streblospio benedicti and Polydora ligni at concentrations up to two larvae per milliliter, but one larva per milliliter or less is to be preferred. At these larval concentrations Dunaliella was added every other day so that the final concentration in the cultures ranged between $0.5\text{-}2 \times 10^4$ cells/ml.

Often associated with the addition of food to polychaete larval cultures is the buildup of bacteria, fungi, or ciliates. Non-

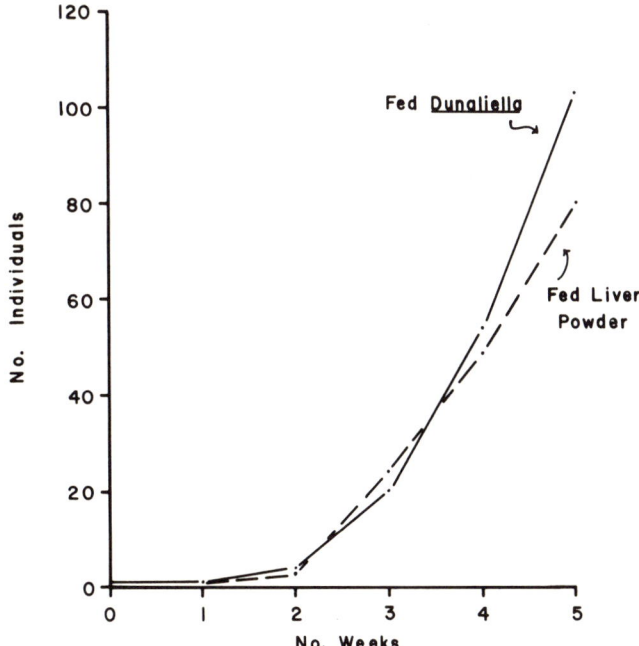

Fig. 7. Population growth of <u>Dinophilus gyrociliatus</u> when fed liver powder vs. <u>Dunaliella</u>.

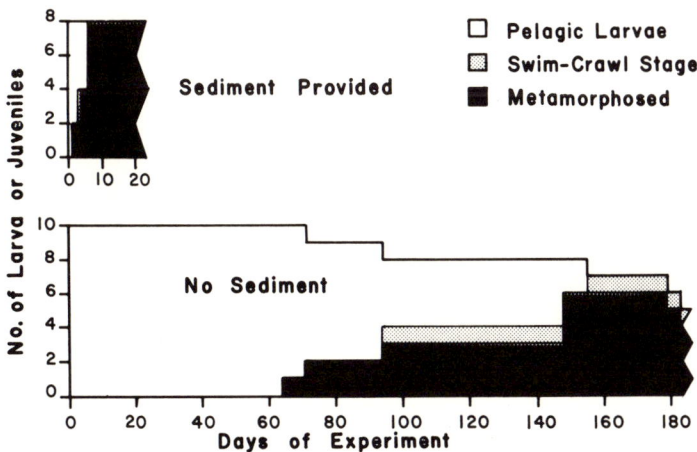

Fig. 8. <u>Spio filicornis</u> larval settlement and metamorphosis.

living food sources such as liver powder and pulverized plant material promote a more rapid buildup of undesirable microorganisms than do living foods. In most instances the buildup of these microorganisms is detrimental to the polychaetes being cultured. A notable exception is the culture of the archiannelid, Dinophilus gyrociliatus (Fig. 7). This organism survives on liver powder and indeed may even utilize bacteria as a source of nourishment. The control of unwanted microorganisms can best be accomplished by 1) adhering to good sterile procedures in the culture of living foods, 2) proper cleanliness of culture vessels, 3) not over-feeding, 4) frequent changes of water and culture vessels, and 5) the use of antibiotics, although to our knowledge, their use has not been adequately evaluated by polychaete culturists.

Settlement and Metamorphosis

In culturing pelagic polychaete larvae, there would appear to be no phase more critical than that during settlement and metamorphosis. The presence of a substrate facilitates the settlement and metamorphosis of larvae. Some species, such as Polydora ligni, may settle and metamorphose indiscriminately on any substrate. However, many species are more selective so that if an appropriate substrate is lacking, settlement and metamorphosis will be delayed or even prevented. A good example of this phenomenon is shown in Figure 8. Eighteen Spio filicornis larvae were isolated from the plankton, eight were placed in a finger bowl containing sediment, and 10 were placed in a similar bowl without sediment. Within six days all larvae in the bowl with sediment had settled and metamorphosed. In the bowl without sediment, the first larva metamorphosed 64 days after the start of the experiment; one larva remained pelagic for 155 days and then assumed the benthic-pelagic, swim-crawl stage and remained in this condition until day 183 when the experiment was terminated. The mortality of pelagic larvae began on day 71 and continued with further delay of metamorphosis. These data, although dramatic, are not atypical of polychaete settlement and metamorphosis. Indeed, it has even been shown that some species are highly specific concerning the nature of the substrata (Day and Wilson, 1934; Wilson, 1952, 1953a, 1953b). Appropriate substrates are also required for culturing species with nonpelagic development. Some of these, such as the epibenthic species Dinophilus gardineri, D. gyrociliatus, Ctenodrilus serratus, Fabricia sabella, and Manayunkia estuarina can readily be cultured in vessels devoid of sediments; some infaunal forms such as Laeonereis culveri, Nereis virens, and Amphicteis floridanus, require sediments similar to those found in the adult habitat. Even differences in grain size of the sediment is sometimes important. For example, the nonpelagic larvae of Laeonereis culveri can be reared successfully in a fine flocculent sediment and remain in the sediment throughout development. On the other hand, if these larvae are placed in a vessel containing only a course-grained sediment, the larvae will intermittently enter and leave the sediment, fail to complete development, and subsequently die. The possibility of gregarious settlement must also be considered for those species whose adults live in dense colonies. Wilson (1968b) has shown that the larvae of Sabellaria preferentially settle and metamorphose on the sandy tubes of previous generations and adjacent to the other recently settled larvae.

ASEXUAL REPRODUCTION

Several species of polychaetes have the ability to reproduce asexually. We won't dwell on the various types of asexual reproduction found in the Polychaeta, for that is not the purpose of this paper. It is our opinion that if adults can be maintained in a satisfactorily healthy condition, asexual reproduction will take place and the progeny can be cultured in the same way as the parent stock. In our laboratory, Ctenodrilus serratus, Dodecacera concharum, Pygospio elegans, and Polydora spp. have successfully reproduced asexually.

The approaches to culturing polychaetes are varied and no one system can be recommended for use with all species. The best system to use is the simplest one that will do the job. The major keys to success are cleanliness, a good quality food, a source of high quality seawater, a lot of common sense, and a great deal of luck.

Contribution No. 50 of the Ira C. Darling Center, Walpole, Maine.

NOTE: Support for this work was furnished in part by NSF Grants G-22549, GB-2179, and GB-4892 and OWRR Grant A-011-Me.

REFERENCES

Åkesson, B. 1967. On the biology and larval morphology of Ophryotrocha puerilis Claparède and Metschnikov (Polychaeta). Ophelia, 4: 111-119.
Bass, N.R. and A.E. Brafield, 1972. The life-cycle of the polychaete Nereis virens. J. Mar. Biol. Assoc., U.K., 52: 701-726.
Blake, J.A., 1969. Reproduction and larval development of Polydora from northern New England (Polychaeta: Spionidae). Ophelia, 7: 1-63.

Costello, D.P., and M.E. Davidson, A. Eggers, M.H. Fox, C. Henley, 1957. Methods of obtaining and handling marine eggs and embryos. Mar. Biol. Lab., Woods Hole, 247 pp.

Dales, R.P., 1950. The reproduction and larval development of Nereis diversicolor O.F. Muller. J. Mar. Biol. Assoc., U.K. 29: 321-360.

Day, J.H. and D.P. Wilson, 1934. On the relation of the substratum to the metamorphosis of Scolecolepis fulginosa (Claparede). J. Mar. Biol. Assoc., U.K., 19: 655-667.

Dean, D., 1969. Comments on pp. 286-289. In: Marine Biology Vol. 5, J.D. Costlow, Jr., (ed.), Gordon and Breach, N.Y. 606 pp.

Dean, D. and P.A. Hatfield, 1963. A method for holding small aquatic invertebrates for observation. Turtox News, 41: 43.

George, J.D., 1966. Reproduction and early development of the spionid polychaete, Scolecolepides viridis (Verrill). Biol. Bull., 130(1): 76-93.

Gilpin-Brown, J.B., 1959. The reproduction and larval development of Nereis fucata (Savigny). J. Mar. Biol. Assoc., U.K., 38: 65-80.

Goerke, H. 1971. Nahrungsaufnahme, Nahrungsaus-nutzung und Wachstum von Nereis virens (Polychaeta, Nereidae). Veroff. Inst. Meersforsch. Bremerh., 13: 51-78.

Grêve, W., 1968. The "planktonkreisel", a new device for culturing zooplankton. Mar. Biol., 1(3): 201-203.

Guerin, J.P., 1970 (1971). Utilisation de nourritures artificielles pour l'elevage de jeunes stades d'invertébrés benthiques. Téthys, 2(3): 557-566.

Hauenschild, C., 1969. Comments on pp. 41-54. In: Marine Biol. Vol. 5, J.D. Costlow, Jr., (ed.), Gordon and Breach, N.Y., 606 pp.

Josephson, R.K. and K.W. Flessa, 1972. Cryolite: A medium for the study of burrowing aquatic organisms. Limnol. Oceanogr., 17(1): 134-135.

Muller, H., 1962. Uber die serualitat des Polychaeten Ophryotrocha puerilis, ihre Determination und ihren Einfluss auf Drusentatigkeit und Kauapparatentwicklung. Z. Morph. Okol. Tiere, 52: 1-32.

Newell, G.E., 1949. The later larval life of Arenicola marina L. J. Mar. Biol. Assoc., U.K., 28: 635-639.

Reish, D.J. and J.L. Barnard., 1960. Field toxicity tests in marine waters utilizing the polychaetous annelid Capitella capitata (Fabricius). Pac. Naturalist, 1(1): 1-8.

Reish, D.J. and T.L. Richards., 1966. A culture method for maintaining large populations of polychaetous annelids in the laboratory. Turtox News, 44(1): 16-17.

Richards, T.L. 1967. Reproduction and development of the polychaete Stauronereis rudolphi, including a summary of development in the superfamily Eunicea. Mar. Biol., 1(2): 124-133.

Sveshnikov, V.A., 1955. Reproduction and development of _Nereis virens_ Sars. (In Russian). Doklady ANSSSR (Zool.), 103(1): 165-167.

Turner, H.J. and J.E. Hanks, 1960. Experimental stimulation of gametogenesis in _Hydroides dianthus_ and _Pecten irradians_ during the winter. Biol. Bull., 119(1): 145-152.

Wilson, D.P. 1932. The development of _Nereis pelagica_ Linnaeus. J. Mar. Biol. Assoc., U.K., 18: 203-217.

_____ 1951. A biological difference between natural sea waters. J. Mar. Biol. Assoc., U.K., 30: 1-20

_____ 1952. The influence of the nature of the substratum on the metamorphosis of the larvae of marine animals, especially the larvae of _Ophelia bicornis_ Savigny. Ann. l'Inst. Oceán., 27: 49-156.

_____ 1953a. The settlement of _Ophelia bicornis_ Savigny larvae. The 1951 experiments. J. Mar. Biol. Assoc., U.K., 31: 413-438.

_____ 1953b. The settlement of _Ophelia bicornis_ Savigny larvae. The 1952 experiments. J. Mar. Biol. Assoc., U.K., 32: 209-233.

_____ 1968a. Some aspects of the development of eggs and larvae of _Sabellaria alveolata_ (L.). J. Mar. Biol. Assoc. U.K. 48: 367-386.

_____ 1968b. The settlement behaviour of the larvae of _Sabellaria alveolata_ (L.). J. Mar. Biol. Assoc., U.K., 48: 387-435.

PROBLEMS ASSOCIATED WITH CULTURE OF MARINE COPEPODS

J.G. Gonzalez, P.P. Yevich, J.H. Gentile, & N.F. Lackie

National Marine Water Quality Laboratory, P.O. Box 277

West Kingston, Rhode Island

In the initial chapter of Perspectives in Marine Biology, Rae (1958) suggested that cultures of marine species suitable for experimentation should be established in the laboratory. He contended that until the counterparts of the fruit fly and the guinea pig are found in the marine environment such work would lack objectivity and realism.

Early investigators, Crawshay (1915), Lebour (1916), Raymond and Gross (1942), and Marshall and Orr (1955) claimed to have success in keeping the calanoid, Calanus, in culture. Murphy (1923) claimed to have success in keeping the cyclopoid, Oithona nana, for two consecutive generations. The above investigators used both filtered and unfiltered natural seawater as a culture medium. They were able to rear these species from eggs to juvenile or adult stages and study their development and feeding habits under laboratory conditions. Prior to these attempts, most of the information gathered was accomplished by following the sequence of generations from preserved seasonal samples.

More recently, numerous workers have cultured other marine zooplankters with varying degrees of success (Provasoli, et al., 1959; Jacobs, 1961; Conover, 1965; Bernard, 1963; Urry, 1965; Johnson, 1966; Zillioux and Wilson, 1966; and others). Most of these investigators also used natural seawater as the culture medium. However, Heinle (1969), Reeve and Cosper (1970), Zillioux (1969), and Zillioux and Lackie (1970) reported using several different synthetic formulations in their culture studies.

One of the major problems encountered in obtaining dependable cultures for laboratory studies appears to be that of finding a suitable medium. The use of natural seawater offers many difficulties. It is not easy to find water samples with exactly the same "chemical quality" in different areas or different times. This makes it difficult to standardize techniques or duplicative results. Laboratories far from coastal regions would not have a supply of seawater with constant characteristics. The introduction of substances from the biological activity of other organisms (red tide, metabolites) and industrial processes adds to the variability of seawater sources. These substances may inhibit or stimulate the growth of experimental animals and it would not be possible to assess their importance or interaction with experimental factors. For these reasons the authors have been working for several years to develop a good artificial seawater formulation.

Davey, et al. (1970) reported on the suitability of artificial and natural seawater for the culture of marine diatoms after trace metals were removed with the sodium form of purified Chelex 100. Kester, et al. (1967) medium was tried for marine zooplankters in our studies using Davey, et al. (1970) technique for removing trace metals.

Methods for preparation and removing trace elements follow:

A. Medium Preparation

Table 1 details components, concentrations, and order of mixing of the Kester (op. cit.) formulation. Note that solutions A, B, and C are prepared separately and then combined while stirring. The following procedure was adopted for removing trace elements.

B. Resin Preparation

The necessity of removing excessive quantities of certain contaminant trace metals from large volumes of artificial seawater has resulted in modifications of the original methodology of Davey, et al. (op. cit.). The modifications are primarily changes in column size and associated problems.

Chelex 100 (50-100 mesh) was purified by mixing 500 cc resin in a Teflon beaker with 10 N NaOH at 60°C constantly for 12 hours. The resin was allowed to settle, rinsed with five bed volumes of high resistance (19 megohm at 20°C) particle-free water, followed by five two-bed volume washes with anhydrous methanol. The resin is then transferred to a 5 cm (I.D.) x 50 cm borosilicate glass column and back-washed (weight + bed volumes/hr.) with five bed volumes of 1N HCl and 1N NaOH interspersed with ten bed volume rinses with deionized water. During this regeneration cycle, the back-washing sizes the resin, permitting removal of the "fines" from the top of the column.

TABLE 1. Formula for 1 kg of 35.00 °/oo artificial seawater

		gm/l	
A	NaCl	23.926	A
	Na_2SO_4	4.008	
	KCl	0.677	
	KBr	0.098	
	H_3BO_3	0.026	
	NaF	0.003	
B	$MgCl_2 \cdot 6H_2O$	10.83	B
	$CaCl_2 \cdot 2H_2O$	1.52	
	$SrCl_2 \cdot 6H_2O$	0.024	
C	$Na_2SiO_3 \cdot 9H_2O$	0.030	C
	$NaHCO_3$	0.196	

Mix A and B in separate containers; combine while stirring. Then add C slowly with constant stirring. pH should come to 8.1-8.2.

After the final deionized rinse, the column is allowed to settle and the supernatant drained to about 2.5 cm above the resin bed. Kester's artificial seawater (pH 8.1 - 8.2) is slowly added to fill the freeboard space and the flow rate is adjusted. The resin is equilibrated with at least five bed volumes of synthetic seawater to reduce the possibility of cationation interaction. Prior to equilibration with seawater the effluent pH is ca. 12.50. There is a rapid pH decrease during equilibration; stabilization occurs around pH 7.5 - 8.0. Occasionally, there is an anomalous pH behavior during equilibration which is characterized by a very slow decline, with stabilization occurring around 9.0 - 9.5. Both algal and zooplanktonic assays as well as anodic stripping data (Davey, pers. comm.) indicate that contaminant metals such as copper and lead are removed to less than 10^{-7} concentrations. Final pH adjustments, when necessary, are made with ultra pure HCl. All collecting tubes and vessels are linear polyethylene.

We determined empirically an optional flow of 13 bed volumes/hr. from radioisotope removal studies. In addition, 1 cc of resin has been found capable of removing contaminant trace metals from two liters of artificial seawater (Davey, pers. comm.). Theoretically, with a 500 cc resin bed, it would be possible to remove the trace metals from artificial seawater effectively at a rate of 6.5 liters/hr. In practice, however, we have never operated to such a capacity, since we rarely exceed 100 liters.

Calanoid copepods, Acartia tonsa, A. clausi, a few harpacticoids, and the gammarid, Corophium isidiosum, were reared successfully for several generations in this medium. Female and male animals were collected from the field and transferred to crystal-

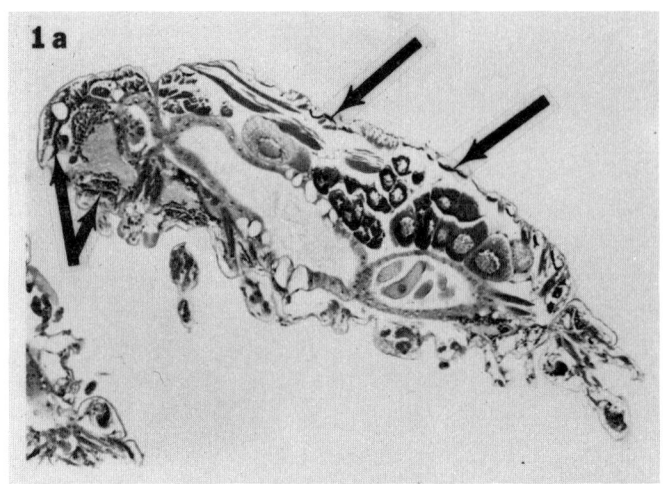

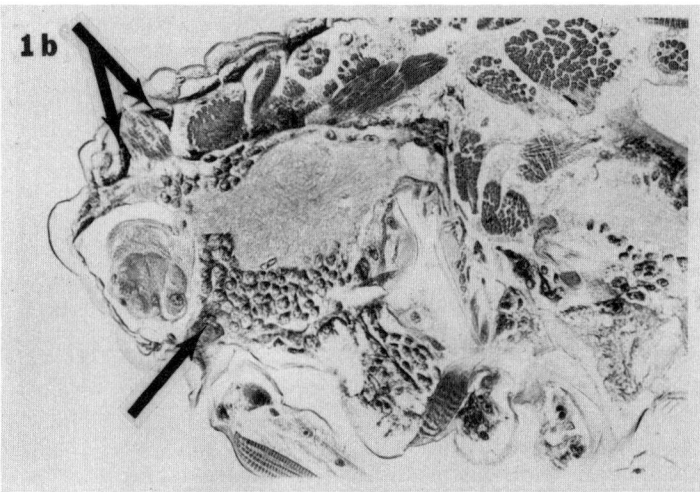

Fig. 1a. Sagital section of a normal healthy <u>Acartia</u> <u>tonsa</u> (Copepoda; Calanoida) collected from the field and fixed immediately for histological examination. Arrows point to unicellular glands; note them especially around circumesophageal ganglia, in the cephalosome area and the rest of the metasome. Hematoxylin and eosin. 190 X.

Fig. 1b. Larger magnification of sagital section of the cephalosome area of normal <u>Acartia</u> <u>tonsa</u> collected from the field. Note numerous unicellular glands at the anterior part of cephalosome and around circumesophageal region. Hematoxylin and eosin. 480 X.

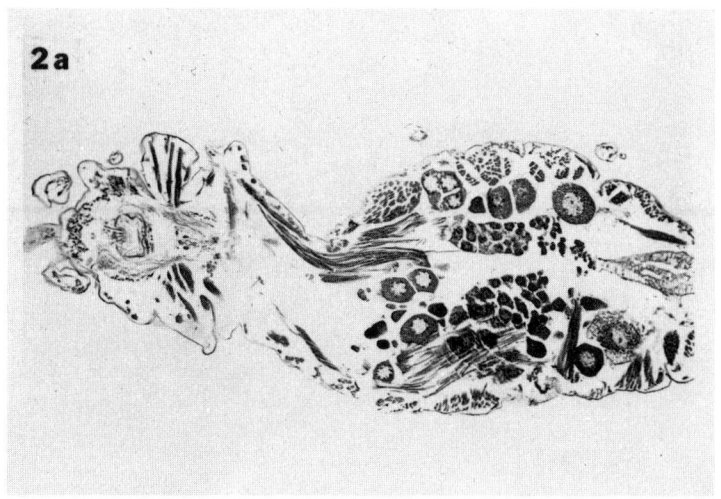

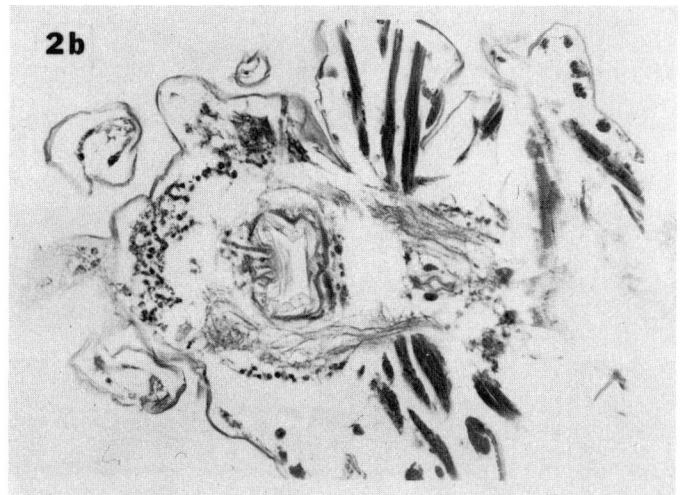

Fig. 2a. Longitudinal section of <u>Acartia tonsa</u> held in the laboratory under starving conditions. Note absence of unicellular glands around circumesophageal region, in the cephalosome area and the rest of the metasome. Hematoxylin and eosin. 190 X

Fig. 2b. Magnification of cephalosome area of same animal of Fig. 2a.

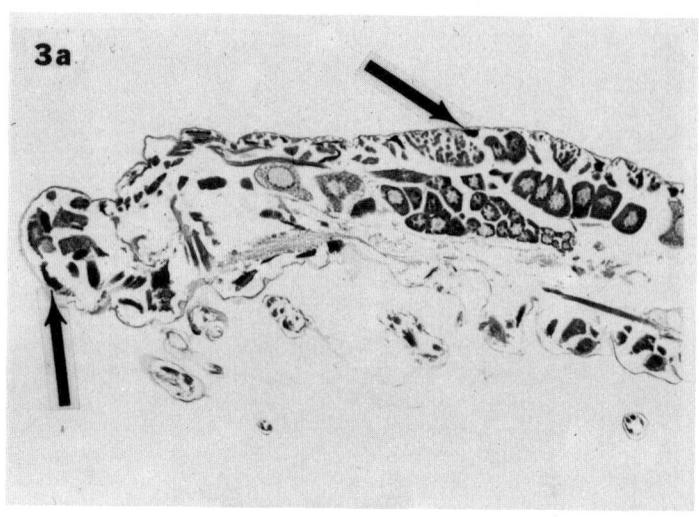

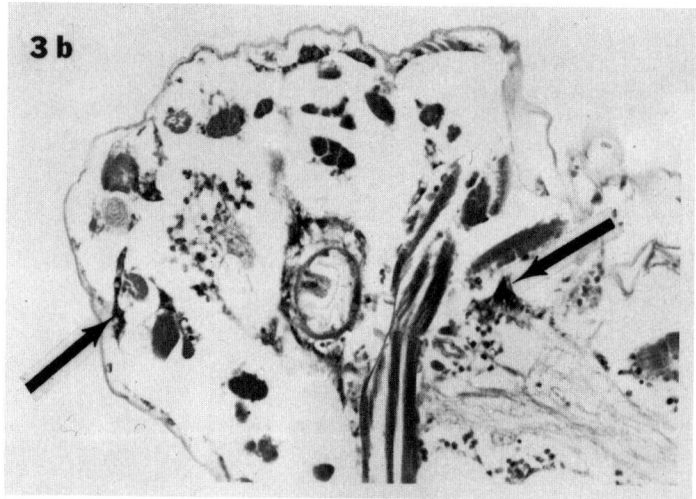

Fig. 3a. Sagital section of <u>Acartia tonsa</u> held in the laboratory for 2 weeks and fed with <u>Isochrysis galbana</u> and <u>Rhodomonas baltica</u>. Note that the number of unicellular glands is considerably less than field animals as shown in Figs. 1a and 1b. Hematoxylin and eosin. 190 X

Fig. 3b. Magnification of longitudinal section of the cephalosome area of <u>Acartia tonsa</u> held and fed under laboratory conditions for 2 weeks. Note again the scarcity of unicellular glands. Hematoxylin and eosin. 480 X.

izing dishes 100 x 190 mm and fed cultures of Rhodomonas baltica and Isochrysis galbana. In all cases, animals were observed to grow well in the new medium. Fecundity was high, survival of nauplii was excellent, and it was possible to maintain successive generations of all the species in the laboratory. However, further experiments on field, starved, and laboratory cultured Acartia tonsa (Copepoda; Calanoida) fixed for comparative histological examinations* showed a consistent dissimilarity from field animals. The condition was evident in copepods reared in filtered and unfiltered seawater, as well as other artificial media.

Figures 1a and 1b are photomicrographs of histological sections of normal healthy Acartia tonsa collected from Narragansett Bay, Rhode Island. The presence of unicellular glands (shown by the arrows) around the circumesophageal region, ganglia, and the rest of the metasome is consistent and characteristic of field animals. The absence of these glands in starved A. tonsa, processed histologically in the same way as field copepods, is shown in Figures 2a and 2b. On the other hand, A. tonsa held and fed in the laboratory showed a decrease in the number of glands as illustrated in Figures 3a and 3b.

We have hypothesized that the observed response of this copepod may be related to nutritional deficiencies. Since the modified Kester formulation is a defined medium, it will be feasible to determine other factors associated with their growth or establish a diet which will provide animals compatible with field populations.

---

* The copepods for this research were filtered from culture dishes (animals cultured in the laboratory) or from Dewar bottles (when field animals were used) and fixed immediately in Dietrich solution. Barszcz and Yevich (ms. in preparation) technique for zooplankton was followed in processing the specimens for histological examination.

REFERENCES

Bernard, M., 1963. Observations sur la biologie en aquarium de Euterpina acutifrons, copépode pelagique. Commission Internationale pour l'exploration Scientifique de la Mer Mediterranée Monaco. Rapports et process Verbaux des Reunions, 17: 545.

Conover, R.J. 1965. Notes on the molting cycle, development of of sexual characters and sex ratio in Calanus hyperboreus. Crustaceana, 8: 308-320.

Crawshay, L.R., 1915. Notes on experiments in the keeping of plankton animals under artificial conditions. J. Mar. Biol. Assoc. U.K., 10: 555-576.

Davey, E.W., J.H. Gentile, S.J. Erickson, and P. Betzer, 1970. Removal of trace elements from marine culture media. Limnol. Oceanog., 15: 486-488.

Heinle, D.R., 1969. Effects on the population dynamics of estuarine copepods. Ph.D. Thesis, University of Maryland. 132 pp.

Jakobs, J., 1961. Laboratory cultivation of the marine copepod Pseudodiaptomus caronatus Williams. Limnol. Oceanog., 6: 443-446.

Johnson, M.W., 1967. Some observations on the hatching of Tortanus discaudatus eggs subjected to low temperatures. Limnol. Oceanog. 12: 405-410.

Kester, D.R., I.W. Duedall, D.N. Connors, and R.M. Pytokowicz, 1967. Preparation of artificial seawater. Limnol. Oceanog., 12: 176-179.

Lebour, M.V. 1916. Stages in the life history of Calanus finmarchicus (Gunnerus), experimentally reared by Mr. L.R. Crawshay in the Plymouth Laboratory. J. Mar. Biol. Assoc. U.K., 11: 1-17.

Marshall, S.M. and A.P. Orr, 1955. The Biology of a Marine Copepod. Oliver and Boyd, London, 188 pp.

Murphy, H.E. 1923. The life cycle of Oithona nana, reared experimentally. University of California Publication in Zoology, 22: 449-454.

Provasoli, L., K. Shiraishi, and J.R. Lance, 1959. Nutritional idiosyncrasies of Artemia salina and Tigriopus in monoxenic culture. Ann. N.Y. Acad. Sci., 77: 250-262.

Rae, K.M., 1958. Parameters of the marine environment. In: Perspectives in Marine Biology. S.S. Buzzati-Traverso (ed.). 3-16 pp.

Raymond, J.E.G. and F. Gross, 1942. On the feeding and breeding of Calanus finmarchicus under laboratory conditions. Proc. Roy. Soc. Edinburgh, Sec. B, 61: 267-287.

Reeve, M.R. and E. Cosper, 1970. The acute effects of heated effluents on the copepod Acartia tonsa from a subtropical bay and some problems of effects on living resources and fishing. Rome, Italy. pp 9-18, Dec. 1970.

Urry, D.L., 1965. Observations on the relationship between the food and survival of Pseudocalanus elongatus in the laboratory. J. Mar. Biol. Assoc. U.K., 45: 49-58.

Zillioux, E.J. and D.F. Wilson, 1966. Culture of a planktonic copepod through multiple generations. Science, 151: 996-998.
Zillioux, E.J., 1969. A continuous recirculation culture medium for planktonic copepods. Mar. Biol., 4: 215-218.
Zillioux, E.J. and N.F. Lackie, 1970. Advances in the continuous culture of planktonic copepods. Helgoländer wiss. Meers., 20: 325-332.

CULTURE TECHNIQUES FOR DECAPOD CRUSTACEAN LARVAE

Morris H. Roberts, Jr.

Director of Research, Aquatic Sciences, Inc.

2624 M.W. 2nd Avenue, Boca Raton, Florida

The history of larval culture for decapod crustaceans dates from the early 1900's. Significant work in the first half of the century included the studies of Lebour (1927) and Hart (1935, 1937). Their methods were adequate for rather hardy species but were not suitable for producing large numbers of larvae of a specific stage as required for physiological ecology studies, bioassay tests, etc.

Costlow and Bookhout (1960) described a methodology for larval culture which has now been widely applied in which larvae are cultured in compartmented plastic boxes (tackle boxes) or small finger bowls with daily transfer to clean vessels containing fresh medium and food. This method is quite laborious and cannot economically produce juveniles for rear-out on a commercial scale nor can it meet all demands for research animals.

The tackle box method has been widely applied to decapod crustaceans and has facilitated studies of descriptive taxonomy, environmental requirements for larvae (salinity, temperature, dissolved oxygen), nutritional requirements (kind and amount of food), behavior, physiology (molt control, amino acid metabolism, osmoregulation, oxygen consumption), and toxicity of various pollutants. A general summary of methodology has been published by Provenzano (1967). However, these papers do not provide details of all techniques involved.

A few other methods have been described but have not gained wide application. Modin and Cox (1967) described a recirculating system which allowed for water treatment and isolation of individual larvae. With this method they reared *Pandalus jordani* which had defied all efforts by the techniques of Costlow. Herrnkind (1968) described an intermediate volume static system which he used to produce sufficient numbers of *Uca* juveniles to perform behavioral studies with a minimum expenditure of time and effort. Yang (pers. comm.) and others have used still larger static systems to produce quantities of juveniles with minimal effort.

Each method has desirable features for certain kinds of studies. For descriptive taxonomy, it is necessary to trace the developmental sequence for each individual. This can best be done by isolating the individuals in plastic trays and examining each animal daily for evidence of molting. Less precise data can be obtained by the culture bowl method with five to ten animals per bowl (160 ml total volume). The culture bowl method can provide sufficient animals for studies such as physiological ecology, toxicity, nutrition, behavior, physiology, and others which require relatively small numbers of animals. The requirement for large numbers of animals of specific larval stages (for research) or juveniles (for mariculture) can be economically met only by large scale culture methods which reduce the manpower requirement. Selection of culture method should include consideration of the type, amount, and precision of data required for the larval period as well as the numbers of organisms required.

CULTURE METHODOLOGY

Before describing the several culture methods most generally employed for decapod larvae, it should be noted that the success of any larval culture procedure depends heavily upon the quality of eggs and maintenance procedures for the embryos prior to hatching. Many failures in culture can be traced to inadequate attention to handling of ovigerous females. If the eggs are subjected to stress, they may hatch prematurely or not at all. Even minor stress not affecting the time of hatching may result in low percent hatch and low viability of the larvae.

Collection and Transport

During collection and all subsequent operations, ovigerous crabs should receive minimal exposure to the air. During transport to the laboratory, the maximum practical water level should be used to maintain the animals. The water should be clean and maintained in aerated tanks. The temperature of these tanks should vary from that ambient at the collection site by no more than $\pm 2°C$.

Egg Incubation

Upon arrival in the laboratory, eggs from each ovigerous female should be staged to estimate when hatching can be expected. To stage the eggs a small aliquot of eggs are carefully dissected from the egg mass with a pair of forceps. The eggs are placed on a slide and examined under a compound microscope. The stages as defined for use at Aquatic Sciences, Inc. are shown in Table 1. These stages are applicable to many species of decapod crustaceans, but may differ in certain details for particular species. When beginning work with a species it is advisable to examine eggs throughout their development and establish meaningful stages with estimates of the time before hatching for each egg stage at the incubation temperature.

If eggs are in Stage 1-4, the crab with its eggs is placed in a large volume tank designated for this purpose (for Callinectes and Menippe, e.g., a 7 foot diameter tank is used at Aquatic Sciences, for smaller species, tanks of 50 gallons are generally suitable). Experience indicates that larger water volumes have greater stability and result in less stress to the eggs. On each subsequent day ovigerous crabs are retrieved from the holding

TABLE 1.

EGG STAGES OF DECAPOD LARVAE
(refers specifically to Callinectes sapidus)

Egg Stage	Description
1	"Blastula" - undifferentiated yolk cells
2	"Gastrula" - yolk cells beginning to localize at one pole of egg
3	Organogenesis - eye placode present, but unpigmented and vague outline of embryo can be seen; over 1/2 egg is yolk
4	Organs well developed; eyes well-formed, pigmented; appendages visible; yolk cells apparent, less than 1/2 egg
5	Eyes well rounded, darkly pigmented; heart beating

tank and a small aliquot of eggs is staged until Stage 5 eggs are observed.

Crabs which arrive in the laboratory with eggs in Stage 5 or which subsequently develop Stage 5 eggs are treated in the following manner. Large numbers of eggs are removed from the pleopods. These eggs are placed in a finger bowl of clean culture medium at the incubation salinity and gently stirred with a glass rod. Large clumps of eggs are teased apart with a pair of forceps and the eggs are carefully pipetted with as little of water as possible into another finger bowl with clean medium at the same salinity. The eggs are examined under a microscope for fragements of tissue or appendages removed from the crab with the eggs and these are carefully removed. Eggs obviously infected with fungi or bearing colonial protozoans are discarded. The transfer of the eggs to clean medium is repeated twice more. At this stage the water overlying the eggs should remain clear. If it does not, the process is repeated. The washed eggs are then suspended in a hatching solution which consists of culture medium containing 100,000 USP units of Penicillin G/l (this solution may also be used to wash the eggs). Twenty milliliters of solution is placed in each compartment of a tackle box with 100-150 eggs. The box is then placed on a shaker table and the shaking rate is adjusted to 60-120 shakes per minute. On subsequent days the box is removed from the shaker table and the eggs are examined. If the eggs have not hatched, they are transferred to freshly prepared media and returned to the shaker table. Eggs which are noted to be infested with fungus or bacteria are removed and discarded. If fungi or bacteria are present, separate pipets should be used for those eggs which are infected to prevent contamination of the "clean" eggs. Pipets and other materials which are exposed to fungi or bacteria are washed in a 10% Chlorox solution followed by 0.1 N sodium thiosulfate rinse and then passed through the normal procedure for washing culture glassware. This procedure is repeated until no more larvae hatch.

Only those larvae which are actively swimming and strongly photopositive are suitable for culture. It has been found that larvae which do not exhibit these characteristic behaviors do not exhibit adequate survival and development in culture.

An alternative procedure for hatching the eggs is to allow the eggs to hatch from the females directly in the large holding tank. Frequently excellent hatches are obtained in this manner. However, some species (e.g. hermit crabs, glass shrimp) have a propensity to eat their own young. Sometimes the eggs will be dropped without hatching. Retrieval of larvae from 50 gallons or more of water is also quite a chore. The chances of obtaining an inadequate hatch or no hatch at all are greatly increased.

Larval Culture

"Tackle Box" Culture: For this procedure, compartmented plastic tackle boxes available in bait shops or the sporting goods section of department stores are used. In my experience the plastic is not toxic to decapod larvae. New boxes may be used after an initial wash with a non-ionic detergent (e.g., Triton X-100). Cleaning with organic solvents such as acetone or ethanol is prohibited because this will craze or dissolve the plastic.

Each compartment of the tackle box is filled with 50 ml of seawater and freshly hatched Artemia nauplii or some other suitable food (see below) is added. For some species the optimal salinity for larval development is reported in the literature. If data is not available, a range of salinities bracketing the salinity at the collection site should be tested. One (or two) newly hatched zoea is placed in each compartment. The boxes are then placed in a location with suitable lighting and temperature control. Photoperiod and light intensity have not to date been shown to have any major effect on development but a "normal" day length is intuitively desirable. Temperature extremes and wide daily fluctuations should be avoided, but rigorous temperature control is not necessary unless this is part of the experimental design.

Data should be collected daily with regard to molting, survival, evidence of feeding, etc. Each day as the data is collected, the larvae are transferred to clean boxes with fresh media and food. The used boxes are emptied and washed. A dilute solution of Chlorox (1-5%) is a good wash solution, provided the box is rinsed in a 0.1 N sodium thiosulfate solution followed by a distilled water rinse.

Exuvia and dead zoeae may be preserved for subsequent examination. These may be placed in a 4 dram vial containing 70% ethanol with 10% glycerol, 1% buffered formalin or any other general preservative.

Bowl Culture: The procedure for bowl culture is basically similar to that for tackle box culture except that a larger number of zoeae are placed in each bowl. In our experience there is no increase in mortality with as many as 40 zoeae per 4" bowl (160 ml) or a comparable density in larger bowls. Above this density, mortality does increase, apparently because of food limitation. Routinely we place 20 zoeae in each 4" bowl to avoid any possibility of deleterious effects from larval density.

The amount of food provided will be described in a separate section. As with tackle box culture, daily transfer is necessary to prevent fouling. A marked increase in mortality may be noted during a second 24 hour period if no transfer is made. This effect is more obvious at higher temperature (above 25°C). The larvae are transferred daily and data recorded. If the stage of the larvae must be known, each larva may be examined microscopically during the transfer. After the larvae are transferred, the culture bowls are emptied and cleaned. The procedure routinely followed at Aquatic Sciences, Inc. is to wash the culture bowls in Bon-Ami, rinse in a clinical dishwasher, and dry in a drying oven. This procedure is adequate provided there has been no disease.

Mass Culture: a. Static Systems: Any large container may be used such as large all-glass aquaria, large battery jars, polyethylene barrels (100 gallons), plastic aquaria, or cylindrical fiberglass tanks. Medium of a suitable salinity and temperature is placed in the container. A unicellular algae is added to 10^5 cells/ml to provide water conditioning (Chlorella or Dunaliella among many other species are suitable). The usual food organisms (see below) are added. Some experience is necessary to determine how much food will be suitable in your culture system. Gentle aeration is provided; violent agitation can mechanically damage the larvae, and even mild aeration rates may cause a "gas bubble disease."

Zoeae are added to the culture vessel immediately upon hatching. An initial density of 15,000/100 gallons seems to be an acceptable number. This compares favorably with the densities used in bowl culture.

In this system no transfer is required nor is quantitative data collected. Every day or two, a sample may be taken to determine the larval stages present. As megalopae are observed these may be removed and cultured separately or the culture may be continued until all larvae have entered a juvenile stage. Algae and food should be added as necessary.

b. Flow-through Systems: An alternative approach to mass culture is a flow-through culture system. This may be especially useful for less tolerant species since there is less build-up of waste products and bacteria. At Aquatic Sciences we have started to test this approach on a relatively small scale (2 liter culture vessels and a 125 l system). Several problems remain to be solved in applying this approach on a larger scale, but these are not insurmountable. There is not sufficient data at the present time to compare this approach with the use of static systems. We are presently testing the effect of flow rate, mechanisms for introducing and removing water, and feeding levels. Ultimately we plan to incorporate systems to control automatically the salinity and temperature in any desired sequence during the culture period.

Water exhausted from the culture system is enriched with nitrogenous wastes. We utilize this "waste" water as the base for medium for algal production after suitable sterilization procedures. Such reuse of the water reduces the cost of a flow-through system.

        c. **Recirculating Sustems**: Modin and Cox (1967) described a recirculating system in which a culture aquarium rested in a water bath such that the temperature could be controlled by an external chiller. Water overflowed from the culture aquarium into a reservoir from which it was pumped through a filter and a UV sterilizing system into a header reservoir and flowed back to the culture aquarium by gravity. Larvae were cultured in compartmented plastic boxes floated in the culture aquarium. Each culture box had the bottom removed and replaced with a fine mesh nylon screen (200 μ "Nytex"). Since this system is quite complex, it has not met with wide-spread application but certainly it has a great deal of merit for the culture of various species. It would be especially useful with highly cannabalistic or fragile forms if only a limited quantity of these larvae must be reared or if large volumes of culture medium are not readily available.

Larval Foods

Most decapod larvae are carnivorous, feeding on small zooplankters. Larvae of a few species such as *Pinnotheres veterum* and *Pinnotheres pisum* (pea crabs) apparently are phytophagous (Atkins, 1955) though other pea crabs are carnivorous (*P. chamum*, unpublished data).

The standard larval food for culture is the nauplius of *Artemia*. Generally the quantity of food is unimportant so long as there are at least 20 nauplii per ml. Of course, excessive amounts (80 nauplii/ml) will lead to oxygen depletion in static systems. This leaves a wide range of acceptable food densities. A little experience will allow the culturist to prepare a stock suspension of nauplii, a few drops of which will provide sufficient nauplii for one day's feeding without actually counting the nauplii.

Larvae of some decapod species are too small to handle *Artemia* nauplii or have mouth parts which are better suited for handling smaller food organisms. For example, the early stages (Zoea I and II) of the hermit crab, *Pagurus longicarpus*, can capture *Artemia* nauplii but often only remove and ingest the appendages, leaving the body of the nauplius behind (Roberts and Pierson, unpublished). *Callinectes sapidus* larvae (Stages I, II and III) apparently cannot capture and ingest any portion of *Artemia* nauplii as shown in repeated tests at Aquatic Sciences with strictly controlled diets.

There is also a possibility that Artemia nauplii are not a complete diet for some species or that the low levels of pesticides in the nauplii lead to the development of morphological anomalies and/or increase mortality (Bookhout and Costlow, 1970).

Alternate foods which have been tested successfully include copepods (Chamberlain, 1962), urchin gastrulae (originally used by Costlow and Bookhout, 1959), Brachionus plicatilis (a marine pelagic rotifer), and the post-trochophore of Arenicola marina. Hart (1935, 1937) reported success with Ostrea larvae as food. My own experience with Crassostrea larvae have been exactly contrary to Hart's with pelecypod larvae clamping themselves to the mouth parts of the crab larvae (Roberts and Pierson, unpublished). Pelecypod larvae may be suitable as food prior to shell deposition.

Several of these foods are readily provided by simple culture techniques. At Aquatic Sciences, urchin larvae are provided by the following procedure: Urchins are placed ventral surface upward on a folded moist paper towel. Electrodes are inserted with a gentle twisting motion. The urchin is then placed over a small culture bowl half filled with seawater at 35 °/oo salinity. The electrodes are connected to a resistance box consisting of a 100 W incandescent bulb connected in series and provided with an on-off switch. The switch is placed in the on position for about 1 minute, then off for about 15 seconds. If gametes (eggs: yellowish cohesive mass, sperm: milky, white substance) are released, the switch is turned on until gamete release ceases. If no gametes are released, the urchin is shocked 2 or 3 times, and if no gametes are released, this urchin is discarded. Once obtained, the gametes are examined microscopically for condition. Eggs should be spherical and yellow; sperm should be motile. All eggs are placed in a large culture bowl with seawater of 35 °/oo salinity. Approximately 1 part sperm suspension is added for 8 parts eggs. The bowl is stirred and allowed to stand for 10 minutes. By this time fertilization should have occurred, as evidenced by presence of a fertilization membrane. The eggs are allowed to settle and as much water as possible is removed without disturbing the eggs. Clean seawater is added, stirred, and again removed after the eggs settle. This removes excess sperm and organic debris. After washing, the fertilized eggs are distributed among several culture bowls and placed in a 25°C incubator. One hour after fertilization the eggs should be examined for proper cleavage. The urchin larvae are then incubated for 12 hours before use as food. At this time they should be motile. Electric shock is not the only procedure to obtain gametes (KCl or an extract of the radial nerve may be injected), but it has proved most reliable for our purpose.

Brachionus plicatilis is readily cultured en masse (e.g., Theilacker and McMaster, 1971). The rotifer is routinely cultured in 20 to 30 gallon aquaria with seawater 30 °/oo salinity. The rotifers are fed Platymonas (green alga) at a level of 10^5 cells/ml daily.

The rotifers also do well on other algae, but have little or no nutritive value for decapod larvae when fed Anacystis (blue-green algae). The rotifers have a doubling time of 0.5 to 1.0 days under these conditions unless the ammonia level is high. In our experience the doubling rate becomes infinite at ammonia levels of 9-10 ppm (no critical experimentation has been done on this observation). Rotifers are harvested by passing the culture through a 64μ screen and resuspending in a small quantity of clean medium. Some algae will remain with the rotifers, but this is not harmful to the decapod larvae.

Arenicola larvae may be obtained by collecting the gelatinous egg strings on a beach. The egg masses are rinsed to remove some debris and placed in large culture bowls with clean seawater. The eggs usually hatch in 24-48 hours. The post-trochophores are photopositive and may be concentrated by a light source. They can be removed with a wide-bore pipet. Several rinses with clean seawater will remove much of the gelatin which adheres to them.

Oysters and other pelecypod larvae can be obtained by techniques described by Loosanoff and Davis (1963). There is no need to redescribe their methods here.

We are presently developing culture methods for several invertebrate species which may have food value for decapod larvae. Protozoans and various polychaete larvae are of particular interest because they are of appropriate size and relatively easy to culture. Barnacle nauplii and copepod nauplii are also desirable, although both have been proven difficult to produce consistently in adequate numbers. Ideally we desire easily cultured foods which can be produced in large quantity with a minimal effort and expense. This frees the culturist from dependence upon collection of materials for food from the field, a laborious, undependable, and expensive process.

Experimentation at Aquatic Sciences has demonstrated that the amount of food as well as quality is important to maximize survival at least for some species. A rotifer density of 40/ml is necessary for Callinectes zoeae in Stages I to III. Thereafter they can be switched to a diet of 25/ml or more to obtain equivalent survival and development. Similar food densities are required for maximal survival of Menippe mercenaria larvae (stone crab).

MEDIUM

The best medium for larval culture for marine species appears to be filtered natural seawater obtained from an unpolluted source and adjusted to the appropriate salinity for the species being reared. A large amount of this water may be collected at a single

time, filtered, and stored in the dark at cool temperature for
periods of up to 1 month prior to use. Initial conditioning of
the water does not appear to be necessary but this has not been
tested rigorously. Such stored water does not retain the pristine
condition of raw natural seawater but at present there is no published data defining precisely the necessary water quality for
decapod larval development.

In lieu of filtered natural seawater it is possible to use
various formulations of artificial seawater. The formulation
found most suitable at Aquatic Sciences is Instant Ocean Artificial Sea Salts (Aquarium Systems, Inc. Eastlake, Ohio). This can
be prepared directly in untreated tap water; however, it is preferable to prepare it in distilled water or water which has been passed through a charcoal filter if the tap water supply contains
significant amounts of residual chlorine or high concentrations
of such heavy metal ions as copper or zinc. This medium will not
give as good results as natural seawater expressed in terms of
percent survival over the period of complete larval development;
however it is more suitable than any other artificial sea salt
which has been tested (Rila mix, e.g., will allow the larvae to
survive for a considerable period of time but the larvae will not
successfully complete even one molt).

DISCUSSION

I have described the several methodologies which have been
successfully used for the culture of decapod larvae. For some
species these methods have not been completely successful; for
example, with the larvae of the spiny lobsters, and the caridean
shrimps, Stenopus, Alpheus and Synalpheus. The reasons for the
lack of success with some species are probably varied, and may
involve consideration of parameters which have not been examined
to date.

No attempt is made here to describe the methodology developed
for the commercially important penaeid shrimp although much of
this technology is identical. Differences in developmental sequence (presence of naupliar stages) and dietary requirements
necessitate some modifications to the procedure. One methodology
for this group has recently been exhaustively described by Tabb,
Yang, Hirono, and Helmen (1972).

The methods described here are also appropriate for fresh
water species such as Macrobrachium, the commercially important
river shrimp; Palaemonetes, a fresh water glass shrimp, and several other species of fresh water crustaceans. The fresh water

used for culture media should be detoxified municipal water or preferably a natural water source that has been properly filtered and stored. Appropriate analogues to the foods described here for marine species should be selected or Artemia may be used directly after washing in fresh water.

It should be obvious that major improvements in techniques are still required. This is especially so with regard to mass culture of larvae such that commercial scale production becomes feasible on an economic basis. Such methods must limit the requirements for technically trained personnel while maximizing the survival and developmental parameters for the species being cultured.

REFERENCES

Atkins, D., 1955. The post-embryonic development of British Pinnotheres. Zool. Soc. London Proc. 124: 687-715.

Bookhout, C.G. and J.D. Costlow, Jr., 1970. Nutritional effects of Artemia from different locations on larval development of crabs. Helgolander wiss. Meers. 20: 435-442.

Chamberlain, N.A., 1962. Ecological studies of the larval development of Rhithropanopeus harrisii (Xanthidae, Brachyura) Chesapeake Bay Instit., John Hopkins Univ., Tech. Rep. 28: 1-47.

Costlow, J.D., Jr. and C.G. Bookhout, 1959. The larval development of Callinectes sapidus Rathbun reared in the laboratory. Biol. Bull. 116(3): 373-396.

___ 1960. A method for developing Brachyuran eggs in vitro. Limnol. Oceanogr. 5: 212-215.

Hart, J.F.L. 1935. Culture methods for Brachyura and Anomura. (from Live Material for the Marine Biologist edited by P.S. Galtsoff). The Collecting Net, 87: 199-200.

___ 1937. Culture methods for Brachyura and Anomura. In: Galtsoff, Paul S., et al., Culture Methods for Invertebrate Animals, Ithaca, Comstock Publishing Co., pp 237-238.

Herrnkind, W.F. 1968. The breeding of Uca pugilator (Bosc) and mass rearing of the larvae with comments on the behavior of the larval and early crab stages (Brachyura, Ocypodidae). Crustaceana, Supp. 2. pp 214-224.

Lebour, M.V., 1927. Studies of the Plymouth Brachyura I. The rearing of crabs in captivity with a description of the larval stages of Inochus dorsettansis, Macropodia longirostris, and Maia squinado. J. Mar. Biol. Assoc. U.K. 14: 795-821.

Loosanoff, V.L. and H.C. Davis, 1963. Rearing of bivalve molluscs. Adv. Mar. Biol. 1: 1-136.

Modin, J.C. and K.W. Cox, 1967. Post-embryonic development of laboratory reared ocean shrimp, Pandalus jordani Rathbun. Crustaceana 13: 197-219.

Provenzano, A.J., Jr., 1967. Recent advances in the laboratory culture of decapod larvae. Symp. Crustacea, Proc. Pt. II, pp 940-945.

Tabb, D.C., W.T. Yang, Y. Hirono and J. Helmen, 1972. A manual for culture of pink shrimp, Penaeus duorarum from eggs to postlarvae suitable for stocking. Sea Grant Special Bull. No. 7: 1-59.

Theilacker, G.H. and M.F. McMaster, 1971. Mass culture of the rotifer Brachionus plicatilis and its evaluation as a food for larval anchovies. Mar. Biol. 10(2): 183-188.

LOBSTER CULTURE

Based on the conference presentation of John T. Hughes

Massachusetts State Lobster Hatchery and Research

Station, Oak Bluffs, Martha's Vineyard, Mass.

COLLECTION OF BROOD STOCK

Most coastal states with a natural lobster population legally regulate the taking of these animals. Therefore, before launching a lobster research project, the state's Department of Natural Resources should be consulted and the necessary permits secured for collecting and holding egg-bearing lobsters and holding sub-legal size lobsters. A commercial lobsterman will also need special permits to collect eggers for research.

Most commercial lobstermen keep their animals alive in a wet tank on board. They can be carried short distances with no special packaging (unless it is an especially hot day) with no significant loss of eggs. In transit over long distances, however, care must be taken to keep the lobsters and their eggs cool and moist. When lobsters are shipped from the hatchery in Massachusetts to California or Hawaii, for example, the same containers are used that the commercial shippers use -- a styrofoam box inside a waxed cardboard box. However, instead of the seaweed used by the commercial shippers, a styrofoam "spaghetti" used in the packing industry is employed for insulation and three packages of reuseable "ice" are included to keep the lobsters cool. Ordinary ice will melt and fresh water is lethal to lobsters.

MAINTENANCE OF BROOD STOCK

Lobsters survive best in seawater of high (30-31 ppt), stable salinity between 2-22°C. In using artificial seawater, the temperature should be adjusted to approximate that of the water from which the lobsters came. Once at ambient temperature it can then be adjusted up or down if done gradually. At the Massachusetts Lobster Hatchery seawater is heated with the air temperature in the building. For example, in the winter ambient temperature is about $2^\circ C$. To condition lobsters to hatch earlier than normal it is necessary to warm the water to about $21^\circ C$. A tank in the hatchery is filled with cold ambient water and warmed by the air temperature in the building. Air is bubbled in to replace the oxygen.

High water quality is the most important consideration. One important precaution is that the water is not taken from, or near, an area where there has been spraying for mosquitoes or other insect pests. Pesticides that kill insects will also kill lobsters. Nitrogen, or gas, disease may also be a problem. The seawater looks normal, but if the pump has an air leak on the suction side, air will be sucked into the pump. The air (mainly nitrogen) is broken up so finely that it stays in an emulsion in the water. It does not effervesce out. The lobsters then get this excess in their blood, causing nitrogen, or gas, disease (the "bends"). Results are fatal.

There are two ways to tell if excess nitrogen is present. First, attach a hose to a petcock near the pump and submerge the end of the hose into a pail of water. If something can be seen coming out the end of the hose, it is air mixed with the water. This is indication of an air leak which would add nitrogen to the water. A second procedure, if gas disease is suspected, is to take an oxygen reading of the water. If the oxygen reading is much higher than it should be for the temperature, the excess reading for oxygen may be nitrogen.

Adult lobsters do not have to be fed for weeks, especially in the winter. In the summer they do not have to be fed for a couple of weeks. They do very well on stored fats. In fact, they do not even lose any weight; tissue used in metabolism is replaced with seawater of virtually the same specific gravity. When fed, a variety of inexpensive, readily available foods, usually shellfish (clams, quahogs, mussels, quarterdecks, periwinkles, or scallop viscera) are used. It is best to avoid fish that are oily. Fish such as mackeral or bluefish foul the tank and deposit a scum on the surface of the water. When possible use a whole animal, such as a periwinkle or quarterdeck. If just the flesh of fish, for example, is fed the lobster receives only muscle tissue. If the whole shellfish is fed, it is given all the tissues and therefore a greater range of nutrients.

Many egg-bearing female lobsters can be maintained in a common tank. Several guides have been employed, such as a maximum of one pound of lobster per gallon of seawater or twenty pounds of lobster per cubic foot in a crowded tank. However, available oxygen is the real guide. One lobster in a tank the size of a small room will die if there is no oxygen, while ten lobsters will live in a bathtub if there is enough oxygen. The amount of oxygen available is dependent on the temperature of the water. Cold water will hold a lot more oxygen than warm water. Also in cold water the lobster is less active and therefore uses less oxygen. As a further precaution, female lobsters are not fed while the eggs are hatching. If fed they are more active and use more oxygen. They also excrete more wastes which in turn use oxygen in decay. Food that is not eaten also decays. There is really no safe rule as to how many lobsters can be held in a certain size tank. Available oxygen is the determinant.

COLLECTING LARVAE

Hatching tanks at the Massachusetts Lobster Hatchery and Research Station are about ten feet long, three-and-a-half feet wide, fourteen inches deep and can hold seventy egg-bearing lobsters under ideal conditions. The water enters one end of the tank and goes out through a two-inch drain on the far end. Separating the drain from the rest of the tank is a fiberglass barrier which extends across the width of the tank close to the drain. This barrier has a V-shaped notch which mates with a similar notch in the fry catcher. The fry catcher, made of wood sides and a screen bottom, is roughly a foot square and four inches deep. As the water flows through the tank it passes through the notches in the barrier and fry catcher and the lobster fry are caught on the screen.

Tanks are fiberglass or wood coated with fiberglass resin. Naturally it is important that any container or tank, whether it is glass, fiberglass, wood, polyethylene, or other material, be seasoned before use with live animals. Run seawater in it for a day or more before use to eliminate toxic materials acquired during storage or manufacture of the tank. This is particularly true of polyethylene rubbish barrels. It has been found that if they are used as purchased new they do have a toxic effect, but if scrubbed and seasoned most are no longer toxic.

CULTURE AND FEEDING OF THE LARVAE

The Massachusetts Lobster Hatchery and Research Station has 100 rearing tanks. These tanks are sixteen inches in diameter, sixteen inches deep, and made of fiberglass. The bottom of the tank has a slight concave shape to it and in the middle of the tank there is a PVC circulator, roughly three inches in diameter. Through the center of the circulator there is a 3/4 inch overflow pipe. The overflow pipe is screened so the lobster fry are not lost. The circulator is similar to half a tennis ball inverted. The water enters the center of the tennis ball and then is forced out through sixteenth-inch holes that surround the circumference.

The whole secret of raising lobsters in large numbers is to eliminate their natural enemies. These include very heavy rainstorms, oil slicks, birds, fish, and other lobsters. In a building they are protected from most of these. Cannibalism, however, can be a serious problem. Cannibalism can be reduced by using circular tanks as described (to eliminate bunching up in corners) and feeding them adult brine shrimp every three hours.

Rearing tanks can be purchased from Glastronics (617-763-2131) for about $17 each, not including the circulator which must be assembled from plumbing parts.

About 3,000 lobster fry are cultured in each rearing tank. They are fed frozen adult brine shrimp every three hours, round-the-clock. It takes about eight pounds of frozen brine shrimp to feed 100 tanks through one day. Newly hatched lobsters resemble mosquito larvae. They are cultured as described above through four molts until they reach fourth stage. At this time they look like minature lobsters and seek shelter on the bottom.

CULTURE OF JUVENILE LOBSTERS

The fourth stage, or juvenile, lobsters are put in separate containers; usually plastic liter refrigerator cartons. Tank size does have something to do with growth. Kept in a small container a lobster will remain small. Presently experiments are under way to determine the tank size necessary for optimum growth without being in excess. A tank that is three times the length, width, and depth of the lobster is certainly adequate, but may be much larger than necessary.

A tank or trough made with a 2 x 6 timber ten feet long and two eight-inch roofers as the sides will suffice through many molts. This long narrow trough is then divided into six-inch sections with plywood partitions. A central screen-covered two-inch hole in each partition allows water to flow through the trough. Juvenile lobsters are fed shellfish or fish.

Shellfish that are small, such as periwinkles, are ideal for lobsters that are three or four inches long, because they can eat the whole thing. When the water temperature drops to 5-10°C lobsters eat very little. Therefore, in the wintertime the juvenile lobsters should be fed once a week. When the water is warm, 20-25°C, they should be fed daily. A piece of food approximately equal in size to one section of the tail fin is suitable.

In standing water aquaria, high water quality is important. Uneaten food should be removed regularly and the oxygen level kept at saturation. Frequency of changing water in such aquaria depends on many factors: dust or soot in the air, evaporation, and organic fouling. Usually water must be changed every week or two. Standing water cultures at the Massachusetts Lobster Hatchery are aerated routinely twice daily -- morning and night. This is done by dipping in an airstone attached to an air pump. Naturally, a small lobster in a large tank with the water surface exposed to the air will not have to be aerated for several days.

Lobsters grow by shedding their exoskeleton. A lobster will usually molt about ten times during the first year. Thereafter molting is much less frequent, often twice the second year and once annually subsequentially.

MATING LOBSTERS IN THE LABORATORY

The only time the female lobster will mate is within approximately 48 hours after she sheds her shell. Best results are obtained within four hours (Hughes and Matthiessen, 1962). They do mate roughly according to size and it is physically impossible for a very large male to mate with a small female. The claws of the male should be banded because some do not know whether to fight or to make love. The cast shell of the female should also be in the mating tank because there is a hormone or pheremone extruded with it that excites the male. The water should be deep enough so that the antennae do not break the surface of the water. In addition, it is important not to disturb them by resting arms or elbows on the tank, walking by, or flashing lights. After they mate, the lobsters can be removed to any desired tank.

At ambient temperatures the eggs will be carried in the ovaries for approximately nine months and then will be extruded. As they are extruded they are fertilized by the stored sperm. To extrude her eggs the female lobster lies on her back, resting on her tail. She balances on her large claws which are extended backwards. She curls her tail up towards her head to touch the last pair of walking legs. Then, as the eggs flow from the ovaries through openings on the second pair of walking legs, they flow across the seminal receptacle, are fertilized, and pass into the pocket or pouch made by the curled tail. At the same time the eggs are extruded there is a cement extruded which will then stick the eggs to the non-plumous hairs of the swimmerettes. It is necessary at this time to have a tank deep enough so that no part of the lobster is out of the water. If the water is too shallow she will extrude the eggs but quickly roll over to get her mouthparts into the water. As she does this the eggs will flow out of the pouch because the cement has not solidified.

VARYING ENVIRONMENTAL PARAMETERS

There are three main abiotic factors: salinity, oxygen, and temperature. Don McLeese at the Fisheries Research Station in St. Andrews (1956) has stated that lobsters can be kept alive in salinities down into the mid-20's provided the temperature is not at either extreme and the oxygen level is high. At the Massachusetts Hatchery lobsters have been kept alive in salinities as high as 48 ppt.

This is also true of the other two variables. One can vary quite a distance from normal as long as the other two are near normal.

Lobsters live in seawater at temperatures down to $-2°C$ and have been kept alive in temperatures up to $32°C$. Optimum growth occurs at $15-20°C$ (Hughes, 1968). When at high temperatures they will need more oxygen, therefore it must be supplied through airstones. It is believed that at high temperatures the lobster is using all its energy just to stay alive and does not grow.

ARTIFICIAL SEAWATER

Artificial seawater is used throughout the world. There are manufactured lobster holding tanks in restaurants that use artificial seawater. The manufacturers of these tanks often compound their own salt mixes and they are readily available. Aquarium Systems, Eastlake, Ohio, for example, produces a mix called "Instant Ocean." There are about six major salts. When mixed with normal tap water an artificial seawater is produced which is about as good as natural seawater for lobster culture. In fact it has some advantages. There is no plankton coming into the system, no shellfish larvae setting in the tanks, and no growths of algae.

In mixing artificial seawater it is essential to make sure that the tap water has not been heavily chlorinated and/or does not contain too much iron or sulphur. More sophisticated techniques are necessary to remove the iron or sulphur, but one simple way to get rid of the chlorine is to bubble air through the water or run it over a few baffles.

There is some lobster culture work being done now at the University of California at Davis, many miles from the coast. Using Aquarium Systems' regular "Instant Ocean" for their seawater they are growing larvae which are doing well.

DISEASES

There are lobster diseases. The Massachusetts hatchery is fortunate, however, in not being plagued by them. There is red tail disease, also called gaffkemia, caused by a micrococcus bacteria which gets into the blood stream of the lobster and will kill it by attacking the blood, breaking down the blood cells. The bacteria, Gaffkya homari, enters through a wound in the lobster, such as caused by pegging a claw as they did in the old days. Some gaffkemia work has been done by Rabin and Hughes (1968) and Jim Stewart et al. in Halifax (1966). Gaffkemia can be eliminated with aureomycin, but this is expensive.

Another is shell disease. This is caused by a bacteria which eats the chitin of the lobster shell. After the disease progresses the lobster looks like he has freckles or eroded spots on his shell.

Particularly with the fry, a disease to watch for is leucothrix. This fungus grows on the lobster, reproduces, and after it progresses the lobster looks like a ball of cotton. If he cannot successfully molt then he will die. After lobsters get to fourth stage or larger this seldom bothers them because it does not grow fast enough to hinder molting.

REFERENCES

Hughes, John T. 1968. Grow your own lobsters commercially. Ocean Industry, 3(12): 46-49.

Hughes, John T. and George C. Matthiessen, 1962. Observations on the biology of the American lobster, Homarus americanus. Limnol. Oceanog. 7(3): 414-421.

McLeese, D.W., 1956. Effects of temperature, salinity, and oxygen on the survival of the American lobster. J. Fish Res. Bd. Can. 13(2): 247-272.

Rabin, Harvey and John T. Hughes, 1968. Studies on Host-Parasite relationships in gaffkemia. J. Invert. Path. 10(2): 335-344.

Stewart, J.E., J.W. Cornick, D.F. Spears, and D. McLeese, 1966. Incidence of Gaffkya homari in natural lobster (Homarus americanus) populations of the Atlantic region of Canada. J. Fish Res. Bd, Canada, 23: 325-1330.

ECHINODERMATA

George D. Ruggieri, S.J.

Osborn Laboratories of Marine Sciences, New York

Aquarium, New York Zoological Society; Brooklyn, N.Y.

There are approximately 6,000 living species of echinoderms. They are found widely distributed throughout the seas and in some abyssal regions may comprise as much as 90 per cent of the sea-floor biomass. All echinoderms are marine and only a very few of them can tolerate even brackish water. Although some synaptid sea cucumbers have been occasionally mistaken for worms, most members of the phylum are easy to recognize. The extremely versatile (respiration, locomotion, nutrition, sensory perception) and efficient tube-feet-water-vascular system is found only in the echinoderms. These animals also possess skin ossicles which give the phylum its name.

Although most echinoderms have bilateral larval stages, radial symmetry is usually a diagnostic feature in the adults. The adults lack a head-end and since they are not bilaterally symmetrical, terms such as anterior, posterior, dorsal and ventral are inappropriate. The two important surfaces in these animals are the oral, which bears the mouth and contains numerous tube feet and the opposite aboral surface. Four of the five living classes of echinoderms move about with the oral surface in contact with the substrate while the crinoids have the oral surface facing upward. The five classes of echinoderms are:

Crinoidea	- sea lilies
Asteroidea	- sea stars
Holothuroidea	- sea cucumbers
Echinoidea	- sea urchins, sea biscuits, sand dollars
Ophiuroidea	- brittle stars, basket stars

Crinoids, if they are found at all in our local waters, are rare. However, a number of representatives of the remaining four classes are present. The sea stars, Asterias forbesi and Asterias vulgaris are plentiful while the small red sea star, Henricia sanguinolenta, is found less frequently and usually in deeper waters. The sea urchin, Arbacia punctulata (Fig. 1), which is ubiquitous along most of our Eastern seaboard, is found in several areas in and around New York. The green urchin, Strongylocentrotus droebachiensis, though common in colder waters of the northern Atlantic and Pacific, is found less frequently and is usually of a smaller size. The common sand dollar, Echinarachnius parma, appears to be less abundant now than in the past. The brittle star, Ophioderma brevispina, which is usually found on sandy bottoms is quite common. Leptosynapta inhaerens, a white sea cucumber ranging in size from four to eight inches is more abundant than the small rose-colored Leptosynapta roseola. Both species are found buried in sand. The more leathery, darkly colored sea cucumber, Thyone briareus, is relatively common in muddy waters.

Although echinoderms are looked upon with disdain by shellfish culturists, these animals have provided embryologists, physiologists and biochemists with research material for the last hundred years. Our knowledge of the mechanics involved in early developmental patterns has been extensively elucidated through the study of sea urchin morphogenesis.

But sea urchins have been destroying the kelp beds in California waters. Kelp, in addition to providing shelter and food for various invertebrates and fishes, is also harvested and processed for use as additives in many medicinal and industrial preparations. In Japan, on the other hand, sea urchins are harvested annually. Sea urchin roe is a prized food item in Japan and many European and South American countries. Attempts are now being made in California to develop new fishery utilizing sea urchin roe as a gourmet food.

PROCUREMENT AND MAINTENANCE OF ADULT SPECIMENS

With the advent of better laboratory facilities, availability of synthetic sea waters, numerous supply houses and quick modes of shipping, even inland laboratories are now able to conduct research on echinoderms. Many of the adult echinoderms are collected by dredging; this is especially true of the deep water forms and those which bury themselves in the sand. Others, like certain sea urchins (Strongylocentrotus purpuratus; Lytechinus variegatus and Diadema antillarum) and sand dollars can be collected in shallow waters at low tide. At the Osborn Laboratories of Marine Sciences the sea urchin *Arbacia punctulata* and the sea stars *Asterias forbesi* and *Asterias vulgaris* are collected by divers.

The maintenance of adult specimens will, of course, depend upon the type of studies that are to be conducted. If the animals are to be kept for any length of time, for example, more than a week, then an open-water system is not only the best but with most species is the only way to maintain them in healthy condition. None of the echinoderms will do well in closed stagnant systems. Short-term use of adult specimens, however, requires very little maintenance. In many cases, merely keeping them cold (approximately $5^\circ C$) and damp will suffice. Some species, for example, *Lytechinus pictus*, are extremely hardy while others like the synaptid sea cucumber found in our waters are sensitive to the point of fragmenting. Ideally the animals should be maintained under conditions which most closely conform to their natural environments. In closed recirculating systems the salinity, temperature, pH and cleanliness of the water are all crucially important. Overcrowding of specimens must be avoided. Not only is crowding detrimental to the health of the adults but if one animal sheds its gametes it is likely to trigger all the others to do likewise. The food, of course, will depend upon the species being studied. Sea urchins do well on a variety of sea weeds. On the few occasions when no sea weed was available we have fed *Arbacia punctulata* spinach and lettuce leaves. Some investigators have kept urchins alive and well on a diet of shrimp. We have maintained sea stars for long periods of time on small pieces of shellfish. Many other echinoderms can be maintained for several weeks without any food at all if they are kept at colder temperatures.

BREEDING SEASONS

The breeding season of the echinoderms found in our waters roughly coincides with the summer months. *Asterias forbesi* and *Asterias vulgaris* usually spawn from May through early July. Sea stars with soft, bulging arms are ripe while small, hard-skinned animals do not, as a rule, contain viable eggs and sperm. The sea urchin, *Arbacia punctulata*, normally spawns from early June through August. We have been able to extend their season and have procured viable eggs and sperm through December and January by using the electrical stimulation method to obtain small quantities of eggs and sperm over a long period of time.

The sand dollar, *Echinarachnius parma*, breeds from April to July. The ophiuroid, *Ophioderma brevispina*, usually spawns from July to August. Attempts at artificial insemination have not been successful in this species nor with any ophiuroid. The delicate

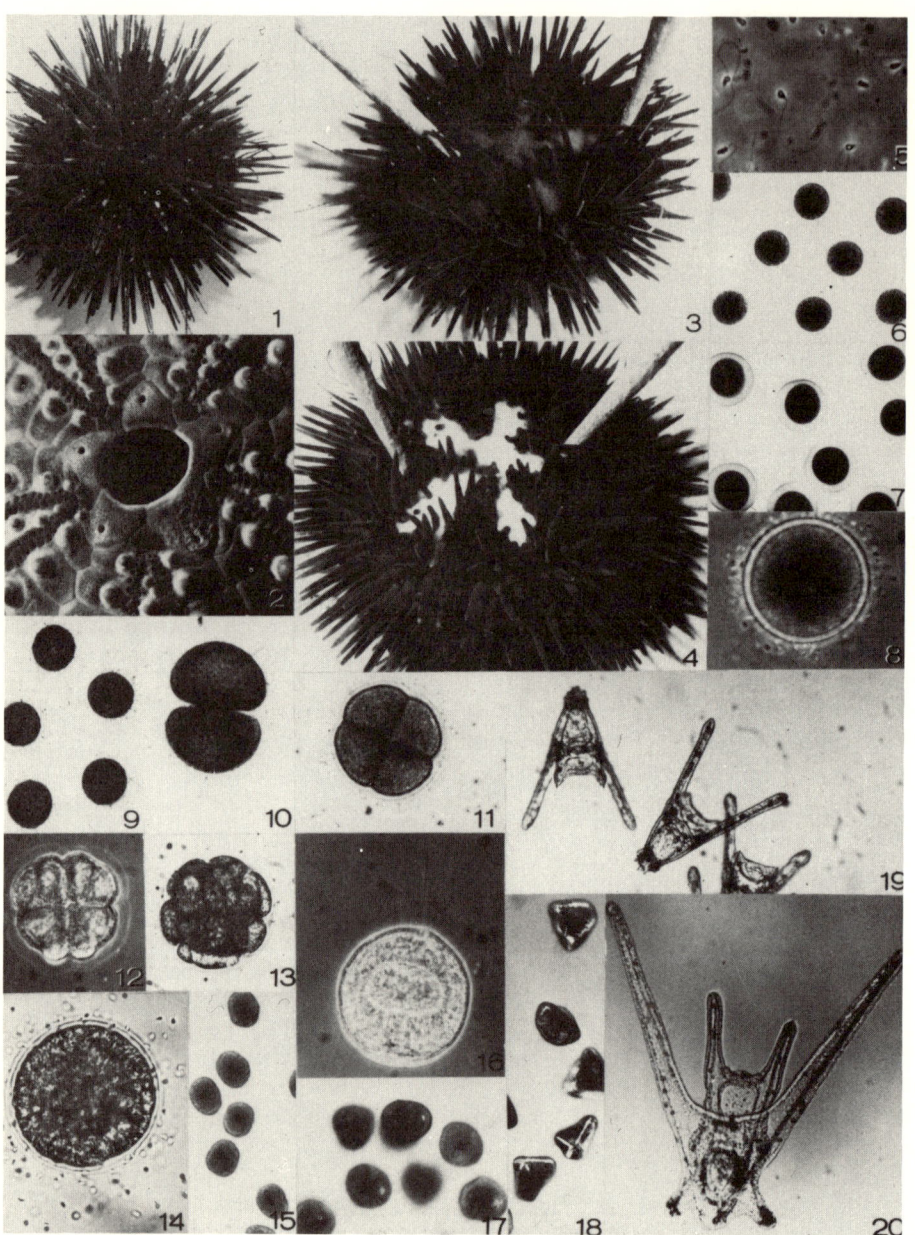

ECHINODERMATA

Fig. 1.   Adult sea urchin, <u>Arbacia</u> <u>punctulata</u> collected from New York waters.

Fig. 2.   Sea urchin with spines removed; note five gonopores through which gametes are extruded.

Fig. 3.   Female sea urchin releasing eggs upon electrical stimulation.

Fig. 4.   Male sea urchin with white threads of sperm.

Fig. 5.   Phase-contrast photomicrograph of sea urchin sperm.

Fig. 6.   Photomicrograph of unfertilized sea urchin eggs; note small but obvious nucleus.

Fig. 7.   Unfertilized sea urchin eggs stained with Janus green to show the jelly coat which surrounds the eggs.

Fig. 8.   Phase-contrast photomicrograph of fertilized egg. Note raised fertilization membrane and supernumery sperm trapped in jelly coat.

Fig. 9.   Fertilized egg in "streak" stage just prior to first division.

Fig. 10. Two-cell stage.

Fig. 11. Second cleavage division (4-cell stage).

Fig. 12. Sixteen-cell stage; note prominent micromeres.

Fig. 13. Fifty-six cell stage. Note micromeres.

Fig. 14. Blastula stage just prior to hatching. Blastula is composed of approximately 1,000 cells (10 divisions).

Fig. 15. Free-swimming blastulae shortly after they dissolved the fertilization membrane.

Fig. 16. Gastrula stage; note long cilia at forward end and area of invagination at the base.

Fig. 17. Prism stage. Photo taken under polarized light to show small spicules.

Fig. 18. Young pluteus larvae. Polarized light, showing skeletal development.

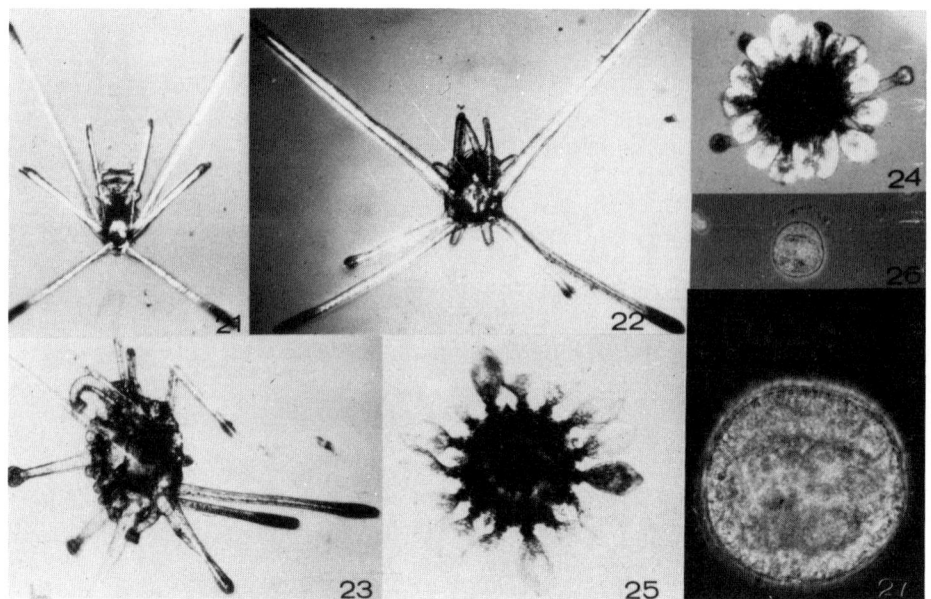

Fig. 19.   Pluteus larvae.

Fig. 20   Some further stages in the development of the larval
thru   form and metamorphosis.
25.

Fig. 26   Low and high power photomicrographs of animalized lar-
&   vae. Note hyper-ciliation, thickened apical plate and
27.   absence of archenteron and skeletal elements.

sea cucumbers, Leptosynapta inhaerens and Leptosynapta roseola, spawn in June and July. Thyone briareus also breeds at this time, but attempts at artificial insemination have not been successful, probably because the ovarian eggs contain a large germinal vesicle, while those shed normally are in metaphase of the first maturation division.

SEXING AND OBTAINING GAMETES

It is impossible, in most echinoderms to determine the sex by external appearance, but the sexes are usually separate. However, our local sea cucumbers, Leptosynapta inhaerens and Leptosynapta *roseola* are hermaphroditic but not self-fertile, whereas the sexes are separate but indistinguishable in external morphology in our local species of echinoderms.

In most echinoderms the only method available to determine the sex prior to obtaining larger quantities of gametes for experimentation is gonadal biopsy. A small sample of the gonads, removed by means of a hypodermic needle, can be analyzed under the microscope without harming the animal or inducing it to shed. After having collected Arbacia punctulata we first sex them by using an electric shock. Figure 2 shows the five gonopores on the aboral surface. Two lead electrodes, delivering between six and ten volts are applied to the test of the urchin. Ripe females exude a pinkish ooze through the gonopores, while the males emit a white thread of sperm (Figures 3 and 4). This electric stimulation method has the great advantage of controlled gamete release. As soon as the current is discontinued the animal stops shedding and can, therefore, be used on other occasions. In sexing the sea urchin we make certain to obtain gametes from each of the five gonopores. This allows us to quickly check for hermaphroditism, which is extremely rare in echinoids. In the last 15 years we have detected three hermaphroditic Arbacia punctulata. We have also described the only known case of in corpore fertilization and development in a sea urchin. (Ruggieri, 1969).

The electrical stimulation method is also used to obtain the eggs and sperm from Arbacia for experimentation. But this method is not effective in procuring gametes from all sea urchins nor does it work with sea stars. The more commonly used method to obtain gametes from sea urchins and sand dollars is the injection of isotonic (0.5 M) potassium chloride into the body cavity. The urchin is placed aboral side down on a beaker full of seawater and the needle of the syringe containing potassium chloride is inserted into the only soft part of the animal on the oral surface. The amount of potassium chloride injected will depend upon the size of the urchin (small individuals, 1 ml; larger specimens 10 ml). The eggs ooze out through the gonopores, stream through the water and are easily collected from the bottom of the beaker. Sperm can be obtained in a "dry" state in a dish without water. This method, like the electrical stimulation technique does not harm the adult but unlike electrical stimulation does not allow for controlled shedding.

Sea stars can be induced to spawn by injection of a radial nerve extract. Extracts prepared from the radial nerve of the sea star, Asterias forbesi, injected into the perivisceral coelom of ripe individuals of the same or different species induces the release of copious amounts of gametes (Chaet and Musick, 1960). An extract of the radial nerve from a sea urchin induces spawning in several species of echinoids but a delayed spawning in asteroids (Cochran and Engelmann, 1972). However, the oldest and still most widely used method, especially with sea stars, is surgical removal of the gonads. After removing an arm from a ripe sea star a mid-dorsal cut along the length of the arm exposes the two large, multi-branched gonads which lie adjacent to the greenish-brown digestive glands. The ovaries are usually pinkish in color and the testes white or beige. The ovaries should be lifted out, rinsed in seawater and placed in a dish or beaker containing seawater. The eggs are released from the proximal end of the ovary. As a general rule eggs released in the first five or ten minutes are best for experimental use. A small section of the testes can be excised and mixed with seawater to make a sperm suspension just prior to fertilization of the eggs.

Although many investigators use either potassium chloride injection or electrical stimulation, surgical removal of the gonads in the sea urchin is also an effective method but, of course, results in the death of the adults. It's important to note, however, that the coelomic fluid of echinoderms has a deleterious effect upon eggs. Therefore, the eggs must be washed several times in seawater prior to use.

Although eggs of a number of echinoderms can be stored for a period of time (24 hours at $5^{\circ}C$ for Strongylocentrotus purpuratus) without losing fertilizability it is advisable to effect fertilization as soon as possible after collection of gametes.

The number of eggs available from a ripe female will depend upon the species and the size of the adult. Even among individuals of the same species the number of eggs will vary considerably. Harvey (1956) reports that a single Arbacia punctulata female may contain as many as eight million eggs. Approximately one-half this number will be exuded after injection of potassium chloride or electrical stimulation.

Sperm is best stored by leaving it in the testes or by collecting it in a "dry" state; that is, in the absence of seawater, until needed. Sperm remain active for only a few hours once they have been diluted in seawater.

After procuring and washing the eggs a small sample should be removed and microscopically examined to determine whether or not the eggs are mature. The mature eggs of Arbacia punctulata are of uniform size (74μ in diameter) and perfectly spherical with a small nucleus (Figure 6). Immature eggs are readily recognized in having a large clear nucleus and prominent nucleolus. Occasionally the eggs may be slightly elongated; this is due to crowding in the ovary. A drop of the sperm suspension is made by adding one drop of the "dry" sperm to ten milliliters of seawater. If the sperm are immotile or move sluggishly they are not likely to fertilize the eggs. The sperm should be vigorously motile or discarded (Figure 5). If both eggs and sperm are viable fertilization is easily effected by adding an appropriate number of sperm to the egg suspension. The tendency, at least in the beginning, is to use too much sperm. Some species of echinoderms are susceptible to polyspermy. Although this is not the case with Arbacia punctulata the excess sperm does decompose and this could have a deleterious effect on the developing embryos. Since the eggs are covered by a jelly coat (Figure 7), which is not ordinarily visible because it has the same refractive index as water, it is easy to get an idea of the number of sperm present. Supernumerary sperm become trapped in the jelly coat (Figures 8 and 9). Ideally, about a dozen sperm should be seen surrounding each egg. If this number reaches several hundred then the number of sperm added was indeed excessive.

Sperm penetration occurs very rapidly. Fertilization is vividly obvious by the lifting or raising of the fertilization membrane (Figure 8). This is seen within a minute after insemination. If the fertilization membranes in all the eggs raise at the same time it is extremely likely that further developmental stages will also be synchronous.

In our experiments a batch of eggs is rejected unless 95 per cent of them exhibit lifting of the fertilization membrane shortly after the addition of sperm.

EARLY DEVELOPMENT

There are a number of excellent texts which describe in detail the time schedule of developing echinoderm embryos, as well as that of other marine and freshwater invertebrates (Costello et al. 1957).

More is probably known about the early development of the sea urchin than about any other organism. There are a number of reasons for this, some of which have already been touched upon. The adults are relatively easy to obtain and maintain, they release their gametes readily and they yield a large number of eggs. Since meiosis is completed in the ovary, the eggs can be fertilized immediately after spawning. Fertilization is easily effected and the embryos usually develop synchronously and with minimum care. The eggs readily take up a large number of chemicals and hence are widely used in physiological and biochemical research.

The rate of development of echinoderm eggs is temperature dependent. Arbacia punctulata eggs are best raised at temperatures between 20°C and 23°C. Temperatures above 30°C are lethal; temperatures slightly lower than 20°C cause a noticeable delay in development. Since the rate of development for most echinoderm eggs has a Q_{10} of slightly more than 2, a 10 degree decrease in temperature would more than double the developmental time schedule.

After insemination and the lifting of the fertilization membrane the egg of Arbacia punctulata divides about ten times in approximately eight hours. The first three cleavages divide the egg into eight blastomeres of equal size. The planes of the first two cleavages are meridional (in the polar axis) (Figures 10 and 11) while that of the third is equatorial.

At the fourth cleavage (16-cell stage) (Figure 12), the four upper cells divide meridionally, forming eight equal mesomeres, while the lower four cells divide equatorially and unequally giving rise to four large macromeres and below these, four small, clear micromeres. In Arbacia punctulata this 16-cell stage offers the first obvious indication of polar differentiation, i.e., an animal pole evident with mesomeres and an opposite vegetal pole clearly marked by the small micromeres. Echinoderm eggs actually undergo polar differentiation while still in the ovary. This polarity, however, is not morphologically evident except in the eggs of a few species whose pigmentation clearly marks off the animal pole from the vegetal pole.

At the fifth division (32-cell stage) the eight mesomeres divide equally and equatorially forming two tiers of cells while the macromeres divide meridionally. At the sixth cleavage (64-cell stage; there are actually only 54 cells at this stage because the eight micromeres fail to divide) (Figure 13), five distinct layers of cells are obvious. These layers have been designated as an. 1, an. 2, veg. 1, veg. 2 and micromeres (Text Fig. I). Layers an. 1, an. 2, and veg. 1 give rise to ectoderm and later to such structures as the apical plate and cilia, and stomodaeum; veg. 2 produces the endoderm and secondary mesenchyme which later differentiates into the coelom, oesophagus, stomach, intestine and anus; the micromeres give rise to primary mesenchyme which differentiates into the skeleton.

Eight hours after fertilization the ciliated blastula rotates within the membrane (Figure 14), releases a hatching-enzyme which dissolves the membrane and emerges as free-swimming (Figure 15). Several hours later gastrulation begins with the invagination of vegetal pole cells (Figure 16). The first signs of bilateral symmetry appear with the flattening of one of the sides of the gastrula; this becomes the oral region. The gastrula transforms in-

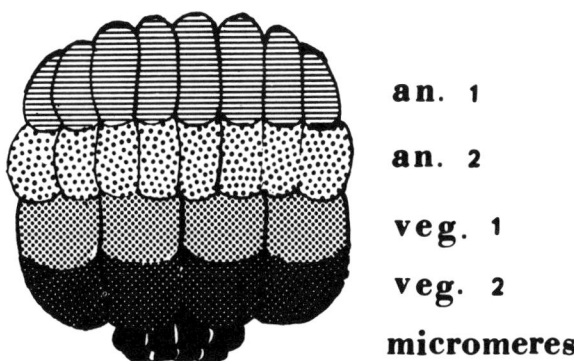

Text Fig. I. Sixth division stage showing the five tiers of cells: an. 1, an. 2, and veg. 1 cells give rise to ectodermal structures, veg. 2 cells give rise to endodermal and secondary mesenchymal structures, micromeres give rise to primary mesenchyme.

to the pluteus larva by degrees with the progressive development of the anal and ventral arms and the overall lengthening of the larva (Figures 17, 18, 19).

Until recently no one had successfully reared sea urchins from one generation to the next. Hinegardner (1969) was able to raise several species of sea urchins from egg to egg. Figures 20 through 25 show further stages in the development of the pluteus larvae and metamorphosis. The entire life cycle takes about six months. Survival through the larval stage and metamorphosis depends in great measure on the control of disease, proper feeding regimen, and on the particular male and female from which the gametes were obtained. Some matings appear to produce a hardier line of urchins than do others. Now it will be possible to apply genetic techniques to the study of sea urchin development. This, together with the other advantages these eggs offer, will make them extremely useful in solving some of the difficult problems involved in the developmental process.

USE OF ECHINODERM EGGS IN RESEARCH

The eggs of some sea urchins are hardier than those of others; and eggs of species such as Lytechinus variegatus are transparent and, therefore, are excellent for studying cell movement.

Arbacia eggs have been extensively studied on the East Coast, probably because of their availability. These eggs, however, are heavily pigmented and are not as hardy as those of some other sea urchins. We have found the gametes of Arbacia punctulata to be especially useful as an assay system in determining the effects of biologically active substances extracted from other marine organisms and various pollutants (Rio et al. 1963).

The types of effects observed when subjecting Arbacia punctulata gametes to these test substances include:

1. immobilation of sperm

2. induction of parthenogenesis

3. delay in cleavage and in overall development of the embryos

4. complete arrest in development and cytolysis

5. induction of multiple karyokinesis without concommitant cytokinesis

6. inhibition of hatching: an interference with the action of the hatching enzyme on the fertilization membrane resulting in an inability of the blastula to emerge as free-swimming

7. fragmentation of developing embryos

8. interference with the synchronous beat of cilia in larval forms

9. inability of the embryo to undergo gastrulation

10. induction of characteristic abnormalities in the developing embryos such as animalization, radialization and vegetalization.

These effects are readily observable within twenty-four hours after the initiation of the experiment.

The most interesting and significant of these effects is the production of abnormal larval forms. Two antagonistic gradients

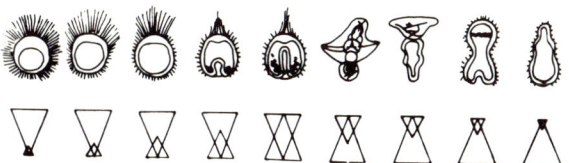

Text Fig. II. Diagram of double gradient system and the effects produced when these gradients become unequal. Various degrees of animalization are depicted left of center; various manifestations of vegetalization are shown right of center. Normal development ensues when the two gradients are equal.

in the sea urchin egg determine the course of ectodermal and endomesodermal differentiations (Text Fig. II). If the two gradients are balanced (as they are in the middle embryo in Text Fig. II), then development is normal. If, however, one is stronger animalization (ectodermization) or vegetalization will occur depending upon whether the animal or vegetal gradient exerts the greater influence. Text Fig. II shows various degrees of animalization to the left of center and of vegetalization to the right. Strongly animalized larvae are characterized by hyper-developed apical tuft, thickening of the apical ectoderm, absence of archenteron, and general suppression of endomesodermal structures (Figures 26 and 27). The degree of extension of the apical ciliary tuft gives an indication of the intensity of the animalization. Vegetalized larvae show excessive development of the endoderm and enlargement of the intestine, which, in strongly vegetalized larvae, does not invaginate, thus forming an exogastrula (i.e., an embryo with the intestine protruding outward). Radialized larvae are elongated along their animal-vegetal axis. The apical ciliary tuft is enlarged and the spicules usually form a ring around the base of the archenteron. Radialization represents a less extreme form of animalization. As indicated in Text Fig. II all stages intermediate between extreme animalization and vegetalization can be obtained. However, only slightly animalized or vegetalized larvae are capable of living for any length of time. Vegetalization can occur either through suppression of animal differentiation or enhancement of the vegetal influence. The same is true of animalization, which can result from a heightened influence of the animal gradient or an interference with the vegetal or perhaps a combination of these factors.

REFERENCES

Chaet, A.B. and R.S. Musick, Jr., 1960. A method for obtaining gametes from Asterias forbesi. Biol. Bull. 119: 292

Cochran, R.C. and F. Engelmann, 1972. Echinoid spawning induced by a radial nerve factor. Science 178: 423-424.

Costello, D.P., M.E. Davidson, A. Eggers, M.H. Fox, and C. Henley, 1957. Methods for Obtaining and Handling Marine Eggs and Embryos. Mar. Biol. Lab., Woods Hole, Mass. 247 pp.

Harvey, E.B., 1956. The American Arbacia and Other Sea Urchins, Princeton University Press, Princeton, New Jersey.

Hinegardner, R.T., 1969. Growth and development of the laboratory cultured sea urchin. Biol. Bull. 137: 465-475.

Kumé, M. and K. Dan, 1957. Invertebrate Embryology, Bai Fu Kan Press, Tokyo (Nolit Publishing House, Belgrade, 1968).

Needham, J.G. (Chm) 1937. Culture Methods for Invertebrate Animals, Dover Publications, Inc., New York, N.Y.

Reverberi, G., 1971. Experimental Embryology of Marine and Fresh-Water Invertebrates, American Elsevier Publication Co., Inc., New York, N.Y.

Rio, G.J., G.D. Ruggieri, M.F. Stempien, Jr., and R.F. Nigrelli, 1963. Saponin-like toxins from the giant sunburst starfish, Pycnopodia helianthoides, from the Pacific Northwest. Amer. Zool. 3(4): 331.

___ 1965. Echinoderm toxins I. Some biochemical and physiological properties of toxins from several species of Asteroidea. Toxicon 3: 147-155.

Ruggieri, G.D., 1969. In corpore fertilization and development in the sea urchin, Arbacia punctulata, Nature 223: 189.

___ 1965. Echinoderm toxins II. Animalizing action in sea urchin development, Toxicon 3: 157-162.

Ruggieri, G.D. and R.F. Nigrelli, 1960. The effects of holothurin, a steroid saponin from the sea cucumber, on the development of the sea urchin, Zoologica 45(1): 1-16.

___ 1962. Effects of Bonellin, a water soluble extract from the proboscis of Bonellia viridis, on sea urchin development, Amer. Zool. 2(4): 552.

___ 1964. Effects of the red web starfish, Patiria miniata, on sea urchin eggs, Amer. Zool. 4(4): 431.

___ 1966a. Effects of extracts of the sea star Acanthaster planci, on the developing sea urchin, Amer. Zool. 6: 380.

___ 1966b. Animalization of Arbacia punctulata larvae by extracts of the sea cucumber, Holothuria edulis, Amer. Zool. 6: 381.

___ 1971. Physiologically active substances from echinoderms. Sym. on Physiol. Active Compounds from Mar. Organisms, St. Petersburg, Fla. (in press).

Ruggieri, G.D., R.F. Nigrelli, and J.J.A. McLaughlin, 1962. Effects of dinoflagellate toxin on development of Arbacia punctulata, Amer. Zool. 2(4): 552-553.

Ruggieri, G.D., R.F. Nigrelli and M.F. Stempien, Jr., 1964. Jelly-precipitating reaction in sea urchin eggs by antibiotic substances from sponges, Amer. Zool. 4(4): 431.

Ruggieri, G.D., M.H. Baslow, S. Jakowska and R.F. Nigrelli, 1961. Developmental aberrations in sea urchin eggs induced by sponge extracts, Amer. Zool. 1(3): 384.

Stempien, M.F., Jr., G.D. Ruggieri, R.F. Nigrelli, and J.T. Cecil, 1969. Physiologically active substances from extracts of marine sponges, Trans. of Food-Drugs from the Sea Symp. (MTS): 295-305.

Wilt, F.H. and N.K. Wessels (eds.) 1967. Methods in Developmental Biology, Thomas Y. Crowell Co., New York, N.Y.

OPISTHOBRANCH CULTURE

David R. Franz

Associate Professor of Biology

Brooklyn College, City University of New York

The snails which are the subject of this chapter, the Opisthobranchia, are less numerous than their ubiquitous cousins, the Prosobranchia. They damage neither man nor his works and transmit no disease. Nevertheless, they are greatly admired by evolutionary-minded zoologists, marine population ecologists, and a small but vigorous and growing band of hearty amateur naturalists, taxonomists and nature photographers. If this seems paradoxical, the question will be answered the first time you discover a nudibranch crawling over a rock, or are confronted with a sacoglossan in a batch of seaweeds. The opisthobranchs are truely the avante-garde of the snail world. They alone of all the many thousands of marine snails have been freed by evolution of the burden imposed by the gastropod shell. Liberated in a sense from this encumbering and conservative influence, they are now clearly in the evolutionary process of striking out in many divergent directions, probing unusual and unorthodox pathways in the solution of all the ancient problems which animals must solve. We find within the Opisthobranchia, a fascinating variety of morphological and physiological specializations, the existence of highly specific predator/prey associations, and the evolution of unique defensive mechanisms. For the curious naturalist, these are the challenge of studying opisthobranchs. But for many people, the matchless delicate symetrical beauty, particularly in juxtaposition with their ferocious carnivorous behavior, is reason enough for interest and study.

The following pages will serve, hopefully, to briefly summarize the art of culturing these animals. The reader must bear in mind that, in contrast to some other groups of organisms discussed in this volume, the culture of opisthobranchs is in its embryonic stages. Indeed, the professional biologists who are interested in this problem are, at present, both very few in number and scattered throughout the world.

KINDS OF OPISTHOBRANCHIA

The dominating characteristic of the subclass Opisthobranchia is the tendency within the group toward the reduction or loss of the gastropod shell. Concurrent with this is a shift in the mantle cavity which in prosobranch snails opens anteriorly. If present at all, the mantle cavity in opisthobranchs open on the right side. Since these evolutionary trends reverse the effects of torsion, the Opisthobranchia are considered to be undergoing "detorsion". The result is a return to almost perfect external bilateral symmetry best exemplified by the Nudibranchia and Sacoglossa.

The subclass is presently undergoing an impressive adaptive radiation so that there is flexibility in the classification of the group at the ordinal level. Table 1 lists and briefly describes the nine major opisthobranch orders. Of these, most of the information on culture is restricted to the three largest: the Cephalaspidea, Sacoglossa and Nudibranchia.

The Cephalaspidea are primarily burrowing snails. They usually possess a thin smooth shell which may be completely internal as in Philine and Navanax. The cephalaspid head is characteristically in the form of a broad shield which is an adaptation for burrowing in soft substrates. Cephalaspids are either predators or deposit-feeders.

The Sacoglossa are herbivorous snails which feed primarily on macroalgae. Sacoglossa seem to be fairly specific, i.e., restricted to one or a few related plant host species. Although some species such as Oxynoe and Lobiger have thin external shells, most sacoglossa are naked. All have specialized mouthparts which are adapted for piercing and sucking macroalgal cells.

The Nudibranchia comprise the naked sea-slugs, nearly all of which are prey-specific predators on other invertebrates.

PROCURING ADULT ANIMALS

The strategy followed in collecting adult opisthobranchs differs according to which group is being collected. Cephalaspid snails are often not visable on the surface of a mud-flat or beach because these animals are adapted to moving just beneath

TABLE 1

Subclass OPISTHOBRANCHIA

Order	Characteristics	Example
Cephalaspidea	Shell thin, often reduced; mantle cavity opens on side of animal; the head forms a large "shield" for burrowing in sand or mud	Haminoea
Notaspidea	Shell internal or absent; single gill on right side of flattened, slug-like body.	Pleurobranchaea
Anaspidea	Crawling or swimming animals with fleshy parapodia; shell reduced and internal; herbivorous	Aplysia
Sacoglossa	Mostly naked snails with a highly modified radula and buccal mass for sucking; herbivorous	Elysia
Nudibranchia	Naked sea slugs with epidermal branchiae; predaceous	Acanthodoris, Aeolidia
Acochlidiacea	Interstitial sand-dwelling shell-less opisthobranchs	Microhedyle
Thecosomata	Planktonic pteropods with coiled shells; herbivorous	Limacina
Gymnosomata	Naked planktonic pteropods; predaceous	Clione
Pyramidellacea	Small, generally parasitic, snails with long coiled shells and a long invaginable proboscis with piercing stylet	Odostomia

the surface. Some species are easily collected by scooping layers of mud from the surface with a spade and gently washing the sediments through a screen. Nested one-quarter inch and one millimeter screens mounted in wood frames are ideal for this. It is advisable to bring the screened sediment back to the laboratory where it can be carefully sorted in an enamel tray. This procedure is not necessary for larger, easily visable species.

Sacoglossa are most easily collected by placing quantities of algae or marine spermatophytes in plastic buckets with seawater. Over a period of several hours (in the shade) sacoglossa tend to leave their host plants and crawl up the sides of the container, or attach to the surface film. It is advisable to handle all opisthobranchs gently with forceps, or in the case of very small specimens, with a pipet. For snorkeling or SCUBA work, a device such as the "Clark Sucker" (Clark, K.B., 1971) or "Acadian SOCK" (Bleakney, J.S., 1969) is enormously useful.

The art of nudibranch collecting is not easily summarized. Some large nudibranchs can be collected in almost completely exposed positions well away from cover or prey organisms. As a general rule, however, nudibranchs are found on or near their prey. Since nudibranchs generally feed on hydroids and other coelenterates, bryozoa or sponges, careful examination of these potential prey animals will often yield good collections, particularly when the adults are large and/or brightly colored. Opisthobranchs produce relatively large, gelatinous egg masses which are frequently attached to the prey, or very close by. The egg masses are often more visable than the cryptically-colored adults, and the presence of egg masses indicates that a very careful scrutiny of the prey colony is in order. For very small nudibranchs, it is necessary to bring clumps of prey (hydroids, bryozoa, etc.) to the laboratory where they can be examined under the dissection microscope. Old pilings, floats, mooring lines and lobster pots are ideal for collecting the clumps of prey in which adult nudibranchs are likely to be found. If an attempt is to be made to culture nudibranchs, it is important to note the prey species, which is usually the organisms on which the nudibranchs are crawling or associated with in nature. As explained below, it may be necessary to provide larvae with the specific prey species to facilitate metamorphosis.

In temperate marine climates, opisthobranchs -- like other invertebrates -- are seasonal in abundance. Seasonal variability is perhaps most striking on the Atlantic coast where the most severe seasonal changes in temperature occur. In southern New England, for example, the meager summer fauna of nudibranchs comprising the "temperate, endemic fauna" (Franz, 1970) disappears in the winter. At that time, and on into the spring, a richer fauna of cold-water species (boreal and boreo-subarctic) reaches maximum abundance. Thus, some knowledge of zoogeography and the breeding ecology of the fauna is helpful in the collection of particular species.

MAINTENANCE OF ADULTS

The major requirement for the maintenance of adults is an adequate supply of seawater and the proper food. Ideally, opisthobranchs should be kept in running seawater but standing seawater will suffice if the water is changed frequently and if the aquarium is gently aerated. Fecal material and sediment should not be permitted to accumulate in the aquarium and it is usually wise to filter the seawater through bolting cloth or a plankton net. Food is usually the limiting factor, particularly in nudibranchs and sacoglossa with specific prey requirements. Colonial prey animals such as hydroids, bryozoa or sponges, cannot be cultured in the quantities necessary to maintain adult nudibranchs. Therefore, successful maintenance hinges on a natural supply of food. In Hawaii, the nudibranchs Phestilla melanobranchia and P. sibogae have been cultured through several generations by continuously providing them with pieces of dendrophyllid and Porites corals collected from the field (Harris, L.G., 1970). Other investigators have had some success in rearing adult nudibranchs on field-collected prey but this requires a knowledge of the food requirements of the predator. Colonial prey, notably bryozoa and hydroids, may themselves be difficult to maintain, even for the short period required to bring them from the field to the laboratory. In this regard, it is particularly important to duplicate field thermal conditions as closely as possible as many invertebrates will not survive severe thermal shock.

Grazing opisthobranchs such as sacoglossa as well as "sea hares" (Anaspidea) can be maintained in the laboratory in good condition as long as the proper plants are provided.

PROCURING AND HANDLING EGGS

Although adult opisthobranchs offer some maintenance problems, a very helpful feature of many opisthobranchs is the tendency for gravid adults to lay quantities of ova in captivity. Eggs are attached to any available substrate including the sides of the container and the surface film and the limit to the quantity of eggs produced is probably the stored energy resources present at the time of capture. Furthermore, since all opisthobranchs are hermaphroditic, there is a very good chance that individuals captured will be fertilized.

To facilitate oviposition, it is useful to provide sacoglossa with a supply of their normal host plant. In the case of Cephalaspidea, it may be necessary to provide a small quantity of natural sediments. The egg masses of Cephalaspidea are attached to the substrate by a fine mucous thread. The adults seem to require the presence of normal sediment in order to produce eggs in the laboratory. Once a snail has oviposited, the egg mass should be removed to clean seawater as the presence of natural sediments may produce reducing conditions in the culture dish.

Opisthobranch eggs are typically enclosed within a thick matrix of clear mucous. The jelly-like egg masses are usually very large relative to the size of the adults which produce them. Under the microscope, an opisthobranch egg can be seen to be enclosed within a fluid-filled capsule. The number of eggs per capsule is variable but fairly constant within a species. Capsules are distributed with a spirally coiled tube or egg string and the string itself is enclosed within one or two membranes. The shape of egg masses and their mode of attachment varies among genera and families, and is very useful in distinguishing species in the field. A classification and description of nudibranch egg masses is provided by Hurst (1967). Estimates of the number of eggs per mass, and measurements of the diameter of uncleaved zygotes are important items of information to be recorded.

Egg masses can be handled, cut with a scissor, even compressed under a coverslip with relatively little damage to the developing embryos. Older egg masses, i.e., those in which hatching is near, are more fragile and rough handling can cause premature release of larvae.

LARVAE AND LARVAL DEVELOPMENT

At the time of oviposition, fertilized ova are usually at the one-celled stage, and nudibranch egg masses are excellent material for observing the early stages of spiral cleavage. Casteel (1904) provides a complete embryological description of the nudibranch <u>Fiona marina</u> and very informative descriptions are provided also by Thompson (1958, 1962) for other nudibranch species.

The development time of the eggs, which is the time between the first cleavage and hatching of the larva, varies from a few days to a few weeks. This depends on the development pattern present in the species and, to a lesser extent, temperature.

In some opisthobranchs, the eggs develop directly. There is no planktonic larval stage, and the animals hatching from the egg mass at the end of the embryonic period resemble to a greater or lesser degree, adult animals except for size. Species with this development pattern generally produce very large eggs (230-400 microns) and the development time may exceed a month (Thompson, 1967).

In most opisthobranchs, the eggs develop into a planktonic larval stage, the veliger. The opisthobranch veliger larva is a microscopic swimming snail complete with shell and operculum. The swimming capability is provided by cilia on the velum, a large bilobed fleshy structure forming the larval head. The larval shell is transparent and sinistrally coiled. Opisthobranch veligers can thus be separated from other gastropod larvae in plankton samples, all of which are dextrally coiled. (Although opisthobranch larval shells are sinistral, the fundamental symmetry of the animal is dextral.) The larval shell of most opisthobranchs is in the form of a simple helicoid spiral of 3/4 to 1 1/2 whorl. In three families of Nudibranchia (Thompson, 1961; Hurst, 1967) a second type of larval shell occurs and consists of an egg-shaped inflated shell in which the number of whorls is not discernable. Diagrams of larval shells and veligers are provided by Thompson (1961, 1967) and Hurst (1967).

Two general patterns of planktonic larval development occur in opisthobranchs: planktotrophic and lecithotrophic. The differences between these are summarized in Table 2. Planktotrophic larvae depend on a planktonic food supply and are unable to complete their development without feeding. Lecithotrophic larvae, on the other hand, are not obliged to feed on plankton. Since the energy needed to permit development without larval feeding is provided in the form of yolk, lecithotrophic ova are considerably larger and more yolky than planktotrophic ova. Although, as indicated in Table 2, both the size and number of ova, as well as several larval morphological characteristics correlate with these developmental patterns, the best evidence is provided by culture data. If the larvae of a given species grow to metamorphosis when provided with phytoplankton, but starved larvae of the same batch do not, this is prima-facie evidence for planktotrophy. If the starved larvae grow and metamorphose in the absence of food, this is prima-facie evidence for lecithotrophy.

PROBLEMS IN LARVAL CULTURE

On the basis of available data, most of which is provided by European zoologists, most opisthobranchs are planktotrophic. Lecithotrophic species are somewhat less common and direct-developing species are least so. In the latter species, the "veliger stage" in embryological development occurs within the egg capsule so that these species provide no problem of larval maintenance.

TABLE 2

TYPES OF PLANKTONIC LARVAL DEVELOPMENT[1]

Planktotrophic	1. Small ova (40-170 microns) 2. Ova produced in large masses (up to 25,000 per mass) 3. Short development time (2-28 days) 4. Eyes frequently lacking in newly-hatched larvae (but probably developing before metamorphosis) 5. Swimming phase greater than 3 days 6. Must feed in the plankton in order to metamorphose 7. Larval shell grows during development (except in species with egg-shaped larval shells) 8. Foot poorly developed at hatching
Lecithotrophic	1. Larger ova (110-250 microns) 2. Ova produced in smaller batches 3. Longer hatching time (4-42 days) 4. Eyes present at hatching 5. Swimming phase short, not exceeding 2 days 6. May feed in plankton but not obligatory 7. No larval shell growth due to detachment of mantle[2] 8. Foot well developed at hatching

[1] Modified after Thompson 1967a

[2] Thompson, 1967b

Almost all of the successful attempts to culture opisthobranch larvae have involved lecithotrophic species (Table 3). Only three planktotrophic species have been successfully cultured through metamorphosis and it is certain that the major difficulty here is providing a suitable plankton food supply.

The general methods applicable to larval culture of opisthobranchs are those described elsewhere in this book and need not be discussed in detail here. Larvae should be reared in small numbers; one to three veligers per ml of seawater in 200 to 500 ml beakers. If possible, the seawater should be membrane filtered (Millipore HA 0.45 microns). It may be necessary to prefilter seawater to remove the bulk of suspended material prior to membrane filtration. Water in the cultures should be changed daily. I have found it convenient to use "Nytex" (Nytex Nylon Monofilament screen cloth. Imported by Tobler, Ernst & Traber, Inc., 71 Murray

TABLE 3

Opisthobranch Species Reared Through Metamorphosis in the Laboratory

Lecithotrophic Species	Reference
Adalaria proxima	Thompson, 1959
Aeolidiella alderi	Tardy, 1962a; 1970
Catriona aurantia	Swennen, 1961
Eubranchus doriae (as Capellinia exigua)	Tardy, 1962b
Phestilla sibogae	Harris, 1970
Tenellia ventilabrum (as T. pallida)	Rasmussen, 1944
Tenellia pallida	Swennen, 1961
Tenellia fuscata	Hendler, 1966 (unpubl)
Tergipes tergipes	Swennen, 1961, Tardy, 1964
Tritonia hombergi	Thompson, 1962

Planktotrophic Species	Reference
Cratena pilata	Castagna & Vogel, 1972 (unpubl.)
Phestilla melanobrachia	Harris, 1970
Tornatina canaliculata	Franz, 1971

St., N.Y., N.Y.) nylon screening for this purpose. The Nytex screen (which is available in a multitude of screen sizes) can be glued to pieces of PVC or nylon pipe cut to any convenient size. Thus, screens of a variety of pore size and diameter can be easily constructed. Larvae retained on the Nytex screen can be washed into petri dishes using a glass pipette. With each daily change of water, it is necessary to carefully wash the culture beakers using a non-detergent soap.

Young opisthobranch larvae are positively phototactic and a problem can arise at the air-water interface. When the veligers come in contact with the surface, they are frequently trapped there by surface tension. Larval mortality can be reduced by keeping cultures in the dark. Harris (1970) reports that cetyl alcohol (n-hecadecyl alc.) flakes were useful in reducing mortality when sprinkled lightly on the surface.

Larvae in culture are very susceptible to bacterial and fungal infections, as well as protozoan infestations (primarily ciliates). Constant attention to cleanliness is the best solution to this problem. Antibiotics such as polymixin and penicillin may be useful in suppressing bacterial growth.

The great stumbling block in the culture of planktotrophic opisthobranch larvae is finding the proper food. Harris used cultures of the green flagellates Dunaliella and Cyclotella but believed them to be less satisfactory than the diatom Phaeodactylum. The latter, however, has a tendency to settle out of standing cultures and the shaking necessary to resuspend them is deleterious to the nudibranch larvae (Harris, 1970). Franz (1971) fed Tornatina (= Acteocina) larvae Isochrysis galbana and/or Monochrysis lutheri at concentrations of 100 cells/ml with success. Castagna and Vogel (pers. comm.) succeeded in growing Cratena larvae on Monochrysis and Phaeodactylum but had the best results with "wild algae", centrifuged and stored in a hot house. This observation is revealing and suggests that opisthobranch larvae may prefer a combination of food organisms, or perhaps just a variety of sizes, for optimum growth.

Methods of procuring and culturing pure strains of nanno-planktonic algae are described elsewhere in this book. The evidence suggests that other planktotrophic opisthobranch species can be cultured successfully using these food organisms. However, none of these are clearly the ideal food for opisthobranchs and further exploration and experimentation with "wild cultures" may yield much better results.

A further problem in larval culture is the induction of metamorphosis. Some species such as Tenellia fuscata and Tornatina canaliculata require no specific settlement stimulus. However, most of the nudibranchs reared so far require the presence of a specific factor to induce metamorphosis. In virtually all cases, the settlement factor is the presence of the nudibranch's specific prey species, although in a single case, an extract or broth made from the prey species also sufficed (Tardy, 1962b). It is interesting that although coelenterate-feeding nudibranchs require the presence of their prey in order to metamorphose, the veligers are sometimes in danger of themselves being eaten if they encounter the tentacles of a whole polyp (Harris, 1970; Hadfield, 1963).

The introduction of prey species to induce metamorphosis should be timed to coincide with the time at which most of the individual larvae are ready to settle. Larvae cannot be induced to metamorphose if they are not morphologically and physiologically ready. Furthermore, Harris has shown that $P.$ sibogae lose most or all of their settling capacity five days after hatching (Harris, 1970). Therefore, timing may be critical. Planktotrophic larvae are ready to metamorphose when they begin to spend most of their time crawling on the bottom of the culture container. By then, the foot is well developed and completely functional. Lecithotrophic larvae are prepared to metamorphose shortly after hatching, but again the behavior of the veligers is the best clue.

Alternate swimming/crawling is indicative of the capacity to metamorphose and at this time, prey individuals or colonies should be introduced to induce metamorphosis.

It is evident from the above discussion that knowledge of the normal prey may be a prerequisite for successful metamorphosis. Since the information is often lacking for any given species, it may be necessary to experiment with available potential prey species from the area where the opisthobranch occurs. This emphasizes the need for careful observations of the opisthobranchs in nature before attempts at larval rearing are initiated.

REFERENCES

Bleakney, J.S., 1969. A simplified vacuum apparatus for collecting small nudibranchs. Veliger 12(1): 142-143.

Casteel, D.B., 1904. The cell-lineage and early larval development of Fiona marina, a nudibranch mollusk. Proc. Acad. Nat. Sci. Phila. 56: 325-490.

Clark, K.B., 1971. The construction of a collecting device for small aquatic organisms and a method for rapid weighing of small invertebrates. Veliger 13(4): 364-367.

Franz, D.R., 1970. Zoogeography of northwest Atlantic opisthobranch molluscs. Mar. Biol. 7(2): 171-180.

___ 1971. Development and metamorphosis of the gastropod Acteocina canaliculata (Say). Trans. Amer. Micros. Soc. 90(2): 174-182.

Hadfield, M.G., 1963. The biology of nudibranch larvae. Oikos 14(1): 85-95.

Harris, L.G., 1970. Studies on the biology of the aeolid nudibranch Phestilla melanobranchia Bergh, 1874. Ph.D. Thesis, University of Calif., Berkely, 293 pp.

Hurst, A., 1967. The egg masses and veligers of thirty northeast Pacific opisthobranchs. Veliger 9(3): 255-288.

Rasmussen, E., 1944. Faunistic and biological notes on marine invertebrates. I. the eggs and larvae of Brachystomia rissoides (Hanl.), Eulimella nitidissima (Mont.), Retusa truncatula (Brug.) and Embletonia pallida (Alder & Hancock) (Gastropoda marina). Vidensk. Medd. Dansk. Naturh. Foren. 107: 207-233.

Swennen, C., 1961. Data on distribution, reproduction and ecology of the nudibranchiate molluscs occurring in the Netherlands. Netherlands J. Sea Res. 1: 191-240.

Tardy, J., 1962a. Cycle biologique et métamorphose d'Eolidina alderi (Gasteropode, Nudibranche). C.R. Acad. Sci. Fr. 244: 3250-3252.

___ 1962b. Observation et experiences sur le métamorphose et la croissance de Capellinia exiqua (Ald. et H.) (Mollusque Nudibranche). C.R. Acad. Sci. Fr. 254: 2242-2244.

___ 1964. Observation sur le development de Tergipes despectus

(Gastéropode, Nudibranche). C.R. Acad. Sci. Fr. 258: 1635-1637.

Tardy, J., 1970. Contribution a l'etude des metamorphoses chez les nudibranches. Ann. Sci. Nat. Zool. 12: 299-370.

Thompson, T.E., 1958. The natural history, embryology, larval biology and post-larval development of Adalaria proxima (Alder and Hancock) (Gastropoda Opisthobranchia). Phila. Trans. Roy. Soc., B. 242: 1-58.

_____ 1961. The importance of the larval shell in the classification of the sacoglossa and the acoela (Gastropoda Opisthobranchia). Proc. Malac. Soc. Lond. 34: 233-238.

_____ 1962. Studies on the ontogeny of Tritonia homberghi Cuvier (Gastropoda Opisthobranchia). Phil. Trans. Roy. Soc. B. 245: 171-218.

_____ 1967a. Direct development in a nudibranch Cadlina laevis, with a discussion of developmental processes in opisthobranchia. J. Mar. Biol. A. U.K. 47: 1-22.

_____ 1967b. Adaptive significance of gastropod torsion. Malacologia 5(3): 423-430.

NEW APPROACHES AND TECHNIQUES FOR STUDYING BIVALVE LARVAE

J.L. Culliney, P.J. Boyle and R.D. Turner

Museum of Comparative Zoology, Harvard University

Cambridge, Massachusetts

Since the appearance of the classic works of Loosanoff and Davis in the 1950's and 1960's, interest in the larvae of bivalve mollusks has increased greatly. Their comprehensive report on culture techniques (Loosanoff and Davis, 1963) is still the most important single work in the field, and should be consulted by anyone planning to work with bivalve larvae. The culturing practices outlined by Walne (1964) should be studied. In more recent years some new terminology and advanced analytical techniques have been introduced by Chanley (1965), Chanley and Van Engel (1969), and Chanley and Andrews (1971). An important study by Pilkington and Fretter (1970) on prosobranch veligers also contains information relevant to feeding of larval bivalves.

The purpose of this paper is to add some refinements to the basic techniques used for culturing and data acquisition, to call attention to some of the problems that may arise and to suggest ways of avoiding them. Our procedures will be useful particularly to investigators beginning small-scale operations for the study of larval systematics, ecology, and behavior.

SPAWNING AND FERTILIZATION

Before spawning, the adults should be cleaned of mud and debris to help control protozoan and bacterial contamination of the gametes and young larvae. Filtered seawater should be used for all operations. Coarse-grade filter paper or the equivalent in bag filters or cartridge type filter tubes will usually suffice. Millipore filtration below one micron is seldom necessary. It should be tried, however, if difficulties arise due to poor water quality resulting from pollution, eutrophication or other causes.

Techniques for inducing spawning have been described previously by Loosanoff and Davis (1963). Thermal stimulation, the most common method, causes ripe bivalves to spawn when the temperature of the seawater is raised a few degrees. Agitation on a laboratory shaker may prove effective with some bivalves (Culliney, 1971).

Preserve and label each pair of parents or spawning group separately so that it is possible to relate them to the larval populations they produced. They then can be retrieved in case of misidentification, or to check some interesting phenotypic feature that appears in the offspring.

Use plenty of sperm to fertilize eggs. In nature, bivalves often spawn together, producing massive concentrations of gametes. The results of Crassostrea spawning shown by Galtsoff (1964, p. 320, fig. 290) illustrates this. Polyspermy is not common except when the temperature is below the optimal level for a species. On numerous occasions we have seen hundreds of spermatozoa attached to the outer egg membrane, but these supernumery sperms do not participate in fertilization, and development is normal. As a rule when fertilizing batches of about 100,000 eggs, contained in about 200 ml of seawater, enough sperm should be added to make the water slightly cloudy.

Eggs and young embryos should not be subjected to excessive temperature and mechanical shock. Unfertilized and newly fertilized eggs may be sieved and washed gently, but once cell division has begun the larvae should not be sieved until they have reached the straight-hinge stage, usually 20 to 30 hours after fertilization. We sieve and wash the zygotes about 15 minutes after adding the sperm, and then rinse them through a slightly larger mesh sieve which traps dust particles and debris, but allows the zygotes to pass into 2-liter jars of filtered seawater at the spawning temperature. Initial population densities should not exceed 50 embryos per ml.

MAINTAINING THE SWIMMING STAGES

As soon as the larvae begin to swim, they should be siphoned into clean culture jars, containing a little seawater so the fragile, shell-less larvae are not damaged on impact with the glass. Adjust the seawater volume in each culture to obtain a larval population density of 10-20 per ml. At the straight-hinge stage

(20-30 hours) the water can be changed completely and the larvae washed on nylon sieves. It is best to have a series of sieves with mesh sizes ranging from 30 to 500 microns in roughly 50 micron intervals. Sieves can be made cheaply from pieces of PVC pipe about 2" in diameter and 4" long. Saw the pieces of pipe in half, place the nylon mesh between the halves, and rejoin them using epoxy or plastic cement. The joints should be sealed around the inside of the sieve with melted paraffin to avoid toxic effects of the cement and to smooth the interior so that larvae are not trapped or injured. Nylon screening can be obtained from numerous dealers and some will send free sample books, on request, containing a series of mesh sizes in pieces large enough to make the sieves. Sieves will last for years if cleaned after each use with a jet of cold tap water (hot water will melt the paraffin), and occasionally resealed with paraffin.

To maintain healthy cultures, the containers and filtered seawater should be changed and the food renewed at regular intervals. The schedule should be determined empirically, based on such factors as temperature, volume of the culture, population density, and size of the larvae. Daily maintenance may be necessary for warm temperate and tropical species; every three days may suffice for cold water species.

Water quality is a persistent problem when rearing small pelagic animals. The high surface-to-volume ratio of the larvae and their obligatory exposure to whatever may be in the water make them highly susceptible to small concentrations of toxic substances or disease-causing agents. Little progress has been made toward identifying the range of factors in natural waters having deleterious effects on larvae. However, Walne (1970) correlated silt content, zinc pollution, and the occurrence of specific algae and bacteria, or their metabolites, with growth fluctuations in *Ostrea edulis* larvae. In Chesapeake Bay (P.E. Chanley, pers. comm.) and in inshore waters at Beaufort, North Carolina, (J.L. Culliney, unpub. obs.) unidentified filterable components inhibit growth and sometimes cause mass mortality of bivalve veligers during the warm months of the year. At present the only remedy for such uncontrollable problems is to obtain and carefully filter offshore oceanic water.

Antibiotics and ultraviolet germicidal lamps have been recommended by previous workers, but we find them of questionable value. Our UV lamp (specified by Loosanoff and Davis, 1963) eliminated bacteria only at flow rates below 10 ml per minute. Even with a flow rate of 8 ml per minute, algal and invertebrate contaminants survived. Good filtration and other "house-keeping" procedures, if strictly followed, will maintain healthy cultures.

Glass is the best material for culture containers. Plastics may absorb toxic substances or provide surface conditions favorable to bacteria. Start culture work with new glassware, and never use it for anything else. Careful cleaning of culture containers is important for consistent success in rearing larvae. Some detergents, including the much used Alconox, contain toxic substances that remain even after careful rinsing (C.G. Bookhout, pers. comm.). The old-fashioned glass-cleaning soap Bon-Ami has proved to be an excellent cleaner which leaves no residual toxicity.

The shape of the container is important for maintaining bivalve larvae. We do not recommend square or rectangular ones because of the "corner effect" which tends to concentrate most of the larvae in the corners of the container. This usually leads to mass mortality, perhaps due to lack of oxygen, the build-up of waste products, or lack of food. A cylindrical jar or bucket is recommended because it offers larvae a more nearly unbounded environment.

It is also advisable to use fairly tall containers as the greater depth of water keeps the swimming larvae well off the bottom. Although relatively little is known about larval diseases (see Guillard, 1959 and Tubiash et al., 1965 and 1970) it appears that diseases and deleterious conditions causing slow or uneven growth are associated with the debris, faeces and pseudofaeces, protozoa, and bacterial films which accumulate on the bottom of culture jars. Therefore, when cleaning the cultures, the larvae should be siphoned off without agitating or sucking up any of the microdebris on the bottom. A few healthy larvae will always be left behind with the dead and diseased specimens on the bottom of the container. Usually these are preserved, labeled, and kept as part of the permanent record. If, however, the culture is small and one cannot afford to lose the few healthy specimens near the bottom, the old culture jar can be refilled and saved. The debris will settle, healthy larvae will swim up off the bottom and they can be siphoned the next day and returned to the main population. Saving the dregs of the main culture in this manner will minimize losses and still avoid the deleterious effects of bottom debris.

When sieving the larvae, combine a mesh size to catch them with one that will allow them to pass through, thus eliminating particles and debris both smaller and larger than the larvae. Have a clean jar of fresh seawater ready so that larvae are not exposed to air for more than a few seconds. Using a range of sieve sizes (about 30 to 500 microns) allows one to catch the larvae on progressively larger meshes as they grow. This aids in the early elimination of protozoans and large amounts of detritus. When selecting the mesh to catch larvae, remember that it must be somewhat smaller than the larvae for they can slip through on the diagonal of the square.

Large numbers of larvae can be concentrated for experimental work by swirling the water containing larvae in a small circular dish (Carriker, 1961). Centripetal forces bring shelled larvae rapidly to the center. It may be necessary to add a small amount of distilled water to inhibit swimming of the most vigorous larvae.

To collect and concentrate veligers of brooding species we tried various existing plankton concentrating devices and found that they either pulled the larvae in with damaging force or the filtering area was so small that the device clogged quickly.

Newly released larvae tend to swim near the top of the water column and can be caught by drawing off and filtering the surface layer. The collector we designed takes advantage of this behavior. The collector (Figs. 1 and 2) is made of 1/8 inch clear plexiglass and proportioned to fit the aquarium on which it is to be hung. Neoprene rubber is used for the seals on the side guides. The large removable filters can be made with nylon mesh glued with silicone aquarium seal on a 1/4 inch plexiglass frame. The flat bottom of the collecting chamber (Fig. 1) allows gentle mixing of the water, eliminating the problem of bottom stagnation.

This larvae collector is especially adapted for retaining the unpredictable spawnings of species which brood their young or the larvae which emerge at uncertain times from egg cases. When using the collector, the adults or egg cases are placed in an aquarium on a water table and the aquarium is tilted to allow overflow only on one edge. The collector is hung on the overflowing edge, and the two side guides are placed on the adjacent sides of the aquarium to prevent spillage at the corners. To trigger the spawning of brooding adults, a supply of warm seawater (a few degrees above ambient) is delivered to the bottom of the aquarium. Water carrying released larvae overflows into the collecting chamber, and it in turn fills and overflows. As the water leaves the collector it passes through the nylon filter which has a mesh size smaller than the larvae being retained. Overflow from the collector flows down the guide track and the outer wall of the aquarium onto the water table (Fig. 1).

After accumulating a large number of larvae, the collector is removed from the aquarium, and the larvae are further concentrated by tipping the collector slowly toward the filter to remove excess water. The contents of the collector can then be poured into a bowl and the larvae adhering to the collector or filter washed

Fig. 2. Larvae collector attached to the aquarium.

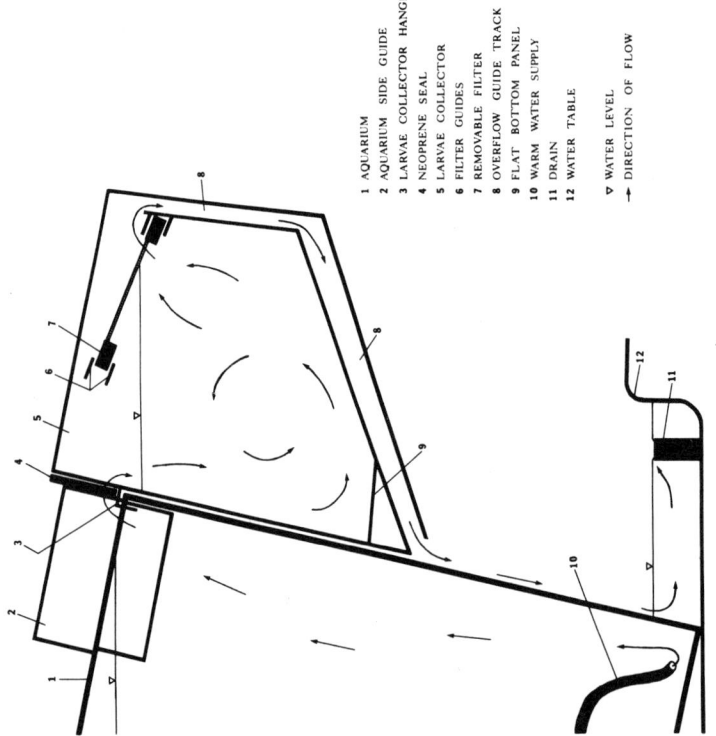

1 AQUARIUM
2 AQUARIUM SIDE GUIDE
3 LARVAE COLLECTOR HANGER
4 NEOPRENE SEAL
5 LARVAE COLLECTOR
6 FILTER GUIDES
7 REMOVABLE FILTER
8 OVERFLOW GUIDE TRACK
9 FLAT BOTTOM PANEL
10 WARM WATER SUPPLY
11 DRAIN
12 WATER TABLE

▽ WATER LEVEL
→ DIRECTION OF FLOW

Fig. 1. Cross sectional view of larvae collector showing flow pattern.

off with a gentle stream of seawater. The larvae may be further concentrated by passing the contents of the bowl through a sieve.

An efficient collector should have a capacity of 1-2 liters, and the surface area of the filter should be 50-100 cm^2. This volume is adequate to hold thousands of veligers without overcrowding, and the large filtering area allows extended operation with a minimum of clogging. Such a system, using unfiltered running seawater and a 240 micron filter functions for about 24 hours before clogging. The accidental introduction and concentration of predators, such as coelenterates, turbellarians, and polychaetes may be a problem if the collector runs for long periods. In this case prefiltering the incoming seawater will be necessary.

The collector was originally designed to concentrate large brooded larvae, but the removable filters, and perhaps prefiltering the incoming water, permits the use of a whole range of mesh sizes, allowing collection of many types of organisms. The collector might also be used to concentrate plankton if a direct line of unfiltered seawater is provided, or to keep organisms collected in a standard plankton tow alive for extended periods.

FEEDING

Isochrysis galbana (Parke) and *Monochrysis lutheri* (Droop) are still the best foods known for larval bivalves. However, the latter species, under bacteria-free conditions, can be toxic to gastropod larvae (Fretter and Montgomery, 1968), and this may be true for bivalves also. We find it advantageous to sieve unicellular algae through a 15 micron mesh immediately prior to feeding in order to remove clumped debris which may interfere with larval activity, stick to their shells, and promote bacterial growth. Food cultures more than one month old, especially if they contain excessive clumped debris, should not be used. Although they may not be strongly toxic, they do not promote as rapid growth as do young healthy food cultures which are approaching their peak density. For basic information on how to culture algae, see Guillard and Ryther (1962) and Loosanoff and Davis (1963). Those interested in biochemical changes and algal senescence, possibly affecting the nutritional value of algae, should consult Fogg (1966).

As the larvae grow larger and the velum becomes a more efficient food gathering organ, the daily food allotment should be reduced. Otherwise mass wastage of food will occur and the resultant pseudofaeces will promote bacterial growth, and many larvae will become trapped by the sticky bottom debris. We generally start with a feeding concentration of 50,000 to 100,000 algal cells per ml of larval culture, reducing this gradually to about 10,000 to 20,000 cells per ml as the larvae approach the pediveliger stage.

Artificial and specially treated foods such as bouillon, milk, pablum, yeast, and dried algal fragments have been tried as laboratory diets for bivalve larvae (see Chanley and Normandin, 1967, for a review of this subject). Most attempts at finding convenient artificial foods have been unsuccessful. However, Hidu and Ukeles (1962) were able to rear clam larvae, Mercenaria mercenaria (Linnaeus), using freeze-dried unicellular algae. This may prove of importance to workers at remote field stations, especially in the tropics where air conditioned cold rooms are essential for culturing algae such as Isochrysis galbana.

EFFECTS OF TEMPERATURE

Along with water quality and food, temperature is one of the most important factors controlling the growth of larvae. Larvae of tropical and temperate zone species grow best at 25-26°C. Although growth rates of many larvae will increase as the temperature is raised to a point slightly over 30°C, bacterial growth may also increase with adverse effects on larvae.

Temperature is especially important at the time of straight-hinge shell formation. Warm water species will often be deformed if the water is too cold at this time, and the larvae may develop incomplete, irregular, lumpy shells or no shells at all. In Bankia gouldi (Bartsch), a warm-temperate to tropical teredinid, temperatures below about 18°C interfere with straight-hinge shell formation. On the other hand, temperatures warmer than optimal can be deleterious to early shell formation in cold water species. A population of the boreal wood-boring pholad Xylophaga atlantica (Richards), reared at 18°C, produced highly deformed straight-hinge valves and suffered mass mortality soon after.

By accident we discovered a simple technique to determine a preferred temperature for rearing larvae of cold water species. Specimens of Xylophaga atlantica, extracted from wood obtained from a depth of 50 fathoms on George's Bank, Mass., spawned at 8-9°C. The fertilized eggs were held at this temperature and the larvae developed normally but did not begin to swim until about 48 hours after fertilization. The tall culture jars were kept in a shallow trough of cold running seawater at 8-9°C so that a vertical temperature gradient existed within them, ranging about 16°C at the surface (exposed to room temperature) to 9°C at the bottom (incoming water temperature). The newly swimming embryos concentrated in a horizontal band, with strikingly sharp edges, about two inches off the bottom. Careful insertion of a thermometer into this layer showed that it was between 11°C and 13°C. Further simple tests were made by raising and lowering the level of cold water in the trough. The band of larvae moved up and down accordingly as the temperature band of 11°C to 13°C shifted with the outside level of the cold water. This seemed to confirm the fact that the larvae were seeking a preferred zone of temperature at an early stage of development. These same larvae went on to form perfect straight-hinge shells, maintaining themselves at 11°C to 13°C. Later, however, as the larvae grew, their precise behavioral regulation disappeared, and they ranged over the whole water column between about 9°C and 16°C.

Precision incubators are not necessary for rearing larvae, except when conducting experiments on temperature physiology. We use a wooden cabinet, heated by a gun-type hair dryer which is controlled by a small thermostat. Thermostats such as those used in poultry breeding or in aquarium heaters are inexpensive and give a level of control of about ± 1°C. For cold-water species, a refrigerator or cold water bath can be adapted to achieve the desired temperature range.

OTHER FACTORS AFFECTING DEVELOPMENT

Salinity seems to have less effect on larval development than the factors discussed above. This is understandable because larvae of inshore and estuarine species probably experience considerable fluctuation in salinity during their tenure in the plankton. However, larvae found in continental shelf waters and beyond are probably adapted to a highly stable salinity regime and may not tolerate much fluctuation.

Studying larval development at low salinities, Culliney (in press) observed that glass-distilled water, used to dilute natural seawater of 30 °/oo, apparently affected the development of the larval shell at 10 °/oo, but not at 15 °/oo. The larvae in the lower salinity developed irregular, ragged-edged valves, possibly a result of a critical reduction of the calcium concentration in the water. Ideally there should be some standard source of fresh water for dilutions, but lacking that, spring or well water, or water from "clean" lakes or streams, when added to seawater, produces a more natural brackish water than dilutions made with glass-distilled water. Distilled water from a metal still should never be used to dilute seawater for cultures; even minute concentrations of many metalic ions are toxic to larvae.

Apparently pH can fluctuate over a fairly wide range with little effect on larvae. Calabrese and Davis (1966) reported normal early embryonic development in *Crassostrea* *virginica* over a range in pH of 6.75 - 8.75, and in *Mercenaria* *mercenaria* from pH 7.00 - 8.75. Larvae of both species survived indefinitely at pH 6.0 - 6.2, but growth rates dropped when pH levels went below 6.75.

DATA ACQUISITION AND RECORDING

Caution should be taken in assigning measurements or other data from cultured larvae to species populations in the field. More than one rearing of larvae from different parents is necessary to assess morphological variability. Differential degrees of inbreeding among populations of bivalves may be an important factor in larval biology within and among species. In addition to genetic considerations, the environment, especially different foods, may determine larval characteristics. The latter point has been well-demonstrated for sabellarid polychaete larvae by Wilson (1968). All of these problems remain to be investigated in bivalves.

Most major works on the systematics of bivalve larvae are based chiefly on shell characters of preserved specimens (Jørgensen, 1946; Sullivan, 1948; Rees, 1950; and Loosanoff et al., 1966). Lebour (1938), however, studied swimming larvae and illustrated them with the velum expanded. Chanley and Andrews (1971), in their key to the bivalve larvae of Virginia, also made use of a variety of larval characters in addition to shell features.

In our work we have found that numerous characters, observable only in living larvae, may be useful in systematics. The presence or absence of a larval eye, the morphology of the velum and foot, the length and distribution of velar and pedal cilia, and the size and shape of the heel or byssal spur of the foot, as well as its position during crawling, are all useful characters. Culliney (1972) described the distinctive crawling behavior of the pediveligers and newly settled dissoconchs of six species of New England bivalves, emphasizing differences in rhythm, the way in which the shell is carried, and the use of muscular and gliding movements. The presence or absence of apical flagella, or their number and length, as well as the shape of the expanded velum, provide clues to the identification of swimming larvae. For example, Xylophaga atlantica can be distinguished from all other known pholad veligers by the enormous size of its velum and the presence of a deep cleft between the lobes anteriorly.

Photography is probably the most efficient way of recording the morphology of living bivalve larvae and cinemicrography the most effective means of studying larval behavior in detail. We are recording the swimming and crawling behavior of veligers on 16mm film for future analysis using graphical techniques and a photoanalyzing projector. Such films yield quantitative and comparative information on types, rates, and rhythms of movements, and perhaps other kinds of morphological coordination and repetitive behavior of species. In addition to adding new characters useful in the identification of larvae, cinemicrography is a valuable tool for the study of functional morphology in these minute bivalves. Such work is expensive (see Turner, 1972 for a list of equipment in our studies). However, any 35mm camera with an attachment for the microscope and perhaps a small electronic flash can provide valuable records of living larvae.

Living larvae can be confined for photography in a small section of glass or plastic tubing, 5 - 10 mm long, set on end in a shallow dish. If necessary, the larvae can be relaxed or slowed down by chilling, degassing the water, lowering the salinity, or by the addition of small quantities of relaxing agents such as magnesium chloride or propylene phenoxitol. The latter is a powerful relaxant and can be toxic. A stock solution can be made by adding several drops to 50 ml of seawater and stirring. When taking up the solution in a pipette avoid the undissolved droplets of propylene phenoxitol. These can remain on the bottom and will serve to keep the stock solution saturated. The larvae to be relaxed should be placed in about 5 ml of clean seawater and the stock solution added drop-wise while continually observing its effects; only a few drops will be needed.

Standard measurements and still photographs of the shell outline and other features of individual specimens are still the best method of recording important taxonomic information. Close-up "portraits" should be supplemented by group photographs of all stages, showing specimens in various positions. When examined with image analyzers, planimetry, or other unforseen techniques, the group pictures may prove to be of great value. Areas of projection, circumference measurements, comparisons of complex shapes and eventually, perhaps, holographic studies in three dimensions may contribute to knowledge of bivalve larvae.

In order to study the hinge structure and the sculpture of the larval shell in detail it is necessary to remove all soft tissue. The best technique we have found for cleaning larval shells is to feed them to small sea anemones. The anemones digest the soft parts of the larvae and egest perfectly clean gaping shells. This process, which works equally well for gastropod veligers, usually takes from 12 - 24 hours at room temperature. Anemones only a few mm long are best because larger ones may leave mucus adhering to the shells. Several species of anemones have been used, and all behave in the same fashion. The larvae should be fed to the anemones by dropping them directly onto the tentacles with a pipette. Care should be taken not to overfeed or the shells will be voided with partially digested organic matter still attached. No more than five large larvae at a time should be given to an anemone under one-half centimeter long. Photographs of the hinge, sculpture, and muscle scars of clean shells may be taken at very high magnification using the scanning electron microscope (Turner and Johnson, 1970; Scheltema, 1971).

Teredinid and pholad veligers are subglobular and very difficult to orient for measurement. By separating the valves of the cleaned shells with fine needles, it is possible to lay the valves flat and measure each separately. This procedure has revealed the fact that teredinid larval shells are slightly inequivalved, and we suspect that the foot, when retracted, is carried in the larger valve. The depth of the valve may be measured using the calibrated fine adjustment of the microscope. First focus on the uppermost surface of the valve and then on the rim where it touches the glass slide. The change in vertical distance shown by the calibrated focusing knob is the measure of the depth.

PRESERVATION OF A REFERENCE COLLECTION

It is important to preserve, for each species reared, a developmental series from the straight-hinge stage to young spat which are sufficiently developed to show adult characters. Before preservation the specimens should be sieved to remove debris.

A preservative for whole larvae, developed by Carriker (1961), has been used by many investigators. Its formula is as follows:

    5 ml commercial formalin

    0.25 g sodium tetraborate

    50 g sucrose

    filtered seawater to make 500 ml

The sucrose is added to enhance color preservation, to act as a clearing agent, and as a safety factor against evaporation. The pH must be checked frequently for Carriker found that it dropped to acidic levels over a period of a few months. In addition, after extended periods the shells may become sticky and a precipitate may cover the shells. For these reasons, this preservative is not recommended for long storage of collections.

A preservative for extended storage recommended in the SCOR/UNESCO Manual for the fixation and preservation of marine zooplankton (in press) is made as follows:

    5 ml formalin buffered to pH 8 with sodium glycero phosphate

    1 ml propylene phenoxitol

    10 ml propylene glycol

    distilled water to make 100 ml of preservative

Before placing specimens in this preservative, they should be washed in distilled water.

When preserving larvae for histological work, 4.5% buffered seawater-formalin can be used if the calcareous parts are to be preserved; otherwise Bouin's or Holland's solutions are recommended.

Cleaned empty valves of larvae to be used for study with the scanning electron microscope should be preserved in 90-95% ethanol (C. Thiriot-Quiévreux, 1972). See also Robertson (1971) on the techniques of preparing larvae for scanning electron microscopy. Alcohol at a strength over 80% will prevent the dissolution of calcium carbonate and will keep shells in good condition, but this is not recommended for preserving whole specimens because the soft parts will become dehydrated and the alcohol-solution acids in the tissues may lower the pH. More dilute alcohol (70-75%) may be used as a preservative for whole larvae, but it must be buffered to pH 8.

Adult specimens are best preserved in 70-80% ethanol buffered to pH 8. Various buffers can be used, but sodium tetraborate (borax), though not readily soluble, is effective. Twenty or thirty grams of borax can be put into a large container of alcohol, shaken vigorously, and allowed to settle, and the clear neutralized alcohol decanted as needed.

The volume of the preservative should be ten times that of the specimens, and the pH should be maintained at 8 or above. If the pH has dropped and the specimens have begun to deteriorate, the process can be arrested by washing them several times in distilled water and replacing them in a properly buffered preservative.

ACKNOWLEDGEMENTS

Our studies, carried out at the Marine Science Institute of Northeastern University, Nahant, Massachusetts, and the Marine Biological Laboratory, Woods Hole, Massachusetts, were supported by ONR Contract Number N000 14-67A-0298-0027 with Harvard University. We are grateful to R. Scheltema, G. Davis, and P. Chanley for reading the manuscript.

REFERENCES

Calabrese, A. and H.C. Davis, 1966. The pH tolerances of embryos and larvae of Mercenaria mercenaria and Crassostrea virginica. Biol. Bull. 131: 427-436.

Carriker, M.R., 1961. Interrelation of functional morphology, behavior, and autecology in early stages of the bivalve Mercenaria mercenaria. J. Elisha Mitchell Sci. Soc. 77: 168-241.

Chanley, P.E. 1965. Larval development of a boring clam, Barnea truncata. Ches. Sci. 6: 162-166.

Chanley, P.E. and R.F. Normandin, 1967. Use of artificial foods for larvae of the hard clam, Mercenaria mercenaria (L.). Proc. Nat. Shellfish Assoc. 57: 31-37

Chanley, P.E. and W.A. Van Engel, 1969. A three-dimensional representation of measurement data. Veliger 12: 78-83.

Chanley, P.E. and J.D. Andrews, 1971. Aids for identification of bivalve larvae of Virginia. Malocologia 11: 45-119.

Culliney, J.L., 1971. Laboratory rearing of the larvae of the mahogany date mussel Lithophaga bisulcata. Bull. Mar. Sci. 21: 591-602.

___ 1972. Cinemicrographic studies of crawling behavior in larval and juvenile bivalves. Amer. Malac. Union Bull. for 1971. p. 29.

___ 1973. Larval development of the shipworms Bankia gouldi (Bartsch) and Teredo navalis L. in response to varying salinity and temperature. Mar. Biol. (in press).

Fogg, G.E., 1966. Algal cultures and phytoplankton ecology. Univ. of Wisconsin Press, Madison, Milwaukee, and London. 126 pp.

Fretter, V. and M.C. Montgomery, 1968. The treatment of food by prosobranch veligers. J. Mar. Biol. Assoc. U.K. 48: 499-520.

Galtsoff, P.S., 1964. The American oyster. Fishery Bull. U.S. Dept. of the Interior (Rish and Wildlife Service). Vol 64: 480 pp.

Guillard, R.R., 1959. Further evidence of the destruction of bivalve larvae by bacteria. Biol. Bull. 117: 258-266.

Guillard, R.R. and J.H. Ryther, 1962. Studies on marine planktonic diatoms I. Cyclotella nana Hustedt and Detonula confervacae (Cleve) Gran. Can. J. Microbiol. 8: 229-239.

Hidu, H. and R. Ukeles, 1962. Dried unicellular algae as food for larvae of the hard shell clam, Mercenaria mercenaria. Proc. Nat. Shellfish. Assoc. 53: 85-101.

Jørgensen, C.B., 1946. Lamellibranchia, p. 277-311 In: G. Thorson (ed.) Reproduction and larval development of Danish marine bottom invertebrates. Medd. Komm. Danmarks Fish. Havunders. Ser. Plankton 4. Copenhagen.

Lebour, M.V., 1938. Notes on the breeding of some lamellibranchs from Plymouth and their larvae. J. Mar. Biol. Assoc. U.K. 23: 119-145.

Loosanoff, V.L. and H.C. Davis, 1963. Rearing of bivalve mollusks, p. 1-136. In: F.S. Russell (ed.) Advances in Marine Biology, Vol. 1. Academic Press, London.

Loosanoff, V.L., H.C. Davis, and P.E. Chanley, 1966. Dimensions and shapes of larvae of some marine bivalve mollusks. Malacologia 4: 351-435.

Pilkington, M.C. and V. Fretter, 1970. Some factors affecting the growth of prosobranch veligers. Helgoländer wiss. Meers. 20: 576-593.

Rees, C.B., 1950. The identification and classification of lamellibranch larvae. Hull Bull. Mar. Ecol. 3: 73-104.

Robertson, R., 1971. Scanning electron microscopy of planktonic larval marine gastropod shells. Veliger 14: 1-12.

Scheltema, R.S., 1971. Dispersal of phytoplanktotrophic shipworm larvae (Bivalvia: Teredinidae) over long distances by ocean currents. Mar. Biol. 11: 5-11.

SCOR/UNESCO Manual for the fixation and preservation of marine zooplankton (in press).

Sullivan, C.M., 1948. Bivalve larvae of Malpeque Bay, P.E.I. Fish. Bd. Canada Bull. 77: 1-36.

Thiriot-Quiévreux, C., 1972. Microstructures de coquilles larvaires de prosobranches au microscope electronique à balayage. Archives de Zoologie Expérimentale et Générale 113: 553-564.

Tubiash, H.S., P.E. Chanley, and E. Leifson, 1965. Bacillary necrosis, a disease of larval and juvenile bivalve mollusks, I. Etiology and Epizootiology. J. Bact. 90: 1036-1044.

Tubiash, H.S., R.R. Colwell, and R. Sakazaki, 1970. Marine vibrios associated with bacillary necrosis, a disease of larval and juvenile bivalve mollusks. J. Bact. 103: 271-272.

Turner, R.D. and A.C. Johnson, 1970. Some problems and techniques in rearing bivalve larvae. Amer. Malac. Union Bull. for 1969. 9-13 pp.

Turner, R.D., 1972. Cine-photomicrography: a tool in biological studies. Amer. Malac. Union Bull. for 1971. p. 30.

Walne, P.R., 1964. The culture of marine bivalve larvae. pp. 197-210. In: K.M. Wilbur and C.M. Yonge (eds.), Physiology of the Mollusca, Vol. 1.

_____ 1970. Present problems in the culture of the larvae of Ostrea edulis. Helgolander wiss. Meers. 20: 514-525.

Wilson, D.P., 1968. Some aspects of the development of eggs and larvae of Sabellaria alveolata. J. Mar. Biol. Assoc. U.K. 48: 387-435.

THE DEVELOPMENT OF METHODS FOR REARING THE COOT CLAM, MULINIA
LATERALIS, AND THREE SPECIES OF COASTAL BIVALVES IN THE LABORATORY

Edwin W. Rhodes, Anthony Calabrese, Wayne D. Cable,
and Warren S. Landers

National Oceanic and Atmospheric Administration
National Marine Fisheries Service
Middle Atlantic Coastal Fisheries Center
Milford Laboratory, Milford, Connecticut

The importance of selecting appropriate biological material for commercial cultivation or for experimentation cannot be over-emphasized. Selection of a species for commercial propagation consists of balancing the economic liabilities of hatchery and field cultivation with various aspects of marketing economics. For experimental use, one can dispose of the commercial aspects in selecting experimental material and concentrate on finding organisms most amenable to controlled culture. Characteristics to be considered should include the availability of adults, the necessity for and ease in conditioning the adults for spawning, spawning methods, the success of larval culture, the growth rate of the immature animals, and the total generation time for the animals. In this presentation, these aspects will be considered for a non-commercial estuarine species, Mulinia lateralis, and three commercial coastal zone bivalves, Spisula solidissima, Arctica islandica, and Placopecten magellanicus.

MULINIA LATERALIS

Availability of Adults

Mulinia lateralis (Say), the coot clam, is found in abundance in favorable environments from Malpeque Bay, Canada, to northeastern Mexico and in the West Indies (Calabrese, 1969a). In Long Island Sound, our study area, Sanders (1956) reported M. Lateralis widely distributed but found that their abundance varied considerably during bimonthly sampling periods. Wass (1965) reported that M. lateralis exhibited a population explosion and then a rapid dieoff in a Virginia estuary. Jackson (1968) studied the size distribution of M. lateralis in an inlet off Long Island Sound and found a very high juvenile mortality, probably resulting from poor adaptability to turbid water and soft substrate. Fluctuations in M. lateralis populations have occasionally made it difficult to find sufficient numbers of adults for experimentation and we have found it desirable to collect large stocks of adults when they are abundant and maintain them in laboratory facilities. Collections can be made readily using an oyster dredge with a small mesh liner.

Maintenance of Juveniles and Adults

Juvenile and adult M. lateralis are best maintained in the laboratory in flowing seawater in trays or tanks with two inches of beach sand substrate. The natural food supply in the flowing seawater is normally sufficient to maintain these organisms in good physiological condition. Supplemental feeding with algal cultures is recommended if such a source is available.

Conditioning and Spawning

Calabrese (1970a) studied the reproductive cycle of M. lateralis in Long Island Sound. Significant numbers of mature M. lateralis can be found in May and June in Long Island Sound and this species normally spawns from July to September. No conditioning of the adults is required to obtain spawn during this period.

It is difficult to obtain ripe M. lateralis from October through December. During this post-spawning period the clams are gametogenically inactive or just beginning gametogenesis and attempts to hasten maturation by increasing the water temperature and supplemental feeding have failed. It may be possible to have M. lateralis available for spawning during this period by placing ripe animals in chilled seawater early in the summer and thus prevent them from spawning. These methods work successfully for other species of clams (Loosanoff and Davis, 1963).

M. lateralis respond well to the conditioning techniques developed by Loosanoff and Davis (1963) for ripening bivalves out of season. M. lateralis are taken from ambient temperatures during the winter and early spring and placed in running water trays. They are then acclimated by increasing the seawater temperature several degrees each day until the conditioning temperature is reached. It requires from 2-5 weeks at 18 to 20°C to condition these animals for spawning.

One advantage to working with M. lateralis is the ease in separating males from females before the spawning attempt is made. The mature female gonad is pink to red-orange in color and the mature male gonad is white. The thin shells of M. lateralis make it possible to see the underlying gonad in the umbone region.

The spawning techniques described by Loosanoff and Davis (1963) are very successful for M. lateralis. Ripe adults are placed in finger bowls and the temperature is raised to about 28°C in a warm water bath. If thermal stimulation alone has not induced spawning in 30 minutes, a sperm suspension prepared from a sacrificed male can be added as an additional stimulant. When the water in the spawning container has become cloudy with eggs or sperm the clams may stop spawning, but can be induced to start again if they are moved to clean water.

Fecundity

The number of eggs released by a spawning female is variable and seems to depend on the size of the animal, as well as on the degree of development of the gonad (Calabrese, 1969a). The smallest number of eggs released by a single female was in the thousands and the largest number was about 7 million. The average egg yield is 3 to 4 million per female.

Rearing Embryos and Larvae

The newly spawned eggs of M. lateralis are often irregular in shape but they become spherical when suspended in seawater for a few minutes. The eggs should be fertilized with motile sperm as soon as possible after they are released. Care should be taken to avoid excess sperm in an egg suspension since polyspermy and abnormal development may ensue.

The fertilized eggs, about 50 microns in diameter, may be separated from feces and other spawning debris by passing them through a sieve coarse enough to pass the eggs easily but fine enough to retain the debris. A sieve with mesh openings of 100 microns works well for M. lateralis eggs. Eggs can be conveniently cultured in one-liter or fifteen-liter cultures using 15μ filtered, ultraviolet-irradiated seawater at a density of 30 eggs/ml. At 20 to 25°C the trochophore stage is attained in 9 hours and development proceeds to the straight-hinge veliger stage in 15 hours (Calabrese, 1969a).

After 48 hours the straight-hinge veliger larvae, now 70 to 75μ in length, are separated from the culture water by pouring or siphoning the contents of the culture container through an appropriate sieve (opening about 40 microns). These 48-hour larvae are redistributed into clean culture water at a density of 15 larvae/ml. Larvae can be successfully reared to metamorphosis in 6 to 8 days at 25°C on an algal diet. The larvae should be screened from the cultures and resuspended in fresh seawater every 48 hours to remove unconsumed food and metabolic products.

A combination of the chrysophytes *Isochrysis galbana* and *Monochrysis lutheri* is an excellent food for *M. lateralis* larvae. These algal species can be grown under carefully controlled conditions in semi-continuous unialgal cultures after the method of Ukeles (1971). These cells are added to the larval cultures at a density of 100-120,000 cells/ml beginning at 48 hours and then every 24 hours.

Most *M. lateralis* larvae undergo metamorphosis and become benthic at a length of 200 to 220 μ. However, the size at metamorphosis is quite variable and individuals from 150 to 245 μ may possess both a foot and a velum and may alternately crawl and swim.

Effects of Various Environmental Parameters

Calabrese (1969a, 1969b, 1970b) has studied the effects of temperature, salinity, and pH on the embryos and larvae of *M. lateralis*. Fertilized eggs develop into normal straight-hinge larvae satisfactorily (70% or more) at temperatures from 15.0 to 25.0°C and at salinities from 22.5 to 30.0 °/oo. The larvae survive and grow satisfactorily (70% or more of the optimum) at temperatures from 20.0 to 27.5°C and at salinities from 20.0 to 27.5 °/oo. The optimum pH range for the culture of *M. lateralis* is from 7.25 to 8.25.

Growth of Juveniles

Recently metamorphosed *M. lateralis* survive and grow best and are most easily handled if they are kept in static cultures with frequent water changes until they are 0.5 mm in length. At 0.5 mm these clams grow more rapidly if placed in running water trays. Some nourishment is supplied by the foods naturally present in the seawater, but laboratory cultured algae are also routinely dripped into these trays to accelerate growth. Rapid growth of juveniles is obtained at 22 to 24°C. Even faster growth of juveniles can be achieved if placed outdoors in boxes of sand kept in tanks of running seawater during the warmer months of the year.

Generation Time

Using the culture methods for M. lateralis outlined above, the egg-to-egg cycle in the laboratory has ranged from 39 to 135 days with the average generation time about 60 days (Calabrese, 1969c). Although fully grown M. lateralis are about 15 to 20mm in length, both male and female laboratory reared M. lateralis have spawned when only 2.7mm in length.

PLACOPECTEN MAGELLANICUS

Availability of Adults

The sea scallop, Placopecten magellanicus (Gmelin), is found from the Gulf of St. Lawrence to Cape Hatteras in areas where maximum water temperatures are 20°C or less (Posgay, 1957). In the commercial fishery the scallops are shucked at sea and no live animals can be obtained at the ports of landing unless prior arrangements are made. We have obtained adult sea scallops successfully by dredge and by diver collection. Adults can be transported in containers with cold, moist toweling with no adverse effects.

Maintenance of Adults

Adult P. magellanicus have been held in the laboratory for over two years in flowing seawater at salinities from 26 to 28 °/oo and temperatures between 0 and 20°C. The upper lethal temperature for sea scallops is 23°C (Posgay, 1953). The adults appear to survive best in shallow tanks with rapid waterflow. Water from Long Island Sound contains sufficient food organisms at these flow rates to provide adequate nutrition for growth and maturation of these animals.

Conditioning and Spawning

Sea scallops can be sexed during most of their annual sexual cycle by holding the valves apart and observing the color of the gonads. Ripe females have brick-red gonads, while ripe males have creamy white gonads. Naidu (1970) gave a good description of the physical appearance of the gonads in relation to the natural sexual cycle.

We have worked with a stock of sea scallops collected near Boothbay Harbor, Maine, that normally spawns in October. This population is sexually mature between July and September, and animals collected during this period can be immediately spawned or held at 10-15°C for later use. Scallops collected between March and June will complete gametogenesis in about 6 weeks when held in the laboratory at 15°C. Ripe P. magellanicus can be held for spawning until November when resorption begins.

Spawning can be best effected in flowing water by raising the temperature gradually to 18°C, and then rapidly cooling the water to 8-10°C. This procedure may have to be repeated, but physiologically ripe scallops should spawn readily. A stripped sperm suspension often hastens the spawning response.

Females have yielded up to 30 million eggs in a single spawning, but as few as 100,000 eggs have been obtained from weakly spawning animals.

Rearing Embryos and Larvae

Fertilized eggs develop into normal larvae at salinities between 25 and 35 °/oo. Development to the straight-hinge stage occurs in about 36 hours at 20°C, in 48 hours at 15°C, and in 72 to 96 hours at 10°C. We have failed to achieve any significant larval growth at salinities from 25 to 35 °/oo and temperatures from 5 to 25°C on a diet of mixed unicellular algae. At 10°C and 30 °/oo salinity larvae have survived for as long as 45 days without any appreciable growth.

ARCTICA ISLANDICA

Availability of Adults

The ocean quahog, Arctica islandica (L.), is widely distributed on the Atlantic Coast of the United States (Merrill and Ropes, 1969). We have had good survival in the laboratory of A. islandica obtained from the Rhode Island fishery.

Maintenance of Adults

Adult A. islandica have been successfully kept in unfiltered flowing seawater trays in the laboratory for over a year. No substrates were used. Temperatures were maintained at 10 to 15°C, and the salinity was 26 to 28 °/oo.

Conditioning and Spawning

Loosanoff (1953) described the reproductive cycle of the ocean quahog from Rhode Island. The spawning period lasts from late June until early October, and animals collected during this period should yield some viable sex products. We have had no success in conditioning ocean quahogs for spawning that were obtained in the fall, and only limited success in conditioning animals obtained in the spring.

We have not been able to induce spawning in A. islandica despite numerous attempts with various stimuli. Normal sperm can be obtained from ripe males by stripping, and viable eggs can be obtained by a combination of stripping and treatment with dilute ammonium hydroxide (Loosanoff and Davis, 1963). The best ammonium hydroxide treatment consists of holding the eggs in a 3.0% solution of $0.1N$ NH_4OH in filtered seawater for 5 to 15 minutes. The eggs are then thoroughly washed and fertilized with stripped sperm.

Rearing Embryos and Larvae

Eggs treated with ammonium hydroxide as described above have been successfully reared in filtered seawater to the straight-hinge stage. Development is best at 10 to $15°C$. Despite numerous attempts, the larvae of A. islandica have been reared to metamorphosis only once. The larval period was 60 days at $10°C$ on a mixed diet of M. lutheri and I. galbana. These larvae began to metamorphose at a length of about 200μ. No information is available on the growth of later stages of A. islandica.

SPISULA SOLIDISSIMA

Availability of Adults

The surf clam, Spisula solidissima (Dillwyn), is the largest clam on the Atlantic Coast of North America and is found from the Gulf of St. Lawrence to Cape Hatteras (Yancey and Welch, 1968). Mature clams are available along the Virginia, Delaware, New Jersey, Long Island (New York), and Rhode Island coasts where they are regularly landed by the commercial fishery. Obtaining clams from the commercial boats is the easiest means of collection, but these clams do not survive well in the laboratory because of damage suffered from the jetting action of the hydraulic dredge in which they are collected. In addition, the daylong exposure on the deck of the vessels prior to arrival at port contributes to the poor condition of the specimens collected in this manner. We prefer to hand gather S. solidissima stocks for spawning from shallow beds in coastal waters. Adults 4 to 5 inches long yield very satisfactory numbers of eggs and survive better and take up less room in the laboratory than larger clams.

Maintenance of Adults

These animals are satisfactorily held in the laboratory in tanks of running seawater. It is not necessary to hold adult S. solidissima in boxes of sand but it is important that the animals be turned ventral side up shortly after being placed in the tanks to allow air bubbles to escape from the shell. Failure to displace these trapped air bubbles will result in gradual stress and subsequent death of the animals. Surf clams will survive in laboratory tanks at temperatures up to 25°C (Loosanoff and Davis, 1963).

Conditioning and Spawning

Surf clams collected from Rhode Island waters from mid-spring to early summer are sexually mature and can be spawned immediately in the laboratory. Ripe S. solidissima can be prevented from spawning by keeping them in flowing refrigerated seawater at 15°C. Mature animals held in the laboratory for spawning in late summer or fall survive and spawn best if maintained at 10°C. Sexually immature surf clams collected in the winter or early spring can be conditioned for spawning in a few weeks by holding them in running water at 15°C.

Surf clams can be artificially induced to spawn by thermal and chemical stimulation. Ripe animals held at 15°C will usually spawn when placed in flowing seawater at 18 to 20°C. Stripped sperm can be added to the warmed animals to further hasten spawning. A more effective spawning technique is to refrigerate mature S. solidissima wrapped in damp toweling overnight, and then thermally stimulate the animals at 18 to 20°C in flowing seawater. This method usually produces spawning of both sexes in one hour or less. Viable sex products can also be obtained from surf clams by stripping, but a smaller percentage of eggs obtained in this manner develops normally when compared with eggs that have been freely spawned.

Rearing Embryos and Larvae

The eggs and larvae of S. solidissima can be reared in standing seawater cultures using the methods described earlier for M. lateralis. We have not as yet clearly defined the optimum salinity and temperature ranges for culturing S. solidissima embryos, but at a temperature of 20°C and salinities between 25 and 30 °/oo development of eggs and growth of the larvae are satisfactory. Under these conditions metamorphosis occurs in an average of 20 days.

As S. solidissima larvae approach metamorphosis, they develop a foot, and then alternate between swimming and crawling behavior. This foot appears when the animal is about 160 microns in length; 80% of the larvae have a foot by the time they are 215 microns long, and at 240 microns, practically all of the larvae will have a well-developed foot (Loosanoff and Davis, 1963). Some larvae metamorphose when they are between 230 and 250 microns in length.

Growth of Juveniles

Belding (1910) measured the growth of juvenile S. solidissima and reported that young surf clams reach a length of one inch in six months and nearly two inches in the first year. He has also reported that surf clams one year old produce mature gametes, but the first important spawning season occurs in its second year.

SUMMARY

The small size, ease in spawning, relatively high fecundity, and short life cycle make Mulinia lateralis an ideal bivalve for experimental biological studies.

Spisula solidissima can be easily spawned, and the larvae successfully cultured, but all of the parameters necessary for rearing the larvae on a routine basis have not been defined. Experimental work is also required before recently metamorphosed surf clams can be successfully grown under controlled conditions.

Placopecten magellanicus can be spawned in season, and the fertilized eggs will develop into straight-hinge larvae. The larvae have not been successfully grown, and this procedure awaits further investigation.

Arctica islandica has not been induced to spawn in the laboratory, but eggs of this species, obtained by stripping, will develop into normal larvae. A. islandica larvae have been reared to metamorphosis only once.

REFERENCES

Belding, D.L., 1910. The growth and habits of the sea clam, **Mactra solidissima**. Ann. Rep. Comm. Fish and Game, Mass., 1909, pp. 26-41.

Calabrese, A., 1969a. The early life history and larval ecology of the coot clam, **Mulinia lateralis** (Say) (Mactridae: Pelecypoda). Ph.D. Thesis Univ. Conn. 101 pp.

─── 1969b. Individual and combined effects of salinity and temperature on embryos and larvae of the coot clam, **Mulinia lateralis** (Say). **Biol. Bull.**, 137: 417-428.

─── 1969c. **Mulinia lateralis**: Molluscan fruit fly? **Proc. Nat. Shellfish Assoc.**, 59: 65-66.

─── 1970a. Reproductive cycle of the coot clam, **Mulinia lateralis** (Say), in Long Island Sound. The Veliger, 12: 265-269.

─── 1970b. The pH tolerance of embryos and larvae of the coot clam, **Malinia lateralis** (Say), The Veliger, 13: 122-126.

Jackson, J.B.C., 1968. Bivalves: Spatial and size-frequency distribution of two intertidal species. **Science**, 161: 479-480.

Loosanoff, V.L., 1953. Reproductive cycle in **Cyprina islandica**. Biol. Bull., 104: 146-155.

Loosanoff, V.L. and H.C. Davis, 1963. Rearing of bivalve mollusks, In: F.S. Russell (ed.), **Adv. Mar. Biol.**, Academic Press, Inc. London, 1:1-136.

Merrill, A.S. and J.W. Ropes, 1969. The general distribution of the surf clam and ocean quahog. **Proc. Nat. Shellfish Assoc.**, 59: 40-45.

Naidu, K.S., 1970. Reproduction and breeding cycle of the giant scallop **Placopecten magellanicus** (Gmelin), in Port au Port Bay, Newfoundland. Can. J. Zool. 48: 1003-1012.

Posgay, J.A., 1953. Sea scallop investigations. In: H.J. Turner, Jr. (ed.), **Sixth Rep. on Invest. of the Shellfisheries of Mass**. Woods Hole Oceanog. Inst., Woods Hole, Mass. pp. 8-24.

─── 1957. The range of the sea scallop. Nautilus, 71: 55-57.

Sanders, H.S., 1956. Oceanography of Long Island Sound, 1952-1954. The biology of marine bottom communities. **Bull. Bingham Oceanog. Coll., Peabody Mus. Nat. Hist.**, Yale Univ. 15: 345-414.

Ukeles, R. 1971. Nutritional requirements in shellfish culture In: **Proc. Conf. Artf. Propagation of Comm. Val. Shellfish**, Univ. of Del., pp. 43-64.

Wass, M.L., 1965. Study of a soft-bottom community in the lower York River, Virginia. **Atl. States Biol. Meet.**, Charlottesville, Va. (Abstract).

Yancey, M. and W.R. Welch, 1968. The Atlantic coast surf clam -- with a partial bibliography. **U.S. Bur. Comm. Fish., Circular** 288, pp. 1-14.

CULTURE OF AMERICAN AND EUROPEAN OYSTERS

Herbert Hidu

Ira C. Darling Center, University of Maine

Walpole, Maine

Over the years much has been written on oyster hatchery rearing techniques. I refer you to Loosanoff and Davis (1963) and Walne (1966) for what have become rather standard methods for conditioning and spawning adult stocks and rearing the free-swimming veliger larvae of American and European oysters (<u>Crassostrea Virginica</u> Gmelin and <u>Ostrea edulis</u> L.)

It would be redundant to repeat these procedures. Rather, I would like to concentrate on variations and shortcuts that we have found to make an oyster hatchery a less complicated affair, something that can be accomplished with a very minimum of equipment and expense.

In recent years many persons have gained experience in rearing oysters in hatcheries. Notable are the experimental efforts of the Mid-Atlantic region in response to the MSX related mortalities (Haskin, Stauber and Mackin, 1966) in the 1950's and 60's and more recently with Federal Sea Grant funded aquaculture efforts on both coasts. I will draw on this experience to tell of some of the new approaches.

CONDITIONING AND SPAWNING

American Oysters

Conditioning regimes for Long Island Sound oysters have been well worked out (Loosanoff and Davis, 1963), i.e., a winter oyster can be brought into spawning condition in 3 to 4 weeks in heated running seawater at 20°C. But it has been suspected (Stauber, 1950) that oysters may exist in a series of physiological races throughout their geographical range with different temperatures necessary to produce gonad maturation. Moreover, it has been felt that southern oysters (south of Delaware Bay) were very difficult to induce to spawn in the hatchery.

Mid-Atlantic research since the 1960's (summarized by Hidu et al., 1969) (Fig. 2) has given us new insights into the conditioning and spawning of Chesapeake area oyster stocks. Price and Maurer (1971) have found that Delaware Bay oysters require 6 to 7 times as long to ripen as Long Island oysters, at temperatures between 12 and 22°C. They have related gonad conditioning to degree days of exposure in equations involving time and temperature. They state that "the cumulative temperature exposure is more significant in ripening Delaware Bay oysters to spawning than exposure to 'high' temperatures per se". If this is generally true, then degree day requirements might be derived for oyster stocks from other regions, thus facilitating spawning procedures.

American oysters achieve natural gonad maturation from 1 to 5 months in the spring and summer, depending on the area, and can be manipulated to spawn in the hatchery during this time. We have observed that Chesapeake stocks repeatedly build up gonad, after spawning, throughout the summer when held in ambient running water from 20-25°C. Thus we have little trouble obtaining spawnable oysters during this period. Occasional Chesapeake oysters have held mature gonads as late as October and November and this may be the case in other areas.

Properly conditioned American oysters are readily spawned by combinations of thermal and chemical (stripped gonad) stimulation (described by Loosanoff and Davis, 1963). However, oysters south of Delaware Bay appear to be more difficult to stimulate and certain modifications of technique are necessary. A heated (25-30°C) running water bath is substituted for the standing water spawning baths described by Loosanoff and Davis (1963). Periodically, the water is stopped, the culture gently drained and stripped gonad is added in dense suspension to stimulate spawning. Spawning oysters from the mass bath are then gently lifted individually to separate water baths filtered to 1 μ^1 for collection of clean eggs and controlled fertilization. The greater effectiveness of running heated baths leads one to suspect that high dissolved oxygen levels are important in stimulating gamete release. The stripping of eggs has been utilized successfully by some laboratories, however, this involves sacrifice of parent stocks and may result in the development of larvae which are not as vigorous as those produced from naturally spawned eggs.

Ostrea edulis

Different techniques are required for conditioning and spawning the European oyster which incubates young larvae (Walne, 1966). In Maine, we have very little trouble obtaining larvae from the introduced Boothbay Harbor stocks (Welch, 1963) between February and October. Early in the year we have the best luck with standing aerated 60 liter vessels containing six oysters. Water is heated to 20-22°C and is renewed daily. Cultured Phaeodactylum sp. or Dunaliella sp. is dripped into cultures commensurate with clearing rates.

Walne has observed release of larvae 30-40 days after placing winter oysters at 21°C and we have observed the same with Maine's Dutch stock of European oysters (Fig. 1). We have never attempted stimulation of spawning since spontaneous spawning regularly occurs as the gonad becomes mature. Spawning of female oysters is indicated by white piles of excess eggs around the shell margins. The progress of the incubating female can be readily monitored by rapidly squeezing the valves together and obtaining several mls of extra-pallial water for microscopic observation of released larvae.

Later in the season, after February, when phytoplankton becomes apparent in most areas, European oysters can be successfully conditioned and spawned by placing them in heated 18-23°C running water baths without supplemental feeding. Of course, a seawater system with heat exchange capacity is a necessary piece of equipment. Periodically oyster extra-pallial fluid is procured as above and incubating individuals placed in standing water for larval release. With some experience in observation the time of release can be predicted within a day or two.

Our natural spawning season in Maine extends from August to September but in warmer areas may extend throughout a longer period. Large quantities of veliger larvae can be obtained by finding incubating oysters by the above technique and isolating them in standing water for subsequent release of veligers.

LARVAL CULTURE

Larvae of both species require feeding of either cultured or natural algae. Feeding natural algae offers great economy, however, it must depend on an area with suitable species and density of natural phytoplankton. Feeding cultured algae involves culturing algae which are then added to larval cultures. Methods are described in Part I of this volume.

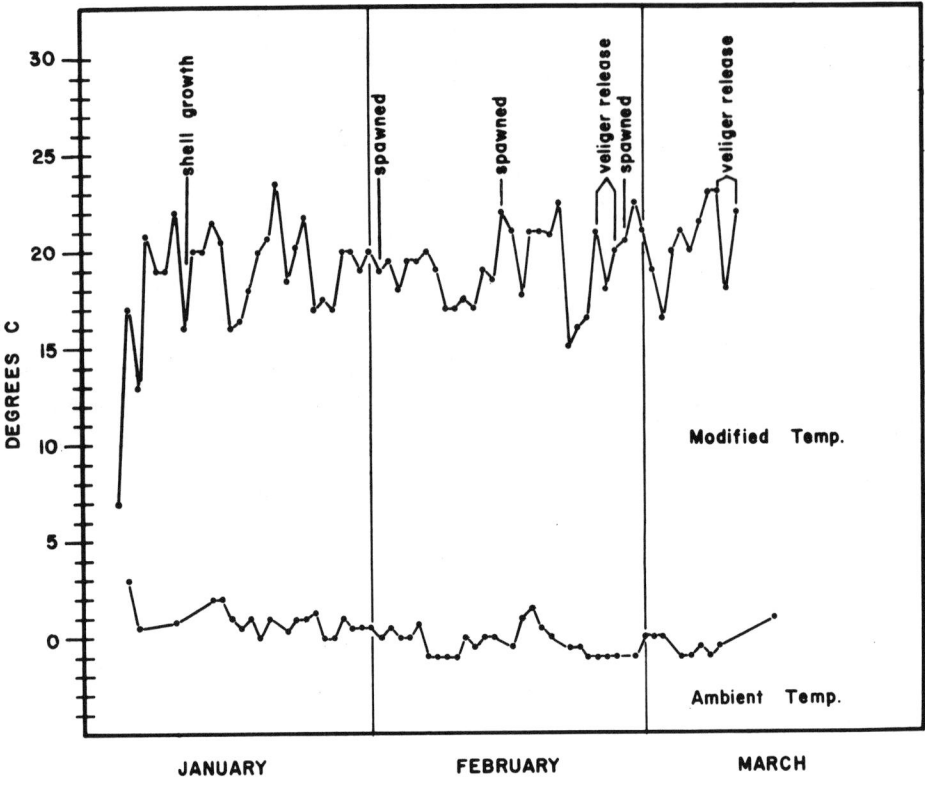

Fig. 1. Spontaneous spawning and larval release in Laboratory populations of European oysters from Boothbay Harbor, Maine. Six animals were held in 60 liters with light aeration and daily changes of water. Cultured Phaeodactylum sp. and Dunaliella sp. are monitored into cultures commensurate with clearing rates.

Feeding natural algae can be extremely valuable especially for rearing large numbers of larvae either commercially or for large scale experiments. The method involves producing an algal bloom by holding centrifuged seawater in sunlight for about 24 hours (Wells, 1920). We have had excellent success using this method in many areas (Fig. 2).

Most larval culture systems are rather simple as in the following procedure that we used with American oysters at the New Jersey Oyster Research Laboratory at Cape May (Fig. 2, No. 1) and the Chesapeake Biological Laboratory at Solomons, Maryland (Fig. 2, No. 6).

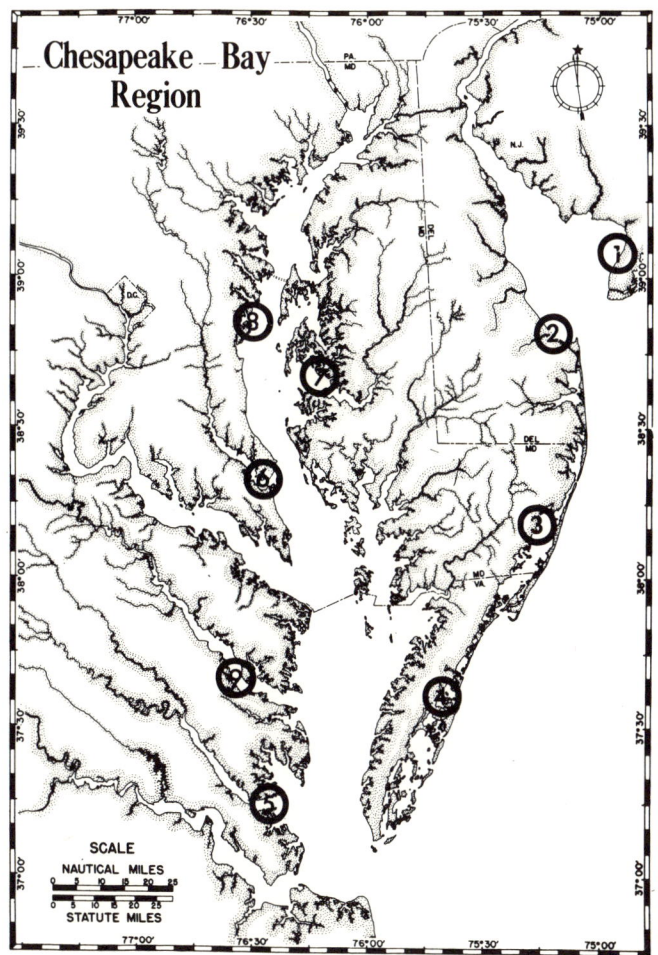

Fig. 2. Laboratories and shellfish hatcheries in the Delaware-Chesapeake Bay region that have participated in "MSX" oyster research and/or hatchery development. No. 1 -- Rutgers, The State University, New Jersey Oyster Research Laboratory on Delaware Bay; No. 2 -- University of Delaware Marine Laboratory at Lewes; No. 3 Snow Hill Field Station of the Natural Resources Institute, University of Maryland on Chincoteague Bay (activities of this station transferred to Solomons, Maryland); No. 4 -- Wachapreague Field Station of the Virginia Institute of Marine Science; No. 5 Virginia Institute of Marine Science at Gloucester Point, Virginia; No. 6 -- Chesapeake Biological Laboratory of the University of Maryland at Solomons, Maryland; No. 7 -- National Marine Fisheries Service Laboratory at Oxford, Maryland; No. 8 -- Chesapeake Bay Oyster Culture Inc., at Shadyside, Maryland; No. 9 -- The Windmill Point Oyster Company (EDA affiliated) at Urbanna, Virginia.

a) Fertilized egg to 2-day veliger larvae -- culture fertilized eggs at densities of 15-30 per ml in polyethylene containers for 2 days at a temperature of 24°C. Laboratory system seawater should be filtered to 1 μ with no contamination by metals. Methods should follow those outlined by Loosanoff and Davis (1963).

b) Two-day larvae to settling larvae

    1) Environmental conditions -- start initially with densities of larvae at 5-10 per ml in polyethylene or other nontoxic culture containers with capacity of at least 15 liters. Large volume cultures of 100 gallons or more have given excellent results. Culture temperatures should range between 24-30°C with some disadvantage to be noted at either extreme. At 24°C larval growth is slower (15 to 20 days to setting); however, there is less chance of loss through starvation and bacterial action. At 30°C larval growth rates are optimal; however, closer control is required to provide sufficient food and eliminate bacterial effects.

    2) Change procedure -- with natural algal feeding change procedures are necessarily more elaborate than with cultured algal systems. Not only do proper amounts of wild algal populations have to be maintained, but there must be control of competitive zooplankton that comes along with the algae. It is difficult to differentially remove all zooplankton, especially copepod instars. Success depends rather upon a proper control of their numbers by differential screening of the larval culture at each change.

    Larval cultures should have water renewed at least once daily. The following are sequences of screening operations that we have followed with good result:

    a) Removal of larvae from cultured water -- use coarsest screen[2] that will retain all oyster larvae so that most small competitors can be eliminated. All screens can be washed into a single container to provide a continuous screening operation. The result of this should be a container of highly concentrated pooled larvae.

    b) Removal of larger competitors, i.e., copepods; re-screen concentrated larvae. Use finest screen

that will just allow passage of all larvae and thus
retain larger competitive organisms. Obviously, competitors which are the same size as the larvae will
remain with the larvae. However, larvae will not be
significantly harmed by these. As competitors, especially copepods, grow larger than the oyster larvae,
they can and must be removed from cultures. Results
of (b) should be pooled culture of larvae which are
as free as possible from competitive zooplankton.

c) Complete removal of all water from pooled larvae
and grading of larvae. All old water must be removed
to lessen bacterial contamination in new cultures.
Larvae should be graded to permit discard of slower
growing individuals which may cause increase of pathogenic bacteria and protozoa. Select a screen series
which will effectively grade the larval population
into 2 to 3 groups. The finest screen in the stack
should collect even the smallest larvae. Examine the
smaller graded samples under a 100x magnification
microscope for signs of mortality and weak larvae.
If any gross differences are apparent between these
and larger size groups, discard or separate the smaller
larvae. Even though slow-growing larvae may eventually set if retained, there is no particular advantage
to their retention. Newly spawned groups can be
raised to setting much quicker without the risk of
disease loss.

d) Renewal of daily food supplies and new culture
conditions; make a stock culture with all healthy
larvae from (c) with about 10-15 liters per million
larvae. (No old culture water should contaminate
this.) Aliquot larvae into culture containers and
renew water supplies. Incoming laboratory culture
water should be filtered with felt bags.[3]

The above procedure has given the most positive results and
has come about through much trial and error. The key to success
is to include enough variation in initial technique so that the
most workable method can be found. Occasionally, an overly dense
algal bloom has been detrimental. A good rule of thumb is to regulate larval densities and algal densities in such a way that larvae noticeably clear the water within the change cycle. Sufficient
algae should be present to show some turbidity over most of the
change cycle. With new cultures (two to six-day larvae with low
clearing rate) larval concentrations can be quite high (up to 20
per ml) with a very modest algal density. With older larvae and
heavier feeding rates, either the density of larvae must be re-

duced or algal concentrations increased. It is difficult to substantially increase natural algal counts except by centrifugation and greenhouse aging of water supply, so that densities of larvae must be adjusted to allow optimal feeding conditions. If algae are too dense, then a fine filtration system[4] is in order. All of the techniques need a bit of experience to be successful. An open eye and much experimentation at each site are necessary ingredients to success.

The same techniques are applicable for European oyster larvae although temperatures as low as 18°C produce satisfactory growth. Large volume cultures (100 gals.) appear to be preferable to small volumes (5 gal.) for unknown reasons. In Maine in 1971 we lost most of our O. edulis larval broods with small volume cultures; in 1972 with large volumes and light aeration we rarely lost larvae. With proper sanitation we have not encountered mortalities that might be attributed to pathogenic organisms and in recent times we have used no antibiotics.

A wide variety of marine areas appear to be suitable for natural algal feeding (Fig. 2) ranging from the upper Chesapeake Bay with salinities of 10 to 15 °/oo (Fig. 2, No. 8) to coastal lagoons such as Chincoteague Bay (Fig. 2, No. 3) with full sea salinities. There have been some exceptions to this rule, for example, Virginia Institute of Marine Science at Gloucester Point, Virginia, (Fig. 2, No. 5) and the National Marine Fisheries Service at Milford, Connecticut, have not utilized natural algal feeding to any great extent. Important factors appear to be the presence or absence of algal species that can be utilized by the larvae. Natural algal feeding techniques have been derived by trial and error and suitable areas for natural algal feeding have also been found by chance. Obviously much must be accomplished experimentally to allow us to understand this situation and predict in advance optimal areas for hatchery location. In the Chesapeake area, most areas appear suitable for natural algal feeding. In Maine, the protected upper estuaries may be the preferred locations.

Natural algal feeding techniques may be useful for a wide variety of algal feeding forms since wild zooplankton such as copepods, annelid and gastropod larvae which occasionally enter as contaminants generally do very well along with our cultured bivalve larvae.

SETTING AND JUVENILE REARING

In setting, we have been aided by the gregarious setting response originally discovered by Cole and Knight-Jones (1949).

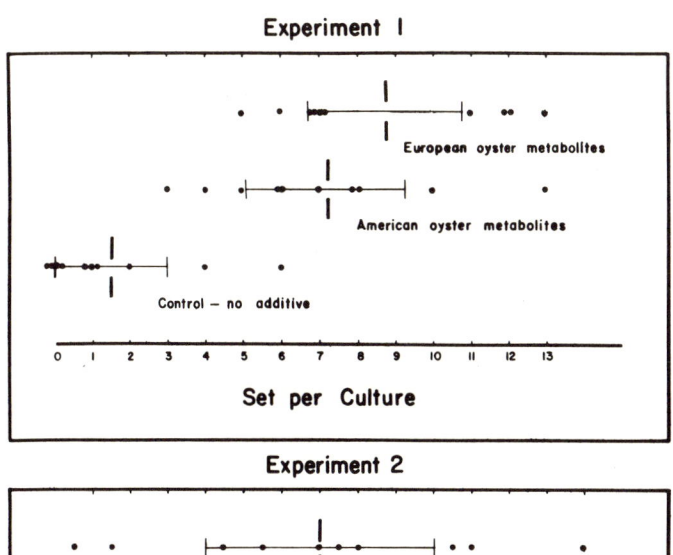

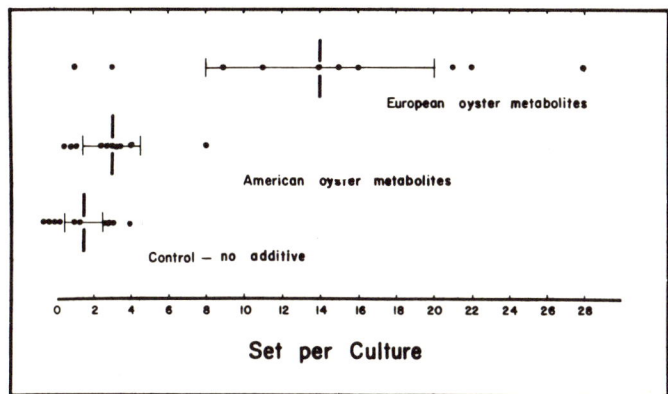

Fig. 3. Setting of the European oyster in response to metabolites of American and European oysters. Vertical bars represent mean values and 95% confidence limits from 10 repetitions within each treatment. Twenty-five ml of adult oyster shell liquor was added to 1 liter cultures containing aliquots of mature larvae and bay scallop shell cultch. Experimental cultures were drained simultaneously in 2 hours and total set counted.

Mature larvae are stimulated to set in the presence of shell liquor or merely water in which adult oysters have pumped. Normally setting will occur sporadically over a period of several days, but by adding adult oyster waste water to setting larvae very heavy sets occur in one-half to two hours. This permits efficient manipulation of cultch surfaces and thus allows optimal setting densities and increases efficiency of conversion of larvae to viable spat. The response appears to be intra-specific (Fig. 3). Metabolites from either species can stimulate the re-

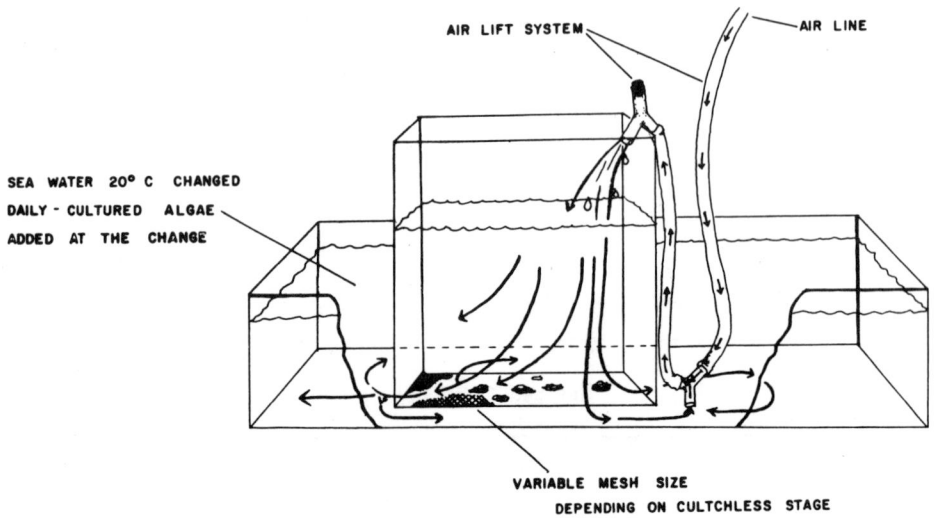

Fig. 4. Air lift circulated closed system found useful for rearing cultchless European oysters to a 1/4 in. size. The entire apparatus is immersed in a constant temperature bath.

sponse in European oysters and possibly the same is true for American oysters. Presently we are at work to identify the biochemical which is responsible for the response.

With the recent development of cultchless oysters we now can use either the traditional cultch or the new cultchless techniques. Cultchless methods are making hatcheries commercially viable and importantly, are giving the researcher a new experimental tool.

The present legal status of cultchless setting is not entirely clear. The idea of obtaining free single hatchery oysters was brought out by Pacific Mariculture of California and at least two techniques are claimed to be proprietary (Pacific Mariculture, 1967; and Long Island Oyster Farms, 1970). However, new spat may be removed by water jets a few hours after setting or set on Mylar sheets may be removed by subsequent flexing (Dupuy, 1972).

Newly removed cultchless oysters must be properly handled to allow acceptable survival and growth. We have had good luck placing the newly freed oysters in screened boxes with closed system water changed daily with the addition of cultured _Isochrysis galbana_. Circulation must be provided either by an airlift system (Fig. 4) or other nontoxic pump. This type of system can be modified to any size, but the essential element is a graded plastic screen series to allow good water and food circulation through the free oysters. Screen size is increased with larger oysters

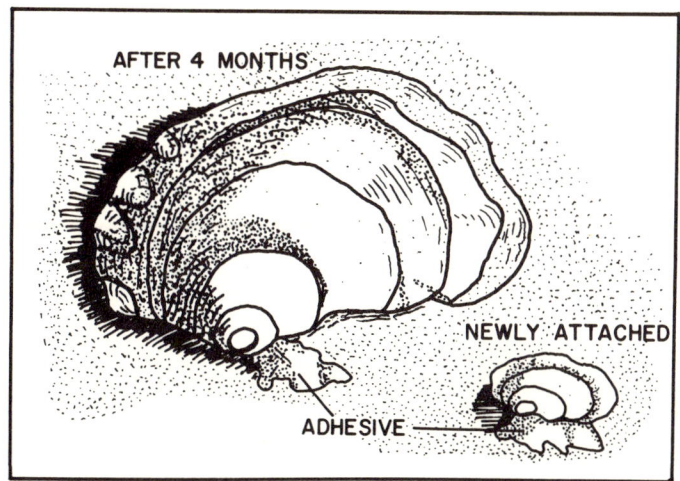

Fig. 5. Growth response of the European oyster when attached to asbestos cement-board panel. New shell growth on the left valve is tightly adhered to the flat surface around most of its margin.

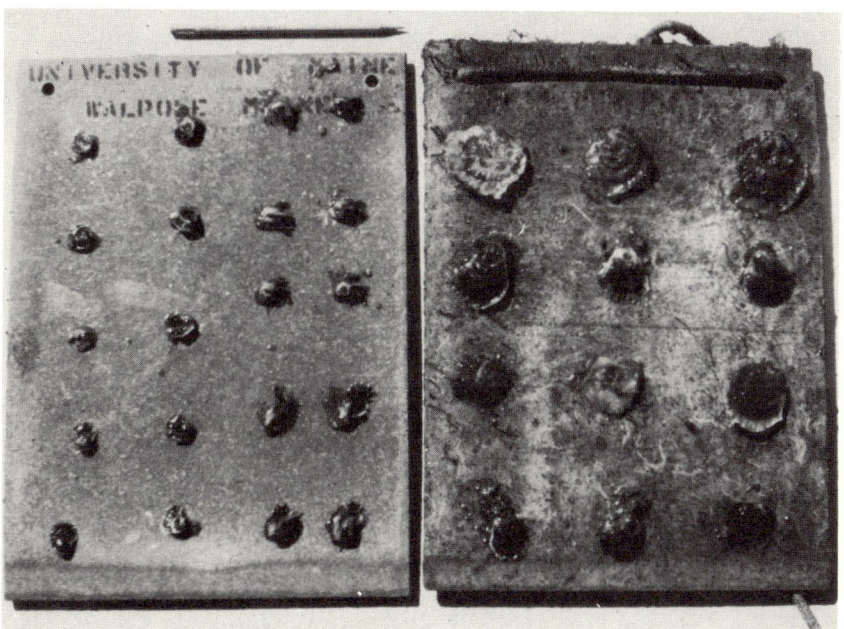

Fig. 6. Experimental oyster panels 2 weeks and 2 months after artificial reattachment. Left panel contains American and Japanese oysters; the right panel European and American oysters. This system has allowed efficient site evaluation for commercial aquaculture in Maine.

to provide easier cleaning. Cultchless oysters can be very easily reared to a 1/4 inch size in the hatchery in this manner and to larger sizes by holding in running water screened boxes. Commercial field nursery techniques are outside the scope of this paper.

A great deal can be done experimentally with cultchless oysters. We have found that if rapid growing free oysters later reattach to a substrate (i.e., asbestos board) they reassume their natural habit of depositing new shell in close proximity to the surface forming a new and permanent attachment (Fig. 5) (Riley, et al., 972). The system has a possible experimental and commercial potential. We have, for example, developed oysters on experimental panels of asbestos board (Fig. 6) to use in determining environmental suitability of new areas. A panel series is now exposed at Maine Yankee Atomic in Wiscasset to monitor growth and survival, nuclide, and heavy metal uptake in the thermally modified waters. Private citizens in Maine also have our panels to evaluate their areas for commercial culture potential.

It is an impossible task to review all newer techniques which help make shellfish hatcheries successful. One should start with proven techniques but retain flexibility of approach to evolve the most economic methods for that area. Each new area has its own peculiarities which must be considered in evolving optimal culture methods.

FOOTNOTES

[1] CUNO MICRO-KLEAN Filters, American Machine and Foundry Co., Meriden, Conn., 06450.
[2] Use either stainless steel sieve series of 325, 250, 200, 150, 100, 80, and 60 mesh per inch (Newark Wire Cloth Co., Newark, N.J.) or hand-manufactured screens of nylon "Nytex" mesh (Tobler, Ernst and Traber, Inc., 71 Murray St., N.Y., N.Y.).
[3] AFCO SNAP-RING Calibrated Filter Bag 5-25 micron opening, AFCO Filter Products Co., Glenville, Conn. This is a method recommended to us by Mr. Theodore Ritchie of the Univ. of Delaware.
[4] Use either fine sand filters or MICRO-KLEAN filters or other similar products.

Contribution No. 51 of the Ira C. Darling Center, Walpole, Maine.

Portions of this work were made possible through Sea Grant Coherent Project SG 1-36099 and NFS Grant GA-28741. Appreciation is expressed to Phyllis Coggins for illustrations.

REFERENCES

Cole, H.A., and E.W. Knight-Jones, 1949. The setting behavior of larvae of the European oyster *Ostrea edulis* (L.) and its influence on methods of cultivation and spat collection. *Fish. Invest. London* Ser. II. 17(3): 39 p.

Dupuy, J.L., and S. Rivkin, 1972. The development of laboratory techniques for the production of cultch-free spat of the oyster *Crassostrea virginica*. *Ches. Sci.*, 13(1): 45-52.

Haskin, H.H., L.A. Stauber and J.A. Mackin, 1966. *Minchinia nelsoni* n. sp. (Haplosporida, Haplosporidiidae: Causative agent of the Delaware Bay epizootic). *Science*, 153(3742): 1414-1416.

Hidu, H., K.G. Drobeck, E.A. Dunnington Jr., W.H. Roosenburg and R.L. Beckett, 1969. Oyster hatcheries for the Chesapeake Bay region. NRI Special Report No. 2, Contrib. No. 382, Natural Resources Institute, Univ. of Maryland, Solomons, Maryland, 18 p.

Long Island Oyster Farms, Inc., 1970. Artificial rearing of oysters. U.S. Patent 3, 495, 573.

Loosanoff, V.L., and H.C. Davis, 1963. Rearing of bivalve mollusks. In: F.S. Russell (ed.), Advances in Marine Biology, Academic Press, London, 1: 1-136

Pacific Mariculture, Inc., 1967. Method and apparatus for growing free oyster spat. U.S. Patent 3, 526, 209.

Price, K.S., Jr., and D. Maurer, 1971. Holding and spawning Delaware Bay oysters (*Crassostrea virginica*) out of season. II Temperature requirements for maturation of gonads. *Nat. Shellfish. Assoc. Proc.*, 61: 29-34.

Riley, J.G., R.J. Rowe, and H. Hidu, 1972. Oysters: Reattachment as a method of rearing cultchless hatchery oysters. *Comm. Fish. Rev.*, 34(5-6): 41-43.

Stauber, L.A., 1950. The problem of physiological species with special reference to oysters and oyster drills. *Ecology* 31: 109-118.

Walne, P.R., 1966. Experiments in the large-scale culture of the larvae of *Ostrea edulis* (L.) *Fish. Invest., London* Ser. II., 15(4): 53 p.

Welch, W.R., 1963. The European oyster, *Ostrea edulis*, in Maine. *Nat. Shellfish. Assoc. Proc.*, 54: 7-23.

Wells, W.F., 1920. Growing oysters artificially. *Conservationist*, 3: 151 p.

LABORATORY CULTIVATION OF ASSORTED BIVALVE MOLLUSKS

Paul Chanley

Shelter Island Oyster Company

Greenport, New York

Probably no more than two or three dozen of the species of marine bivalves have ever been cultured through their entire life cycle. Larval stages of perhaps fifty additional species have been reared in the laboratory, but juveniles have not been cultured. The life cycles of bivalves in the orders Solemyoidea and Septibranchia are not even known. Nonetheless, a considerable body of information on cultivation of marine bivalves is available. Although the bulk of experimental work has been specifically with commercially important species, many of the techniques are applicable to, or adaptable to, other species.

Innumerable investigators have made significant contribution to our knowledge of culture techniques for marine bivalves. Much of what follows is a synopsis of their work. However, to avoid excessive citations, literature references have been omitted. Instead, a list of helpful references, the references from which much of the present report was drawn, is appended. A number of considerations will be treated in the following order: water supply and treatment, collecting and culturing, obtaining reproduction, care of eggs, care of larvae, diseases and problems, and setting.

WATER SUPPLY AND TREATMENT

Some success has been achieved culturing bivalves in the better artificial seawaters, especially Instant Ocean (Aquarium Systems, Eastlake, Ohio). However, none have thus far proven as satisfactory as natural seawater or salt-water well water. Nonetheless, natural seawater as a culture medium has disadvantages. It varies in quality seasonally and some dinoflagelate blooms render it useless for bivalve culture. Sometimes this problem can be overcome by utilizing salt-water wells. Well water is apt to be more consistant in quality, although not all wells are suitable. One well produced water with such a heavy iron concentration a precipitate was formed which concentrated in the guts of larvae. The resulting constipation proved fatal.

Water used in cultivation of bivalves can easily be rendered unfit by pipes, pumps, and fittings. As a general rule avoid all metal. Copper is particularly notorious. Some metals, such as iron and aged lead are relatively harmless. The safest procedure, however, is to avoid contact between larvae, their food or culture water, and metals. Most, but not all, plastics are safe. Usually polyethylene, polypropylene, polyvinyl chloride, and nylon are acceptable. It is a good idea to bioassay any material to be used.

Natural water must be filtered or centrifuged to remove undesirable organisms before it can be used to culture bivalve larvae. A variety of centrifuges, such as cream separators and clarifiers, have been used successfully. The simplest filtration technique involves pouring water through glass wool or cotton wrapped in cheesecloth (cheesecloth should be rinsed thoroughly before use). The most serious disadvantage with this type of filter is the rapidity with which cotton and cheesecloth decompose in salt water. The resultant bacterial load usually has undersirable effects on bivalves.

Many of the cartridge filters are quite satisfactory for filtering water. They are available at most plumbing supply houses in a variety of micron sizes. Many can be cleaned by repeated alternate back flushing and beating. This undoubtedly destroys the micron rating of the filter but many are still acceptable after several cleanings.

We are presently using 10 and 50 μ polypropylene bag filters manufactured by G A F Corp., Glenville Station, Greenwich, Conn. 06830. They have a greater flow capacity and can be cleaned in a washing machine. Washing probably destroys the original micron rating but washed bags still filter satisfactorily.

After water has been filtered it should be sterilized. Larvae can be reared in untreated filtered seawater. However, when larvae are regularly cultured, or whenever it is desirable to obtain a high rate of survival, sterilization of water is highly recommended.

Water is most commonly sterilized by irradiating it with ultraviolet light. We use UV lamps from Atlantic Ultraviolet

Corp., 24-10 40th Ave., Long Island City, New York, and mount them over a trough with a bottom of corrugated fiberglass panelling over which water flows. Several such units are used if greater water flow is necessary. We generally try to keep flow rate at no more than 1 gal/min/light unit. Water depth below the light should be as shallow as possible.

Ozone has also been used to sterilize water. Generally ozonation is more expensive than UV irradiation and ozone must be removed from water before it can be used. Aeration of culture water never does any harm and sometimes may be helpful, but is not usually essential.

COLLECTING AND CULTURING BIVALVES

Larvae

Bivalve larvae can be obtained for cultivation by collecting living larvae from the plankton or by collecting adults and obtaining gametes, larvae or juveniles from them.

Larvae can be collected live in a plankton sample by screening water through a plankton net or other cloth or screen. Large larvae can be collected with a mesh of 100-200 μ but a mesh of 35 μ is none too small for the smallest larvae. They are heavier than most other plankton and can be concentrated in the mid-bottom of a beaker or other container by swirling the water. They can then be picked up with a pipette before they resume swimming. These larvae can be cultured with the same techniques discussed later for larvae obtained from known parents.

Total separation of bivalve larvae from extraneous plankton or separation by species requires careful micromanipulation, dilutions, and re-collections. Such procedures are time consuming and consequently relatively few larvae can be obtained. Other disadvantages in cultivation of planktonic larvae are: 1) It is difficult to be sure that the smallest larval stages of any given species have been collected. 2) Identification by species is difficult and sometimes impossible unless larvae are cultured to identifiable juvenile stages. Most bivalve larvae and early juvenile stages bear little resemblance to the more familiar adult form. 3) Only larvae of species with pelagic stages can be collected from the plankton.

In spite of these disadvantages, the late larval and early juvenile stages of several species have never been obtained except by this technique.

Adults

Usually it is more practical to collect adults with mature gametes and induce them to reproduce in the laboratory. Collection of healthy, sexually mature adult bivalves is complicated by several factors which influence their availability, distribution, and condition. Many smaller species have short life cycles and adults may be available only irregularly or seasonally. Although bivalves may be essentially a sedentary group many species are capable of extensive migration. For example, <u>Donax variabilis</u> and <u>Ensis directus</u> may be entirely absent from beaches from which they were previously collected in great numbers. These variations in abundance may occur seasonally or unpredictably. In most cases clams are available when you can't use them and unavailable when you want them.

Generally the optimum time for collection is just before or early in the natural spawning season. There are exceptions. <u>Tagelus plebeius</u> in Virginia contain apparently mature gametes from June until fall. Nonetheless repeated attempts to spawn them in the laboratory were futile until September. Estimates of spawning seasons of several species in north and mid-Atlantic states are given in Table I. Spawning seasons vary geographically. <u>Mercenaria mercenaria</u> develop gametes in the fall in Connecticut and can be spawned in the laboratory from late December into the normal spawning period which begins in June. On Long Island, in Virginia, and North Carolina gametogenesis does not begin until Spring and laboratory spawning cannot be induced outside the natural season, without special conditioning.

OBTAINING REPRODUCTION

After sexually mature adults have been obtained they may be maintained in the laboratory until they reproduce; mature gametes can be stripped from them, or they may be induced to spawn. Marine bivalves have at least four distinctly different types of larval development: Direct, Protobranch, Pandoracean, and Veliger. The type of development and whether or not larvae are incubated determine which culture techniques are appropriate.

Direct Development

Species that have direct development incubate eggs and larvae, usually within the parent, throughout development. Typical larval stages and structures are often modified or absent and juvenile clams are the first free-living stage. Eggs are large (over 150 μ) and few in number (sometimes less than 100). This type of develop-

TABLE I

Estimates of Periods When Some Marine
Bivalves Will Have Mature Gametes[1]

Species	Season
Argopecten irradians	Spring through summer
Amygdalum papyria	May through July
Anadara ovalis	Summer
Anadara transversa	Summer
Anomia simplex	May through summer
Barnea truncata	April through November
Brachidontes recurvus	Spring and summer
Crassostrea virginica	April through November
Cyrtopleura costata	April through November
Donax variabilis	June through November
Ensis directus	February through May
Gemma gemma	March through September
Laevicardium mortoni	March through October
Lyonsia hyalina	March through June
Mercenaria mercenaria	May through October
northern specimens	January through July
Modiolus demissus	May through September
Montacuta percompressa	March through June
Mulinia lateralis	March through October
Mya arenaria	March through May; September, October
northern specimens	Summer
Mytilus edulis	Late winter through early fall
Noetia ponderosa	June through October
Petricola pholadiformis	March through September
Pitar morrhuana	May through August
Rangia cuneata	May through September
Spisula solidissima	March through June
Tagelus plebeius	June through September
Tellina agilis	April through August
Teredo navalis	June through October

[1] Varies geographically

ment occurs most frequently in species that are small as adults or that typically inhabit regions of low species diversity. Eggs or larvae of this type will not continue development when stripped from the adult unless their removal approximates the time of natural release. The only satisfactory technique for culturing bivalves with this type of development is to maintain adults in the laboratory until they reproduce.

A common species with direct development that can be cultured easily is Gemma gemma. Adult G. gemma can be maintained, with or without a sand substrate, in about 50 ml of seawater/adult providing water is changed at least three times weekly and a few cc of a unicellular algal culture added daily. Juveniles, after their release, can be cultured with the same methods used for adults.

Incubation of Larvae

A modified technique is necessary for species that incubate their larvae for only part of their development. Some species, such as Montacuta percompressa, M. floridana, and Teredo navalis, release young veliger larvae that are pelagic throughout most of their development. Others, such as several species of Ostrea, incubate larvae for about half the developmental period. Still others, such as many Teredinids and Ostreids, are incubated through most development and have only a few hours pelagic existance.

These species have a veliger type larva, so named because of the large bilobed organ larvae use for swimming. Veliger larvae are planktotrophic and increase in size during larval development. Typically young larvae have a straight hinge line and are D-shaped. They develop a swelling or umbo, characteristic of the species, that obscures the hinge line with growth. Typical larval shapes are sometimes modified especially if the incubation period is long. Number, and the size of eggs is dependent on the size of the adult and the length of the incubation period. Most species that incubate veliger larvae produce from 10,000 to two million or more eggs.

Except for T. navalis and possibly a few other species with short incubation periods, eggs or zygotes do not develop when removed from the parent prematurely. Therefore adults must be maintained until larvae are released. Sometimes a few adults can be maintained in a large aquarium providing water is changed frequently and quantities of algae are added often. Usually unless the adult is small they should be kept in running salt water with continuous addition of unicellular algae. If the incubation period is known, bivalves can be transferred to tanks of standing water a day or two before the anticipated release of larvae. When larvae are released they should be collected and cultured separately from the adults. Otherwise many will be ingested. O. edulis and O. lurida lose a few white eggs when spawning and these can be seen on the bottom around the shell margin. Their larvae are usually released about eight days after spawning.

The commensal clam Montacuta percompressa is so small that it can be maintained in one liter of water, which is changed three times weekly. They should be fed about 50 cc of algae daily. The presence of the host, Leptosynapta inhaerens, is desirable but apparently not necessary. The male is a parasite within the female. The female forms a bulbous brood pouch with a posterior extension of the mantle. Fertilized eggs are white and can readily be seen circulating in the brood pouch. As the larval shell develops larvae become brown. After four-five days they are released as straight-hinge larvae and can be cultured as typical veliger larvae. Care must be taken not to disturb females incubating eggs or larvae during water changes or feeding because they readily abort them. Larvae have been grown through the free-swimming stage in the laboratory but juveniles, though remaining alive for some time, have not grown. Consequently nothing is known about the development of sexual dimorphism or the initial association with the host.

Stripping and Spawning

Most familiar marine bivalves release gametes into the water where fertilization and larval development occur. When sexually mature individuals of species that reproduce in this manner are available gametes can either be stripped directly from the adults or adults can be induced to spawn.

Stripping: Stripping is desirable when it is impossible or requires too much time to induce spawning. It is also desirable when it is necessary to cross one female with several males. If sperm from more than one male are used during spawning it is impossible to tell the male parent. Even though a spawning female may be moved from one container to another, sperm and eggs are carried along in the mantle cavity and on the shell.

Disadvantages of stripping are that a relatively small number of eggs can be obtained and only a low percentage of these develop normally. Stripping destroys the parent, thus preventing subsequent breeding or back crosses. Eggs of many species are not fertilizable when stripped.

Males can be stripped by inserting a pipette into the gonad and withdrawing the sperm. They are then mixed with salt water to make a sperm suspension. Spermatozoa thus obtained swim actively within a few minutes after their removal from the gonad if they are mature and healthy. Spermatozoa are seldom satisfactory if they do not begin to swim in salt water.

Greater care must be taken in stripping eggs for they are much larger and are very susceptible to injury. Eggs can be removed from the gonad by removing overlying tissues and then washing the exposed ovarian tubules with a stream of salt water. Eggs that are washed free can be collected with the wash water. It is frequently desirable to make criss-cross incisions in the gonad to expose more eggs to the spray of water. Eggs that do not readily wash from the gonad are usually immature and will not develop. Fully ripe eggs begin to ooze from ovaries as soon as the gonad is cut.

During the stripping process care should be taken to avoid entering or damaging the digestive diverticulae. It is also desirable to selectively screen eggs before and after fertilization to eliminate extraneous tissue and excess sperm.

Stripped eggs of many species are unfertilizable because of the failure of the nuclear membrane to break down. This can be overcome sometimes by keeping eggs in a 3% salt water solution of .1 molar ammonium hydroxide for 15 to 20 minutes before the addition of sperm.

Spawning: Several techniques have been employed to induce sexually mature bivalves to spawn in the laboratory. Electric shock, injection (or otherwise treating adults) with KCl, NaOH, and NH_3OH have been reported as inducing spawning. Mytilids have been spawned by pricking or otherwise injuring adductor muscles, especially by wedging the shell open. Kraft mill effluent has also stimulated Mytilid spawning. Mature Mytilid gametes have been obtained by excising mantle tissue containing the gonad and placing it in filtered seawater.

In my experience the most successful technique for inducing spawning involves rapidly increasing the water temperature and adding a suspension of sperm stripped from a male of the same species. Spawned sperm is even better. This technique has always worked as well or better than any other I have tried. Frequently increasing temperature is the only stimulus required. The temperature should be raised as rapidly and as high as possible but not to the point where animals become sluggish or cease pumping. This point varies with season or conditioning. When animals are pumping actively, stripped sperm can be added to the incurrent water of each individual by pipette or merely placed into the water. Sometimes, if the animals fail to respond, cooling and reheating the water will induce spawning. Changing the water also frequently induces spawning even though the new water may be at a lower temperature. Aeration is also sometimes helpful. A suspension of eggs instead of, or in addition to, sperm may also induce spawning, especially of males. However, as a general spawning stimulant eggs are inferior to sperm.

Some species respond more readily if refrigerated briefly (1/2 to 1 hour) prior to spawning. This technique is particularly helpful in summer when the ambient temperature is so close to the maximum tolerable temperature that it cannot be increased enough to stimulate spawning. Cooling in an ice water bath can also be used to lower the water temperature of animals to be spawned. It is usually not necessary to lower water temperature below 15°C in this manner. The response of bivalves to laboratory spawning stimuli varies considerably from species to species and also within a species from one geographic area to another (Table II).

More sperm is usually released than is necessary to fertilize eggs. Therefore only a small percentage of the available sperm should be used. Check eggs microscopically for fertilization. At 24-26°C many multiple cell stages should be present 1 1/2 hours after spawning unless egg yolk is over 100 μ in diameter. If cell division has not begun the addition of extra sperm is necessary. Hermaphroditic species produce far too much sperm in laboratory spawnings. It is therefore desirable to have a special "egg" dish in which they can be placed when spawning as female. They should be removed from the egg dish as soon as they begin releasing primarily sperm. Except for the danger of polyspermy and decomposition of excess sperm autofertilization appears to have no detrimental effects, at least in the first generation. Many bivalves have a variety of organisms living in or on their shells. Accordingly these animals will respond to the same spawning stimuli as the bivalves. Eggs of these animals should be removed because other invertebrate larvae compete with bivalves for food.

Conditioning

It is not always convenient or possible to collect adults during the natural spawning period. The spawning period of most bivalves can be altered by temperature manipulation.

The spawning season can be delayed by either refrigerating adults just prior to the beginning of the spawning season or by shipping the animals to a more northern locale for storage in naturally colder water. The latter is more satisfactory but involves risk of introducing associated exotic species or diseases to the area. Many states prohibit the placing of shellfish from other areas in waters under their jurisidiction without special permit.

TABLE II

Relative Response of Some Marine Bivalves
to Laboratory Spawning Stimuli[1]

Easy to Spawn	Moderately Difficult to Spawn	Difficult or Impossible to Spawn
Argopecten irradians[2]	Amygdalum papyria	Anomia simplex
Bankia gouldi	Anadara ovalis	Arca zebra
Barnea truncata	Anadara transversa	Astarte castanea
Crassostrea virginica (Northern)	Donax variabilis	Atrina rigida
Cyrtopleura costata	Noetia ponderosa	Brachidontes recurvus
	Rangia cuneata	Chione intapurpurea
Ensis directus	Tapes philippinarum	Chione cancellata
Gemma gemma[3]		Congeria leucophaeta
Laevicardium mortoni[2]		Crassostrea virginica (Southern)
Lyonsia hyalina[2,4]		Cyprina islandica
Mercenaria mercenaria		
		Dosinia discus
Montacuta percompressa[5,6]		Lima pellucida
Montacuta floridana[5,6]		Lithophaga bisulcata
Mulinia lateralis		Macoma balthica
Ostrea edulis[6]		Macoma tenta
Petricola pholadiformis		
		Macrocallista maculata
Spisula solidissima		Modiolus demissus
Tellina agilis		Mya arenaria
Teredo navalis[6]		Mytilus edulis
		Pandora gouldiana
		Pitar morrhuana
		Placopecten magellanicus
		Solemya velum
		Solen viridis
		Tagelus plebeius

[1] This subjective evaluation, in many cases, is based on few observations. Occasionally "difficult or impossible to spawn" species have spawned readily and profusely. Conversely some of the "easy to spawn" species, on occasion, have been difficult to spawn. Relative ease of spawning varies with season and geographically. Usually spawning can be induced more readily early in the spawning season and in the northern part of the geographic range.

[2] Functional hermaphrodite.

[3] Direct development. Larvae are incubated and released as small clams.

[4] Lecithotrophic larvae. Complete development in 3 - 5 days.

[5] Males live within females.

[6] Incubate veliger larvae for part of their development.

Refrigeration, either dry or wet, involves a risk of damaging gametes or adults. Many species do not survive dry refrigeration. Even the hardiest should not be refrigerated dry over 2-3 weeks. They should be covered by a damp cloth to reduce dehydration. Bivalves, refrigerated in water, should not be overcrowded. No more than half the bottom area of trays should be covered with shellfish. Water should be changed at least weekly and only with water that is of the same temperature. Otherwise they are apt to spawn. If possible, water should be aerated. Bivalves refrigerated in water can be kept for a longer period of time than those refrigerated dry, but should probably not be kept longer than 6-8 weeks. Again, some species are hardier than others. Species whose distribution is limited to southern waters do not survive well under refrigeration. Also, animals are usually less hardy in summer than in spring or fall.

Spawning should be delayed at least a day or two after shellfish are removed from refrigeration because eggs spawned immediately either do not develop or have a high percentage of abnormalities. This can be accomplished with hardy species by keeping them dry. Refrigerated water in which bivalves are placed also can be allowed to rise gradually but kept at temperatures just below the level inducing spawning.

A variation of the cooling technique can be used to provide mature specimens of early spawning species after the normal spawning season. Individuals can be spawned out early in the season or collected immediately after the peak of the spawning period when most have relatively empty gonads. Animals can then be kept refrigerated for at least a month and may develop mature gonads usually in about a month after being returned to natural water in late summer or early fall.

The most satisfactory conditioning technique is to maintain animals in running water at controlled temperatures and with an abundant supply of algae. Unfortunately this is expensive and few laboratories have the facilities. Without abundant water and algae, bivalves are starved not conditioned by elevating water temperature. In northern areas of the east coast natural plankton may not be rich enough to support filter feeders in winter if temperature is increased. Many, but not all, species require a rest period after the normal spawning period prior to developing gametes. Consequently a period of 1 or 2 months of inactivity at low temperatures should be permitted or induced. After this period gametogenesis can be induced by elevating water temperature to summer levels. Time required for gametogenesis varies with temperature, species, and within a species from one geographic area to another.

CULTURING FREE-LIVING LARVAE

There are three types of free-living bivalve larvae. The Protobranch type of the Nuculoidea, develop within a barrel-shaped ciliated test. Development is lecithotrophic (feeding on egg yolk). Consequently eggs are large (150 μ or more) and usually few in number. Larvae do not require additional food. The larval period is brief, usually 3-5 days and there is little or no increase in major dimensions during the larval period. Larvae have a bean shape within the test. At the end of the larval period the ciliated test ruptures and is cast off, releasing the juvenile clam. Since the work of Drew near the turn of the century there are no reports of work with Protobranch larvae. Apparently they would be relatively easy to cultivate, since they require no food and have a short larval period. Nucula delplimodonta incubates this type of larvae in a sand grain nest attached to the shell of the female. In the laboratory juvenile and adult stages would require a unicellular algal food.

Another type of larva, the Pandoracean larva, is common to the genera Pandora and Lyonsia. Pandoracean larvae resemble Protobranch larvae except the cilliated test that encloses the latter is modified to form a true velum. Larvae develop from large eggs and are lecithotrophic, bean shaped, and briefly planktonic. They require no food until the end of the pelagic period and can be cultured at concentrations given for veliger larvae.

The majority of estuarine bivalves have pelagic veliger larvae. Veliger larvae are lecithotrophic only until the end of the trochophore stage. This is usually 18-30 hours after gametes are released. At this stage they become planktotrophic, feeding on plankton, and unicellular algae can be added to the culture. Shape and growth pattern of this type of larvae are discussed previously. Some species increase 25 times or more in volume during pelagic stages. Others have a relatively short pelagic period and increase only about four times in volume. Eggs are usually small (40-100 μ) and numerous (up to 100,000,000 or more). The pelagic period is rarely less than a week and sometimes in excess of a month. In addition to specific differences, the rate of growth and development varies considerably with environmental influences. At the end of the free-swimming period a foot develops, the velum disintegrates and larvae "set" on the bottom. At this point they still do not resemble the adult.

Care of Eggs

Concentration of eggs should be reduced as soon as possible after spawning. Numbers can be estimated by thoroughly mixing the egg suspension and counting the eggs in a 1 cc sample. This is a rough estimation and replicate samples are desirable. The heaviest concentration for culturing eggs should not exceed 50 eggs/ml if eggs are 60 μ or less in diameter, or 30 eggs/ml if eggs are 60-100 μ in diameter.

Optimum water temperature for eggs of estuarine species is about 24-28°C. Temperatures below 20°C should be avoided, especially if spawning occurred near 30°C. Temperature requirements vary with species. West coast, marine and northern species undoubtedly have lower requirements.

Salinity should approximate salinity at which gametogenesis occurred. Many bivalves thrive from 5-25 °/oo. Nonetheless development will not be normal if gametogenesis occurred near one extreme and eggs are exposed to the other.

Care of Larvae

Larvae of different species vary considerably in ease of culture. Some species such as Crassostrea virginica, Barnea truncata, and Cyrtopleura costata grow satisfactorily only when fed certain species of algae. Others such as Mercenaria mercenaria will grow when fed a variety of foods, some even non-algal. Some species seem to be especially prone to hatchery diseases while others are more resistant.

Densities at which larvae can be grown vary with the size of the larvae. Other culture conditions also influence optimum density at which larvae can be grown. Usually there is a wide range of acceptable concentrations with rate of growth decreasing as concentration increases. Generally near-optimum growth can be expected if larval concentrations are kept below the concentrations shown in Table III.

TABLE III

Maximum Concentrations For Optimum Larval Growth

Larval Length	Number/cc
50 - 100 μ	15
100 - 200 μ	8
200 - 300 μ	5
300 μ +	1

The most common source of food for bivalve larvae in the laboratory is cultured unicellular algae. Algae vary considerably in their food value. Dinoflagellates are worthless or even toxic. Diatoms and Cryptomonads are generally acceptable for large larvae and juveniles of most species providing they are small enough. Other algae vary in food value depending on the species. The most widely used and most generally acceptable species are Isochrysis galbana and Monochrysis lutheri. Any bivalve that has been successfully cultured can be grown when fed either of these two species.

The concentration of algae fed must be controlled, although there is usually a wide range of acceptable concentrations. Underfeeding results in reduced larval growth but overfeeding can be fatal. Some species of algae have narrower acceptable concentration limits than others. Algal concentrations are determined by cell counts using a haemocytometer, or by centrifuging to determine the "wet packed cell volume." Optimum concentration for cell counts depends on the size of the algal species being fed. Packed cell volume optimum concentrations are approximately 10×10^{-3} mm^3 per ml of larval culture per day. Diatom cultures regularly have higher packed cell volumes than do other algal cultures.

Because there is a wide range of acceptable concentrations for most algae and because algal density is relatively constant we regularly feed our larvae 1 part algal culture to 40 parts of larval culture daily. This is tripled at the end of the free-swimming period. Although it is doubtful that such a rough approximation consistantly results in optimum growth of larvae we do get acceptable growth rates and the nuisance of cell counts and centrifuging is avoided.

Unicellular algae vary in food value depending on several factors. First, growth phase of the algal culture is important. Algal cultures rapidly increasing in cell concentrations are of more value than older cultures. Then, media in which algae are grown also seems to effect their food value. Ammonia may stimulate algal growth and produce very dense cultures, but these cultures may be inferior as larval foods. Glucose in the media may increase the food value. Also, bacterial contaminants in algal cultures can decrease larval growth or even cause larval mortality. Other bacteria, however, apparently enhance the food value of the algae. Much more needs to be known about the role of bacteria in bivalve nutrition.

The procedures for growing algae are complex, well documented, and covered in Part I by R.R.L. Guillard. However, there are a few shortcuts that have been used in bivalve culture that are worth mentioning:

1) Wild algal culture can frequently be started by the addition of 1 gram of commercial fertilizer per 4 liters of filtered seawater. Sometimes it is necessary to innoculate the culture with a hardy form such as Chlorella sp. Food value of such cultures is variable and success varies seasonally and with the type of fertilizer.

2) In several areas seawater has been filtered or centrifuged and then aerated in sunlight at high temperatures (30°C) for 24 hours. This water, without further treatment or additional food, is used as culture water for bivalve larvae. Sometimes it is also innoculated with desirable algae prior to the 24 hour storage period. Again results are variable. Record rates of growth of oyster larvae have been obtained with this technique. Water is changed daily and larval concentrations kept at a minimum.

3) A wide variety of processed or artificial foods have been used to grow bivalve larvae. Processed or artificial foods that have shown the most promise include frozen finely ground Ulva lactuca (sea lettuce) and frozen or freeze dried unicellular algae. A wide variety of other substances have been tested and only particulate foods appear to be utilized. Some species do not utilize any processed foods and even the best processed foods have not proven satisfactory for long term use.

4) Running water culture of bivalve larvae has not proven successful. Screens or filters small enough to retain larvae clog too frequently to permit continual water exchange. Therefore, most larval culture has occurred in standing water. When cultured algae are used as foods and fed daily, water changes are usually made every other day. We routinely change Monday, Wednesday, and Friday. Water changes less frequently are usually accompanied by less success. However, less frequent changes are occasionally possible and I have reared larvae through the entire free-swimming period without changing water.

A series of screens to hold larvae during water changes is necessary. We have found the most economical to be made of "PE" cloth available from Tobler, Ernst and Traber, Inc. (P.O. Box 112 420 Saw Mill River Road, Elmsford, N.Y. 10523). Nytex, also available from the same company, is good but cannot be acid-cleaned. We make screens by gluing the cloth to cut sections of plastic pipe and securing it by gluing a cut section of pipe union to the pipe and cloth. PE screens can be cleaned by washing with fresh water. After continued use they tend to clog and must be briefly dipped in concentrated NaOH, rinsed in water, dipped in concentrated HCl, and rinsed in water again. A 35 μ screen is most useful though it will not hold the smallest eggs and larvae. It is also desirable to have several larger sizes for selective screening and to reduce wear on the more expensive smaller micron screens. For larval and early juvenile culture we find it convenient to have screens of the following approximate micron ratings: 35 μ (or smaller), 70 μ, 100 μ, 150 μ, 200 μ, 300 μ, 400 μ, 700 μ and 1,000 μ. The greater the variety the more precise selective screening is possible.

During water changes we remove culture water through a screen by siphon, leaving an inch or so of water in the container and this last bit of water is then washed out of the container and screened separately. This is usually the dregs of the culture. It contains all manner of junk and, more importantly, sick and dying larvae. The larvae screened down by the siphon (suspension larvae) should be checked and a sample measured to determine rate of growth. The remaining larvae (bottom deposits) should be carefully examined for diseases and other problems and then either discarded or cultured separately.

5) Larvae of estuarine species can adapt to almost any salinity in which the adult can survive providing extreme changes are made gradually. The optimum temperature for cultivation of most east coast estuarine species is 24-27°C. Probably lower temperatures are necessary for species whose normal habitat has lower summer temperatures. Development is more rapid at 30°C but there is an increased risk of epizootics. Temperatures in excess of 30°C frequently result in slower growth and higher mortality. Minimum temperature for growth varies considerably by species. Larvae of most species survive temporary cooling down to almost 0°C, but growth is negligible below 20°C and activity decreases with temperature.

DISEASES AND PROBLEMS

The first sign of difficulty in bivalve larval cultures is a cessation of growth. Healthy larvae grow continuously and sometimes at incredible rates, providing all conditions are optimum. If basic conditions, temperature, density, pH, salinity, etc. are adequate a cessation or reduction in larval growth is the first indiction of trouble. If the problem is not corrected larvae cease swimming and increasingly larger numbers occur in the bottom deposits. Shortly thereafter massive mortalities will occur. Usually this pattern of events is caused by a change in water or food quality. It is sometimes traceable to dinoflagellate blooms and toxins from new equipment, in water lines, or culture systems.

Sometimes the problems are caused by pathogenic organisms. Increasing mortality without an accompanying decrease in the rate of growth of survivors is evidence that the problem is caused by disease rather than poor water quality or some other physical factor. More often diseases also reduce rate of growth. Very little is known about diseases of bivalve larvae. Only rarely have Koch's postulates been fulfilled to identify the causative organisms. Occasionally pathogens may become routinely epidemic. Several years ago we were able to rear clam larvae without problems until they were ready to set at which time healthy cultures were routinely anihilated within 48 hours by a fungus disease. Most pathogens decimate cultures within hours. Mortalities of virtually 100% have been obtained in less than 14 hours when cultures were inoculated with a bacterial pathogen. Obviously prevention is better than cure. Nonetheless, there is some advance warning of impending epidemics. Careful examination of recently dead larvae in bottom deposits can indicate the presence of a pathogen and treatment can be started before massive mortalities occur. Even then mortality is frequently heavy.

Recently dead larvae will still have meat between the valves but internal structures will be indistinct because of decomposition. The only motion inside the shell is that of bacterial and protozoan scavengers. Larvae decompose quickly leaving only empty shells which persist for several days. Dead larvae with tissue still between the shells usually have died during the previous 24 hours. Moribund larvae frequently have gaping shells with the velum slightly extended. They do not swim but may rotate slowly on one valve as a result of feebly moving velar cilia.

Bacterial Pathogens

Relatively clean culture water with strings of mucus to which algal foods are attached is evidence of bacterial problems. Other cultural symptoms indicating bacterial problems include an increase in protozoan population and minute hair-like growths on larval shells. Some species of Pseudomonas stain culture vessels pink. Occasionally larval shells are also stained pink. These symptoms are usually accompanied by a reduction in rate of growth and eventually slow steady mortality. Epidemic mortalities are rare.

Several species or strains of the genus Vibrio have been demonstrated to cause massive larval mortalities. These bacteria, when present, can be seen swarming around the shells of quiescent larvae in bottom deposits by careful examination at 100X. Frequently empty valves are a dull brown. Epidemic mortalities are a strong possibility.

Most marine bacteria and all known bivalve larval bacterial pathogens are gram negative. Therefore antibiotics effective against gram negative bacteria are more useful than those controlling gram positive bacteria. We have found streptomycin or aureomycin at 50-200 mg/liter of culture a satisfactory treatment. We use a veterinary preparation called Vet Strep (manufactured by Merck Chemical Division, Rahway, New Jersey). Bi Strep (John D. Copanos and Co., Inc., 610 Robinhood Rd., Baltimore, Md. 21225) is also relatively inexpensive and probably even better since it is a combination of streptomycin sulfate and dihydrostreptomycin sulfate.

More expensive but even more effective is chloramphenicol (Rachelle Laboratories, 71 Grand Ave., Palisades Park, N.J. 07650). It is used at 50 mg/l.

Sulmet (sulfamethozine) has also been used to control bacteria in larval cultures but in our experience Sulmet and penicillin are relatively worthless in controlling larval pathogens.

Protozoan Predators

Most protozoa are probably secondary invaders. A large granular ciliate has been observed in and apparently feeding on living larvae. This ciliate has never become abundant and has not been suspected as causing catastrophic larval mortalities.

A small slow-moving amoeba, on the other hand, has been associated with extremely heavy mortalities. Infected larvae appear to have a series of clear bumps around the edge of the mantle. These are the amoebae which will sometimes leave the larvae and crawl around the slide if left under the microscope for lengthy periods.

Condylostoma, a large protozoan, feeds on bivalve larvae. It can be controlled by selective screening.

There is no satisfactory treatment for protozoa. We have dipped screens, holding larvae, into fresh water briefly to make healthy larvae close their valves, then dipped the screens into a very weak formalin solution (less than 1%). The formalin solution should be strong enough to kill protozoa but not kill larvae. This does kill most protozoa but enough usually survive between the valves of dead larvae to quickly repopulate the culture.

Fungus Diseases

At least one fungus disease of bivalve larvae has been described. It is Sirolpidium zoophthorum. The thallus of this organism resembles a contorted worm in early stages. It is of the same color as larval tissue and cannot be seen, at this stage, except in empty shells. The thallus produces many branches which extend as finger-like masses toward the margin of the shell. Sometimes branches will extend beyond the shell margin. Zoospores are released and infect healthy larvae. Sirolpidium is frequently endemic in cultures but can quickly become epidemic, destroying most larvae in just a few hours. Another fungus commonly seen in larval cultures is some form of Labrynthula. These infestations look like branching strings of connected colorless wheat kernels. They have not, however, been associated with heavy larval mortalities and may not be pathogenic. The most effective treatment for Sirolpidium is to prevent its occurrence by ultraviolet irradiation of seawater. Sirolpidium infected cultures can also be treated by keeping them at 34-35°C for 48 hours. Such treatment is rough on larvae and must be accompanied by addition of antibiotics.

COMPETITORS AND PREDATORS

Copepods are common weeds in bivalve cultures. They can be controlled readily by insecticides. We prefer TEPP, available from most farm supply stores. It is effective at 1 or 2 ppm and disintegrates in the water within 24 hours.

Annelid larvae occasionally prey on bivalves. They can usually be removed by selective screening.

As cultures grow older Vorticella-like protozoa frequently form colonies on shells of larvae and juveniles. These colonies sometimes become so thick a small clam resembles a ball of fuzz. The formalin treatment, or similar treatment using brine instead of formalin, has been used with some success. Sometimes the Vorticella colonies can be knocked off shells with a strong stream of water.

SETTING

As the free-swimming period ends larvae alternately crawl on the bottom and swim. Eventually the velum disintegrates and swimming is no longer possible as larvae become juveniles. Bivalves are extremely active as early juveniles and even species that are sessile as adults move about at least briefly. Most possess a byssus thread and are capable of climbing on the sides of setting trays. Many will escape if placed in running water at this stage, even though they cannot swim.

At the end of the free-swimming stage it is necessary to reduce larval concentrations; use a culture container with greater bottom area per unit volume and triple the amount of food. Our larval culture containers are tall plastic cylinders (garbage pails). Our setting trays are 4' long, 2' wide, and only 3" deep.

Most bivalves will set without the presence of a specific substrate. The rate of setting and percent successfully setting can be increased in some species by treating the substrate or providing a special substrate. The same effect can frequently be achieved by rapidly increasing water temperature 3-4°C. Generally it is not desirable to provide a sand substrate until juveniles are large enough to be separated from the sand by screening.

Juveniles can be kept in setting trays and cared for much as larvae until they are 400 μ in length. Then it is desirable to provide them with running water to which a unicellular algae is added, continuously if possible. Species normally burrowing in sand can be placed in sand at about 1 mm in length. For optimum growth beyond this point it is desirable to plant them overboard. However, they must be protected by trays, cages or screens or they will be consumed by predators. All sizes of crabs and fish are especially destructive. Juveniles should be checked regularly. Fouling organisms and excessive silt should then be removed.

When dealing with only a few specimens or small species it is possible to keep bivalves in standing water much longer and in some instances through their entire life cycle.

Perhaps the most important rule of thumb for culturing marine bivalves is to remember that, though they are often encased in a rock-like shell they are living organisms. They may be hardier than some animals but cannot survive mistreatment and only thrive when given good care. No one would think of calculating the number of chickens that could be packed into a chicken coop and then put that many chickens in a coop without providing additional food. Yet this type of thinking has occurred repeatedly in shellfish culture. It has not worked and will not work. Successful cultivation requires that bivalves be provided with ample supplies of a suitable medium, space, oxygen and suitable food. Some provision must be made to dispose of waste products. Failure to provide any of these basic necessities of all living things inevitably results in failure to cultivate.

REFERENCES

Allen, J.A., 1961. The development of _Pandora inaequivalvis_ (L.) J. Embryol. Exp. Morphol. 9(2): 252-268.
Brown, C., 1973. The effects of some selected bacteria on embryos and larvae of the American oyster, _Crassostrea virginica_. J. Invert. Path. 21(3): 215-223.

Carriker, M.R., 1950. Notes on the killing and preservation of bivalve larvae in fluids. Nautilus 64(1): 14-17.

Chanley, P., 1968. Larval development in the class Bivalvia. Proc. Symp. on Mollusca - Part III. India 475-481.

Chanley, P. and M. Chanley, 1970. Larval development of the commensal clam, Montacuta percompressa Dall. Proc. Malac. Soc. London 39(1): 59-67.

Chanley, P. and R.F. Normandin, 1967. Use of artificial foods for larvae of the hard clam, Mercenaria mercenaria (L.). Proc. Nat. Shellfish. Assoc. 57: 31-37.

Davis, H.C. and R.R. Guillard, 1958. Relative value of ten genera of microorganisms as foods for oyster and clam larvae. Fish. Bull. U.S. 58(136): 292-304.

Drew, G.A., 1899. Some observations on the habits, anatomy, and embryology of members of the protobranchia. Anat. Anz. 15(24): 493-514.

Galtsoff, P.S., F.E. Lutz, P.S. Welch, and J.G. Needham (eds.), 1937. Culture Methods for Invertebrate Animals. 1959, Dover, N.Y. 590 pp.

Imai, T., M. Hatamaka, and R. Sato, 1949. On the artificial propogation of Japanese common oyster, Ostrea gigas Thunb. by non-colored naked flagellates. Bull. Inst. Agric. Res., Tohoku Univ., 1(1): 33-46.

Jørgensen, C.B., 1946. Reproduction and larval development of Danish marine bottom invertebrates. 9. Lamellibranchia. Medd. Komm. Havundersog. Kbh. Ser. (d): Plankton, 4: 277-311.

Loosanoff, V.L. and H.C. Davis, 1963. Rearing of bivalve mollusks. 1-136 In: Advances in Marine Biology. F.S. Russel, (ed.). Acad. Press, London.

Loosanoff, V.L. and J.B. Engle, 1942. Use of complete fertilizers in cultivation of microorganisms. Science 95: 487-488

Price, K.S. and D.L. Maurer (eds.), 1971. Proc. Conf. Art. Prop. of Comm. Val. Shellfish - Oysters - 1969. College of Marine Studies, Univ. of Del. 212 pp.

Prytherch, H.F., 1934. The role of copper in the setting, metamorphosis, and distribution of the American oyster, Ostrea virginica. Ecol. Monogr. 4: 47-104.

Selmer, G.P., 1967. Functional morphology and ecological life history of the gem clam, Gemma gemma (Eulamellibranchia: Veneridae). Malacologia 5(2): 137-223.

Smith, W.L. (ed.), 1969. Proc. Conf. Shellfish Culture held at Suffolk County Community College 1968. The Regional Mar. Res. Council, Hauppauge, N.Y. 106 pp.

Stickney, A.P., 1964. Salinity, temperature, and food requirements of soft-shell clam larvae in laboratory culture. Ecology 45: 283-291.

Tubiash, H.S., P. Chanley, and E. Liefson, 1965. Bacillary necrosis. a disease of larval and juvenile bivalve mollusks. J. Bacteriol. 90: 1036-1044.

Vishniac, H.S., 1955. The morphology and nutrition of a new species of <u>Sirolpidium</u>. <u>Mycologia</u> 47(5): 633-645.

Wells, W.F., 1920-1926. Early oyster culture investigations by the New York State Conservation Commission, New York State Conserv. Dept., Div. Mar. and Coastal Res. 1969.

CONTRIBUTORS

Marie B. Abbott
Gray Museum, Marine Biology Laboratory
Woods Hole, Massachusetts

P. J. Boyle
Museum of Comparative Zoology
Harvard University
Cambridge, Massachusetts

A. Harry Brenowitz
Adelphi University Institute of Marine Science
Garden City, New York

Elsa Brunn
Adelphi University Institute of Marine Science
Garden City, New York

Wayne D. Cable
National Oceanic and Atmospheric Administration
National Marine Fisheries Service,
Milford, Connecticut

Anthony Calabrese
National Oceanic and Atmospheric Administration
National Marine Fisheries Service,
Milford, Connecticut

David G. Cargo
Chesapeake Biological Laboratory
Natural Resources Institute
University of Maryland
Solomons, Maryland

Joseph M. Cassin
Adelphi University Institute of Marine Science
Garden City, New York

Patricia E. Cassin
Queensborough Community College
City University of New York
Bayside, New York

Paul Chanley
Shelter Island Oyster Company
Greenport, New York

J.L. Culliney
Museum of Comparative Zoology
Harvard University
Cambridge, Massachusetts

Anthony D'Agostino
St. John's University
Jamaica, New York
New York Ocean Science Laboratory
Montauk, New York

David Dean
Ira C. Darling Center for Research,
Teaching and Service
University of Maine
Walpole, Maine

David R. Franz
Ecology Section
University of Connecticut
Storrs, Connecticut

Kenneth Frenke
Adelphi University Institute of Marine Science
Garden City, New York

J.H. Gentile
National Marine Water Quality Laboratory
P.O. Box 277
West Kingston, Rhode Island

J.G. Gonzalez
National Marine Water Quality Laboratory
P.O. Box 277
West Kingston, Rhode Island

Robert R.L. Guillard
Biology Department
Woods Hole Oceanographic Institution
Woods Hole, Massachusetts

Herbert Hidu
Ira C. Darling Center
University of Maine
Walpole, Maine

CONTRIBUTORS

John M. King
Director of Research
Aquarium Systems, Inc.
Eastlake, Ohio

N.F. Lackie
National Marine Water Quality Laboratory
P.O. Box 277
West Kingston, Rhode Island

Warren S. Landers
National Oceanic and Atmospheric Administration
National Marine Fisheries Service
Milford, Connecticut

John J. Lee
Biology Department
City College of the City University of New York
New York, New York
New York Institute of Technology
Wheatley Road
Old Westbury, New York

Michael Mazurkiewicz
Ira C. Darling Center for Research,
Teaching and Service
University of Maine
Walpole, Maine

William A. Muller
Biology Department
City College of the City University of New York
New York, New York
New York Institute of Technology
Wheatley Road
Old Westbury, New York

Helen M. McCammon
Environmental Protection Agency
Boston, Massachusetts

Michael Priano
Adelphi University Institute of Marine Science
Garden City, New York

Edwin W. Rhodes
National Oceanic and Atmospheric Administration
National Marine Fisheries Service
Milford, Connecticut

Morris H. Roberts, Jr.
Director of Research
Aquatic Sciences, Inc.
2624 N.W. 2nd Avenue
Boca Raton, Florida

George D. Ruggieri, S.J.
Osborn Laboratories of Marine Sciences
New York Aquarium
New York Zoological Society
Brooklyn, New York

Haskell S. Tubiash
U.S. Department of Commerce
National Oceanic and Atmospheric Administration
National Marine Fisheries Service
Oxford, Maryland

R.D. Turner
Museum of Comparative Zoology
Harvard University
Cambridge, Massachusetss

Heidi G. Wetherall
Adelphi University Institute of Marine Science
Garden City, New York

Neil T. Wetherall
Adelphi University Institute of Marine Science
Garden City, New York

P.P. Yevich
National Marine Water Quality Laboratory
P.O. Box 277
West Kingston, Rhode Island

ANIMAL AND PLANT INDEX

Acanthodoris, 247
Acartia clausi, 201
   A. tonsa, 201, 202, 203, 204, 205
Acaulis ilonae, 139, 141
Achromobacter, 70, 73, 78, 111
Acteocina (Tornatina), 254
Adalaria proxima, 253
Aeolidia, 247
Aeolidiella alderi, 253
Aeromonas, 78
Aequipecten (see Argopecten)
Aetea Anguina, 160
Aglaophamus, 181
Alcyonidium polyoum, 162
Alpheus, 218
Allogromia laticollaris, 91, 93, 100
Ammonia beccarii, 91
Amphiblestrum flemingii, 160
Amphicteis floridanus, 194
Amphinema dinema, 139
   A. rugosum, 139
Amphiporus formidabilis, 113, 129
Amphora, 92, 93, 94, 96, 101
Amygdalum papyria, 301, 306
Anadara ovalis, 301, 306
   A. transversa, 301, 306
Anaspidea, 247, 249
Anomia simplex, 301, 306
Aplysia, 247
Arbacia, 240
   A. punctulata, 230, 231, 232-238, 240
Arca zebra, 306
Arctica islandica, 273, 278-279, 281
Arenicola, 217
   A. marina, 216
Argopecten (Aequipecten) irradians, 62, 65, 301, 306
Artemia, 65, 122, 123, 124, 125, 138, 139, 140, 141, 146, 147, 149, 150, 151, 152, 213, 215, 216, 219, 224
   A. salina, 122, 127, 141

Ascophyllum, 159
Astarte castanea, 306
Asterias forbesi, 230, 231, 235
   A. vulgaris, 230, 231
Asteroidea, 229, 230
Atrina rigida, 306
Aurelia, 147, 149, 150, 151
   A. aurita, 139, 148-150
Australorbis glabratus, 127
Bacillus, 70, 78
Balanus eburneus, 114, 115, 128
Bankia gouldi, 264
Barnea gouldi, 306
   B. truncata, 301, 306, 309
Bougianvillia, 137
   B. carolinensis, 139, 141
   B. muscus, 139
Bowerbankia gracilis, 160, 162, 163
Brachidontes recurrus, 301, 306
Brachiomonas submarina, 30
Brachionus pilictilis, 216
Bugula, 167
   B. avicularia, 163
   B. simplex, 160
   B. stolonifera, 163
   B. turrita, 160
Busycon, 65
Caenorhabditis elegans, 126
Calanus, 129, 199
   C. finmarchicus, 114
Callinectes, 211, 217
   C. sapidus, 215
Callopora aurita, 160, 162
Campanularia, 138
   C. flexuosa, 139, 141
Capellinia exigue (Eubranchus doriae), 253
Capitella capitata, 181, 182, 187, 192
Cassiopea, 151, 152
Catriona aurantia, 253
Cellepora americana, 161
Cephalaspidea, 246, 247, 250
Cerebratulus californiensis, 129
Cereus pedunculatus, 137
Chaetoceros calcitrans, 30

C. simplex, 30
Chione cancellata, 306
C. intapurpurea, 306
Chironex, 145, 146
Chlamydomonas, 93, 101
C. coccoides, 30
Chlorella, 10, 214, 311
C. autotrophica, 30
Chlorococcum, 10, 30, 94, 101
Chlorococcus, 92
Chlorohydra viridissima, 112, 129
Chondrus, 159
Chroomonas salina, 30
Chrysoara, 145, 147, 148, 149, 150, 151
C. hysoscella, 147
C. quinquecirrha, 139, 140, 146, 147
Cladophora, 10
Clava multicornis, 139, 140
Clione, 247
Clytia attenuata, 138, 139, 141
Coccolithus huxleyi, 30
Condylostoma, 314
Congeria leucophaeta, 306
Conopeum reticulum, 162, 163, 168, 169
Cordylophora lacustris, 138, 139, 141
Corophium isidiosum, 201
Corynebacterium, 70, 73
Craspedacusta sowerbii, 139
Crassostrea, 216, 248
C. virginica, 62, 65, 113, 169, 258, 265, 283-284, 301, 306, 309
Cratena pilata, 253, 254
Crepidula fornicata, 65, 114
Cribrilina punctata, 160, 162
Crisia eburnea, 160
Crisosphaera carterae, 30
Cryptomonas, 30
Cryptosula pallasiana, 161, 162
Ctenodrilus, 181, 187
C. serratus, 187, 194, 195
Cyanea, 145, 150, 152
C. capillata, 150-152
Cyclotella, 254
C. nana, 30

Cylindrotheca closterium, 93, 101
Cynea, 145
Cyprina islandica, 306
Cyrtopleura costata, 301, 306, 309
Cytophaga, 70
Daphnia, 124
Dendrobeania murrayana, 160
Diadema antillarum, 230
Dicrateria inornata, 30
Dinophilus, 181
D. gyrocitiatus, 193, 194
D. gardineri, 194
Ditylum brightwellii, 30
Dodecacera concharum, 195
Donax variabilis, 300, 301, 306
Dosinia discus, 306
Dugesia, 126
D. dorotocephala, 112
D. lugubris, 129
Dunaliella, 19, 96, 101, 192, 193, 214, 254, 285, 286
D. parva, 93, 94
D. salina, 93, 94, 101
D. tertiolecta, 30, 128
Echinarachnius parma, 230, 231
Electra "crustulenta" (Membranipora lacroixii; Canopeum reticulum), 162, 163, 168, 169
E. hastingsae (E. monostachys), 162
E. Pilosa, 158, 160, 163, 164, 168
Elphidium, 91
Elysia, 247
Enis directus, 300, 301, 306
Enteromorpha, 10, 102, 104
Eubranchus doriae, 253
Euterpina acutifrons, 114, 115
Fabricia, 181
F. Sabella, 185, 194
Farella repens, 158, 163
Fiona marina, 250
Flavobacterium (Flavobacter), 70, 73, 78, 93, 111
Flustrellida hispida, 160, 166
Fragilaria construens, 93
Fucus, 159

ANIMAL AND PLANT INDEX

Gaffkemia, 227
Gaffkya homari, 227
Gemma gemma, 301, 302, 306
Glycera, 181
   G. dibranchiata, 178, 184
Gymnolaemata, 156, 160
Haminoea, 247
Henricia sanguinolenta, 230
Hexamita inflata, 112
Hippoporina americana, 161
   H. porosa, 161
Hippothoa, 167
   H. hyalina, 160, 164
Homarus americanus, 9, 11, 221-227
Hydra, 139
   H. littoralis, 141
   H. oligactis, 112, 129
Hydractina echinata, 139, 141
Hydroides dianthus, 182
   H. hexagonus, 183
Isochrysis, 19, 93, 192
   I. galbana, 30, 204, 205, 254, 263, 264, 276, 279, 292, 310
Labrynthula, 315
Laeonereis culveri, 181, 183, 186, 194
Laevicardium mortoni, 301, 306
Laminaria, 159, 186, 194
Lecane inermis, 113
Lepidonotus squamatus, 182, 183
Leptosynapta inhaerens, 230, 234, 303
   L. Roseola, 230, 234
Leucothrix, 227
Lima hyalina, 306
   L. pellucida, 306
Limacina, 247
Lineus vegetus, 113, 129
Lithophaga bisulcata, 306
Lobiger, 246
Lyonsia, 308
   L. hyalina, 301
Lytechinus pictus, 231
   L. variegatus, 230, 239
Macandrevia, 18
Macoma balthica, 306
   M. tenta, 306
Macrobranchium, 218

Macrocallista maculata, 306
Macrocystis, 138
Magellania venosa, 17, 19
Manayunkia (a)estuarina, 185, 194
Membranipora (membraniporids), 18, 164
   M. lacroixii (Electra "crustulenta"), 162, 168, 169
   M. membranacea, 169
   M. tenius, 162, 168, 169
Menippe, 211
   M. mercenaria, 217
Mercenaria mercenaria, 62, 65, 66, 114, 115, 264, 265, 300, 301, 306, 309
Merga galleri, 139
Micrococcus, 70, 73, 78, 93
   M. cryophila, 73
Microhedyle, 247
Micromonas pusilla, 30
Microporella ciliata, 161, 162
Milliamania fusca, 91
Minchinia nelsoni, 70
Modiolus demissus, 301, 306
Monochrysis, 30, 93
   M. lutheri, 30, 254, 263, 276, 279, 310
Montacuta percompressa, 301, 302, 303, 306
   M. floridana, 302, 306
Mulinia lateralis, 273-277, 281, 301, 306
Mya arenaria, 113, 301, 306
Mytilus edulis, 113, 301, 306
Nannochloris, 30, 92, 93, 94, 101
   N. atomus, 30
Nassarius reticulatus, 114
Navanax, 246
Navicula diversistriata, 93
Nereis, 148, 181
   N. arenaceodenta, 187
   N. succinea, 178, 182, 183, 190
   N. virens, 180, 182, 186, 194
Nitzschia acicularis, 93, 101
   N. brevirostris, 93, 96
   N. hungarica, 94
Noetia ponderosa, 301
Notosaria nigricans, 19
Nucula delplimodonta, 308

Obelia, 139
Odostomia, 247
Oithona nana, 199
Oncaea, 114, 115
Ophelia bicornis, 191
Ophioderma brevispina, 230, 231
Oscillatoria, 10
Ostrea, 216, 302
   O. edulis, 62, 65, 113, 115, 169, 259, 283, 285, 290, 303, 306
   O. lurida, 303
Oxynoe, 246
Oxyrrhis, 163
Pagurus, 4
   P. longicarpus, 215
Palaemonetes, 218
Pandalus jordani, 210
Pandora, 308
   P. gouldiana, 306
Parasmittina nitida, 161
Penaeus aztecus, 114
Percursaria, 10
Petricola pholadiformis, 301, 306
Phaeodactylum, 19, 192, 285, 286
   P. tricornutum, 30, 101
Phestillia melanobranchia, 249, 253
   P. sibogae, 249, 253, 254
Philine, 246
Photobacter, 78
Phycodrys, 159
Phylactolaemata, 156
Phyllodoce mucosa, 185
Pinctada fucata, 7
   P. maxima, 113
Pinnotheres chamum, 215
   P. pisum, 215
   P. veterum, 215
Pitar morrhuana, 301, 306
Placopecten magellanicus, 273, 277-278, 281, 306
Platymonas, 30, 115
   P. maculata, 115
   P. suecica, 30
Platynereis dumerilii, 183
Podarke obscura, 183
Podocoryne carnea, 139, 141
Polydora, 181, 185, 195

P. ligni, 181, 183, 190, 191, 192, 194
Potamilla, 181
Protohydra leuckarti, 139
Pseudomonas, 70, 73, 76, 78, 93, 111, 313
Pygospio elegans, 195
Pyramimonas grossii, 30
Quinqueloculina lata, 91, 93
Rangia cuneata, 301, 306
Rhabditis marina, 91, 92, 95
Rhodomonas, 30
   R. baltica, 204, 205
Rhodospirillum, 78
Rhopilema, 145, 151
   R. verrilli, 152-153
Rosalina leei, 91, 94, 100
Rotaliella, 91
Sabellaria, 195
Sacoglossia, 246, 247, 248, 249, 250
Sagartia troglodytes (Cereus pedunculatus), 137
Sagitta, 141
Sarsia tubulosa, 139
Schizoporella unicornis, 161, 162, 168, 170
Scolecolepides viridis, 182, 184
Sirolpidium zoophthorum, 314-315
Skeletonema costatum, 30
Solemya velum, 306
Solen viridis, 306
Spio filicornis, 188, 193, 194
Spiroloculina hyalina, 91, 92, 93, 100
Spirorbis, 185
   S. borealis, 192
Spisula solidissima, 273, 279-281, 301, 306
Stauridiosarsia japonica, 139
Staurocladia portmani, 140, 141
Staurocoryne filiformis, 139
Stenopus, 218
Streblospio benedicti, 181, 182, 183, 185, 186, 192
Streptococcus, 75, 80, 83, 84
Strongylocentrotus droebachiensis, 230
   S. purpuratus, 11, 230, 236

ANIMAL AND PLANT INDEX

Synalpheus, 218
Syncoryne exima, 137, 139
Tagelus plebeius, 300, 301, 306
Tegella unicornis, 160
Tellina, 18
   T. agilis, 301, 306
Temora stylifera, 114, 115
Tenellia fuscata, 253, 254
   T. pallida, 253
   T. ventilabrum (T. pallida), 253
Terebratalia transversa, 23, 24, 25
Terebratulina septentrionalis, 19
Teredo navalis, 62, 65, 301, 302, 306
Tergipes tergipes, 253
Tetraselmis suecica, 30
Thalassiosira fluviatilis, 30
   T. pseudonana, 30, 34, 37
Thyone briareus, 230, 234
Tigriopus, 138
   T. japonicus, 115, 127
Tintinnopsis, 112
Tornatina canaliculata, 253, 254
Tornitina (Acteocina), 254
Tritonia hombergi, 253
Trochammina inflata, 91
Tubularia crocea, 112, 129, 139
Uca, 210
Udothea, 140
Ulva, 10
   U. lactuca, 311
Vibrio, 70, 78, 313
   V. alginolyticus, 66
   V. anguillarum, 62, 63, 64, 65, 66, 67, 68
Victorella pavida, 162, 169
Vorticella, 315
Waltonia inconspicua, 19
Xylophaga atlantica, 264, 266
Zoobotryon, 156

GENERAL INDEX

Acidification, 7, 10
Acochlidiacea, 247
Acrylic acid, 104
Actidione (cycloheximide), 113, 127
Activated carbon, 10, 11
Activated charcoal, 47
Aeration, 8, 35, 37, 38, 44, 147, 163
Aerators, 39, 40, 187
Aerosporin (Polymyxin B sulfate), 92
Agar, 62, 102, 119, 120, 121
Agnotobiotic, 88, 89, 90, 98, 109, 116
Air-lift, 20
Alanine, 97, 103
Albamycin (Novobiocin sodium), 92, 99, 112, 116, 117, 118
Alcohol, 74, 127, 213, 253, 269
Alconox, 260
Alfalfa, powdered, 182, 192
Algae
  benthic, 10
  competitors, 88
  culture, 29-60
  deleterious effects, 259
  food organisms, 101, 182, 217, 276, 285-286, 288-290, 309-311
  fouling organisms, 22
  media for isolation, 102
  tolorance to antibiotics, 122-125
Aluminum, 6, 39
Aluminum chloride ($AlCl_3$), 54
Amictic eggs, 127
Amino acids, 10, 19, 56, 80, 97, 103, 104, 111, 129
  alanine, 97
  arginine, 97, 103
  asparagine, 97, 104
  aspartic acid, 97
  cysteine, 80
  glutamate, 80, 103, 104
  glutamic acid, 97
  glycine, 99, 103
  histidine, 93, 103
  hydroxyproline, 97
  isoloucine, 97
  lysine, 97, 103
  methionine, 80, 97, 103
  norleucine, 97
  phenylalanine, 97
  proline, 80
  serine, 97
Ammonia, 4, 8, 9, 110, 217
Ammonifiers, 110
Ammonium chloride (NH_4Cl), 49, 51, 54
Ammonium hydroxide (NH_4OH), 279, 304
Ammonium ion, 46, 49, 56
Amoeba, 314
Amphipods, 159
Amphotericin-B (Fungizone), 92, 119
Anacystis, 217
Annelids, 81, 141, 290, 315
Anthozoa, 137
Antibiotics, 20, 22, 109-133, 259, 313, 315
  Actidione (cycloheximide), 113, 127
  Aerosporin (Polymyxin B sulfate), 92
  Albamycin (Novobiocin sodium), 92, 99, 112, 116, 117, 118
  Amphotericin B (Fungizone), 92, 119
  Aureomycin (Chlortetracycline), 112, 125, 314
  bacitracin, 98, 117
  Bistrep, 69, 314
  Candicidin, 117, 118, 119, 123, 125
  carbomycin, 117, 118, 119
  chloramphenicol, 69, 92, 98, 112, 114, 116, 117, 118, 120, 121, 122, 123, 125, 127, 129, 314

Chloromycetin (see chloramphenicol), 20, 92, 125
chlortetracycline (Aureomycin), 112, 117, 118, 119, 125, 127
Colistin, 117, 118, 119
Coly-mycin (Sodium colistimethate), 92
Combi-Strep, 69, 111
cycloheximide (Actidione), 113, 127
dihydrostreptomycin (SO_4), 69, 92, 99, 111, 112, 114, 116, 125, 127, 314
erythromycin (erythromycin lactobionate; Erythrocin), 92, 98, 99, 112, 117, 118, 119, 121, 122
Filipin, 112
Fumidil B^2, 112
Fungizone, 92
Kanamycin, 69, 116, 117, 118, 120, 121, 122, 123
Mycostatin, 92, 112
Nalidixin, 123
Naramycin, 123
Neomycin, 69, 98, 112, 113, 116, 117, 118, 120, 121, 122, 123, 125, 126, 127
Novobiocin sodium (Albamycin), 92, 99, 112, 116, 117, 118, 121, 122
Nystatin, 117, 118, 119, 123, 126
oleandomycin, 117
oxytetracycline (Terramycin), 117, 118, 119, 125
Penicillin, 20, 69, 92, 99, 112, 113, 114, 115, 116, 117, 118, 119, 121, 122, 123, 124, 125, 126, 127, 129, 212, 253, 314
Polymixin B, 69, 92, 98, 116, 117, 118, 123, 125, 127, 253
sodium colistimethate (Coly-mycin), 92
Sulmet (sulfamerazine), 111, 113, 314

Streptomycin (Streptomycin sulfate), 20, 69, 111, 112, 113, 114, 115, 116, 117, 118, 120, 121, 122, 123, 125, 126, 127, 129, 314
Terramycin (oxytetracycline), 125
tetracycline (Achromycin), 69, 112, 114, 117, 118, 121-123, 125, 127, 129
Trichomycin, 117, 118, 119, 123
Trypaflavin, 112
Vancomycin, 112
Vet-Strep, 69, 314
Antisera, 66
Arenicolids, 181
Arginine, 97, 103
Arthropoda (arthropods), 30, 114
Articulata, 16
Articulate brachiopods, 16, 17, 18, 19, 23, 26
Artificial foods, 311
Artificial seawater, 56, 57, 218, 221, 226-227, 297
Aseptic, 24, 91
Asparagine, 97, 104
Aspartate, 103
ASP medium, 56
ASPM base, 56
Aureomycin (chlortetracycline), 112, 125, 314
Autoclave, 39, 42, 44, 47, 54, 58, 59, 104
Axenic, 124, 128
Bacillariophyceae, 30
Bacillary necrosis, 61-71
Bacitracin, 99, 117, 118, 119, 125
Bacteria
 acidifiers, 7, 110
 aerobic, 9
 ammonifiers, 110
 competitors, 53, 54
 control of, 10, 11, 109-133
 deleterious, 88, 129, 192-194, 259-260
 denitrifiers, 5
 favorable, 110
 gram negative, 110, 116

gram positive, 116
media for isolation of, 102
natural populations of, 73-86, 98, 119
nitrifiers, 4, 8, 9, 10
nutrients, 23, 92-93, 99, 115, 310
pathogenic, 22, 29, 61-71, 111, 148, 212, 227, 253, 289, 313-314
seasonal populations, 111
Bactopeptone, 45
Barnacles, 15, 22, 26, 148, 217
Barophily, 74
Bay scallop, 62
Beef heart, 15-27
Bicarbonate, 4, 6, 38, 56
Bile esculine azide broth, 80
Biological filters, 5, 8, 9
Biotin, 20, 47, 48, 52, 53, 54
Bistrep, 69, 314
Bivalves, 19
Bivalves, mollusks, 61, 70, 113, 257-318
Blastula, 211, 238
Blood cells, red, 11
Blue green algae, 217
Bolivina, 91
Bon-Ami, 214, 260
Borax, 269
Boric acid, 6, 55, 56, 149
Boron, 55, 56
Boron stock solution, 56
Borosilicate (glass), 29, 48, 56, 200
Bouillon, 264
Bouin's solution, 269
Brachiopoda (brachiopods), 16, 19, 22, 23, 26
Brachiopods, articulate, 16, 17, 18, 19, 23, 26
Brachiopods, inarticulate, 19
Brachyuran, 115
Brine shrimp (see Artemia)
Brittle stars, 229, 230, 231
Bromide, 6
Bryozoa, 115-176
Buffer, 10, 11, 53, 56, 269
Bugulids, 158, 166

Calanoida, 199, 201, 205
Calcite, 9
Calcium, 6, 7, 8, 9, 50, 55, 138, 265
Calcium carbonate, 9, 39, 155, 269
Calcium chloride, 40, 46, 53, 54, 55, 149
Calcium sulfate, 148
Candicidin, 117, 118, 119, 123, 125
Cannibalism, 224
Carbohydrates, 10, 97
Carbomycin, 117, 118, 119
Carbon, 37
activated, 10, 11
Carbon Dioxide, 31, 37, 38, 58
Carbonate, 4, 9, 37
Carboys, plastic, 47
Caridean shrimp, 218
Carriker's solution, 268
Cells, red blood, 11
Cetyl alcohol (n-hecadecyl alcohol), 253
Chaetognath, 141
Charcoal, 40
activated, 47
Cheilostomata (cheilostome), 156, 158, 160, 162, 163, 164
Chelex-100, 200
Chemoautotrophic bacteria, 4
Chitin, 155, 158, 227
Chitons, 159
Chloramphenicol, 69, 92, 98, 112, 114, 116, 117, 118, 120, 121, 122, 123, 125, 127, 129, 314
Chloride, 6, 55
Chlorine, 218, 226
Chloromycetin (see chloramphenicol), 20, 92, 114, 125
Chlorophyceae, 30
Chlorophyta, 91, 92, 93, 125
Chlorox, 127, 212, 213
Chlortetracycline (Aureomycin), 117, 118, 119, 125, 127
Chrysophyta, 30, 125, 276
Ciliates, 88, 98, 192, 314
Cirratulid, 187
Cirripedia, 114
Citrate, 46, 50, 52

Citric acid ($C_6H_5O_7$), 52
Clark's artificial seawater medium, 54
Clams, hard, 62, 63, 66, 69, 73, 111
Cnidaria (coelenterata, coelenterates), 22, 26, 112, 126, 137-143, 145-154, 263
Cobalt, 6
Cobalt chloride hexahydrate, 48, 51
Coelenterata (see Cnidaria)
Coliform bacteria, 75, 79, 80, 81, 82, 83, 84
Colistin, 117, 118, 119
Coly-mycin (sodium colistimethate), 92
Competitors, 61, 289
Copepoda (copepods), 22, 114, 115, 129, 138, 141, 119-207, 216, 217, 288, 289, 290, 315
Copper, 6, 218, 298
Copper sulfate pentahydrate, 48, 51
Coral, 9, 249
Coronatae, 145
Coryneform bacteria, 78
Crinoidea (crinoids), 19, 23, 26, 229, 230
Crustacea, 114, 115, 124, 145, 148, 209-220
Cryolite, 181
Cryptomonads, 310
Cryptophyceae, 30
Ctenophores, 141, 147, 148, 151
Ctenostomata (ctenostomes), 156, 158, 160, 162, 163, 164
Cubomedusae, 145
Cultchless setting, 292, 294
Cyanophyta, 125
Cycloheximide (Actidione), 113, 127
Cyclopoid, 199
Cyclostomata, 156, 164, 166
Cyphonautes, 164, 165, 168, 169
Cysteine, 80

Dayho sea salt, 57
Decapoda, 114
Decomposition, 7
Denitrification, 4
Denitrifying bacteria, 5
Density, 312
Detergents, 192, 213, 260
Detritus, 73, 74, 88, 101, 110, 181, 182, 192, 260
Diatoms, 29, 30, 37, 43, 48, 49, 56, 88, 91, 92, 93, 98, 151, 254, 310
Dietrich solution, 205
Dihydrostreptomycin sulfate, 69, 92, 99, 111, 112, 114, 116, 125, 127, 314
Dinoflagellates, 44, 298, 310, 312
Disease, 22, 148, 150, 227, 260, 312-315
Dolomite, 9
Echinodermata, 229-243
Echinoidea (echinoids), 163, 229, 230, 231
Echiuroids, 23, 163
Ectoprocta (ectoproct), 15, 18, 19, 22, 26, 156
EDTA (ethylenediaminetetraacetic acid), 6, 11, 38, 46, 48, 49, 50, 51, 52, 54, 115,
Electrinids, 164
Enrichment media, 48, 49, 50
Enterics, 70
Entoprocta, 156
Enzymes, 74
Ephyrae, 147, 150, 151, 152
Epiphytes, 104
Erdschreiber, 91, 102
Erythromycin (Erythromycin lactobinate; Erythrocin), 92, 98, 99, 112, 117, 118, 119, 121, 122
Ethanol, 269
Ethylenediamine tetraacetic acid disodium salt, 48, 49, 52
Ethylenediamine tetraacetic acid ferric sodium salt, 54
Euglenophyta, 125
Eutrophication, 163, 258

Fecal, 75, 79, 80, 81, 82, 84
Feeding rate, 93, 95
Fernbach flask, 32, 38, 39, 42
Ferric chloride, 48, 49, 52
Ferric citrate, 52
Ferric hydroxide, 10
Ferric sequestrene, 49, 51, 52
Filipin, 112
Filter, biological, 5, 8, 9
Filters
  autotrophic, 10
  biological, 5, 8, 9
  charcoal, 40
  glass fiber, 47
  material, 7, 9, 11
  methods, 47, 76, 215, 257, 263, 294, 311
  Millipore, 39, 47, 91, 179, 252
  paper, 47
  Seitz-Filter, 140
  sterilization by, 39
  substrate, 20
  under gravel, 146
Filtration, 75, 76, 178, 186, 311
Fish, 4, 19, 30, 141, 148
Flagellates, 30, 44, 110, 115, 254
Fluorescent light, 33, 89, 183
Foam towers, 10
Foods
  alfalfa, powdered, 182, 192
  algae, freeze dried, 264, 311
  Artemia, 146, 151, 213, 215, 216, 224
  bacteria, 93, 115, 310
  beef heart, 26
  bouillon, 264
  brine shrimp, frozen, 140
  Chlorella, 10, 214, 311
  crustaceans, 146
  ctenophores, 148
  Cyclotella, 254
  detritus, 73, 110, 181, 192
  diatoms, 192. 310
  Dunaliella, 10, 19, 192, 193, 254, 285
  egg yolk, 139
  glucose, 19, 23, 97, 310
  infusoria (salt water), 147
  Isochrysis, 19, 192, 205, 254, 263, 276, 292, 310
  liver, 22, 193, 194
  macroalgae, 138, 192
  milk, 264
  Monochrysis, 254, 263, 276, 310
  nannoplankton, 163
  pablum, 264
  Phaeodactylum, 19, 192, 254, 285, 286
  polychaetes, 148
  protozoa, 146
  rotifers, 146, 217
  sea weeds, 231
  shrimp, 231
  small fish, 148
  spinach, 182, 231
  Ulva, 311
  Udothea, 140
  urchin (grastrulae), 216
  yeast, 264
Foot-candles, 33, 89
Foraminifera, 88, 90, 91, 93, 94, 95, 98, 100, 163
Formalin, 268, 314, 315
Fossils, 16
Fouling communities, 157
Fumidil, 112
Fungicides, 22, 119
  Amphotericin-B, 119
  Candicidin, 119
  Fungizone (Amphotericin-B), 92
  Nystatin, 119
Fungizone (Amphotericin-B), 92
Fungus (fungi), 22, 40, 110, 116, 117, 119, 192, 212, 227, 313, 314-315
Gammarids, 201
Gas bubble disease, 222
Gastropoda (gastropods), 114, 163, 245-256, 263, 290
Gelbstoff, 7
Germicides, 127
Glass shrimp, 212
Glucose, 19, 23, 24, 25, 26, 310

Glucose fermenters, 110
Glutamate, 80, 103, 104
Glutamic acid, 97
Glycerids, 181, 184
Glycerol, 104
Glycine, 99, 103
Glycylglycine, 53, 56
Gnotabiotic, 98
Gymnosomata, 247
Haptophyceae, 30
Halophilic, 74
Halophobic, 74
Halotolerant, 74
Haptophyceae, 30
Hard clams, 62, 63
Harpacticoid copepods, 115, 141, 201
Heart, beef, 15
Hermaphroditism, 183
Hermit crabs, 212, 215
Heterotrophic bacteria, 73
Histidine, 97, 103
Holland's solution, 269
Holothuroidea, 229, 230
Hopkins tubes, 144
Hydranths, 138, 140, 141
Hydrocarbons, 192
Hydrochloric acid (HCl), 54, 311
Hydrogen peroxide, 187
Hydroids, 129, 137, 138, 140, 141
Hydroxyproline, 97
Illumination, 33, 34, 35
Inarticulate brachiopods, 19
Infusoria, 147
Insecticide, 315
Instant Ocean, 6, 20, 57, 141, 218, 226-227, 297
Iodide, 6, 10, 149
Iodine, 10, 149
Ion exchange, 10
Iron, 6, 10, 49, 50, 54, 226, 298
Iron chloride hexahydrate, 48, 49
Isoloucine, 97
Japanese pearl oyster, 7
Jellyfish, 145-154
Kanamycin, 69, 116, 117, 118, 120, 121, 122, 123

Kelp, 159, 186, 194, 230
Kester's artificial seawater, 201
Kraft mill effluent, 304
Lead, 298
Lecithotrophic larvae, 164, 166, 192, 251, 252, 253
Light, 31, 32, 33, 34, 35, 36, 43, 44, 122, 146, 149, 166, 183, 190, 191, 213, 225, 286
Light meter, 33
Lipolytic bacteria, 78, 111
Lithium, 6
Lithium chloride (LiCl), 54
Liver, 19, 22, 193, 194
Lobster, american, 11, 221-227
Lobster, spiny, 218
Lophophore, 16, 17, 19, 23, 25, 26, 153, 155
Lysine, 97, 103
M-16655, 112
Macroalgae, 159, 186, 194, 230
Magnesium, 6, 7, 9, 11, 50, 55, 138
Magnesium chloride, 48, 51, 53, 54, 149, 179
Magnesium sulphate, 53, 54, 179
Manganese, 6
Manganese chloride, 48, 51, 53, 54
Mannitol, 104
Medium (media), 46, 47, 49, 52, 54, 102-104, 119, 120, 122, 126, 127, 297, 316
Medusae, 139, 140, 141, 147, 148, 151, 152
Membrane filtration, 76, 90, 91, 179, 252
Meroplankton, 164
Merthiolate, 127
Mesophily, 73
Metabolites, 9, 259, 291
Metals, 10, 46, 48, 50, 51, 52, 122, 138, 182, 192, 265, 288
Methamine-mandalate, 123, 125
Methionine, 80, 97, 103
Microherbivores, 98
Mictic eggs, 127
Milk, 264

GENERAL INDEX

Mollusca (mollusks), 16, 30, 61, 70, 113, 114, 245-318
Mollusc shells, 9
Molybdate, 6
Mortality, 7
MSX, 283, 287
Mycostatin, 92, 112
Mylar sheet, 292
Mytilids, 304
Nalidixin, 123
Nannoplankton, 163
Nansen bottle, 74
Naramycin, 123
Nauplii, 122, 123, 139, 215, 217
Necrosis, granular, 65
Nematocysts, 150
Nematodes, 88, 89, 90, 91, 92, 93, 94, 95, 98, 126
Nermerteans, 129
Neomycin, 69, 98, 112, 113, 116, 117, 118, 120, 121, 122, 123, 125, 126, 127
Nereids, 181, 182, 184
Neutral red, 148
Niskin sampler, 75
Nitrate, 4, 5, 6, 7, 8, 9, 10, 46, 58, 80
Nitric acid, 4
Nitrification, 9
Nitrifiers, 4, 8, 9, 10
Nitrite, 4, 5, 8, 9, 80
Nitrogen, 4, 5, 8, 55, 57, 222
Norleucine, 97
Notaspidea, 247
Novobiocin sodium (Albamycin), 92, 99, 112, 116, 117, 118, 121, 122
Nuculoidea, 308
Nudibranchia (nudibranchs), 159, 246, 247, 248, 249
Nylon, 298
Nystatin, 117, 118, 119, 123, 126
Nytex, cloth, 215, 252, 311
Oil spills, 11
Oleandomycin, 117, 118, 119
Ophiuroidea (ophiuroids), 163, 229, 230, 231

Opisthobranchia (opisthobranchs), 245, 256
Oxidation, 7
Oxygen, 191, 223, 226, 260, 316
Oxytetracycline (terramycin), 117, 118, 119, 125
Oysters, 62, 63, 64, 70, 73, 111, 113, 168, 170, 217, 283-295
Oxone, 10, 11, 69, 299
Pablum, 264
Pandoracean larva, 300, 308
Parasites, 10, 61
Parthenogentic, 240
Pea crabs, 215
Pelecypoda (pelecypods), 15, 16, 18, 19, 22, 26, 145, 217, 257-318
Penaeid shrimp, 218
Penicillin, 20, 69, 92, 99, 112, 113, 114, 115, 116, 117, 118, 119, 121, 122, 123, 124, 125, 126, 127, 129, 212, 253, 314
Pesticides, 100, 192, 216, 222
Ph
  aeration, 37-38
  alkaline, 115
  antibiotic activity range, 118, 119
  ASPM base, 53-54
  bacterial taxonomic character, 79
  Clark's artificial seawater, 54-55
  control of, 9, 10, 31, 231, 269
  decrease, 4, 8, 268
  effect of, 7, 100, 116, 124, 276
  Kester's artificial seawater, 201
  natural factor, 99, 149, 265, 312
  related to autoclaving, 49, 59
  trace metals stock solutions, 51-52
Phenol, 127, 192
Phenylalanine, 97
Pheremone, 183, 225
Pholads, 264, 266, 268
Phoronids, 19

Phosphates, 6, 7, 8, 9, 10, 46, 58
Phosphorus, 56, 58
Photography, 266-267
Photoperiod, 34, 90
Phytoplankton, 15, 19, 26, 29-60, 182
Planaria, 126
Plankton, 19, 110, 151, 163, 184, 194, 226, 261, 263, 299
Planktonkreisel, 187, 188
Planulae, 146, 148, 151
Platyhelminthes, 112
Plectolophous lophophore, 16, 17, 25
Pleurobranchaea, 247
Podocysts, 147, 151
Pogonophores, 15, 23
Pollutants (pollution), 192, 258
Polychaetes, 148, 177-197, 217, 263
Polydorids, 182
Polymyxin-B, 69, 92, 98, 116, 117, 118, 123, 125, 127, 253
Polyplacophora, 159
Polyps, 140, 146, 147, 148, 149, 150, 151, 152
Polyspermy, 237, 305
Polyzoa, 156
Potassium, 6, 7, 8, 55, 138
Potassium bromide (KBr), 54, 149
Potassium chloride (KCl), 53, 54, 149, 235, 236, 304
Potassium iodide (KI), 54,
Potassium orthophosphate mono, 54, 55
Prasinophyceae, 30
Prawns, 73
Predators, 61
Preservative, SCOR/UNESCO, 268
Preservatives, 268, 269
Pressure, 73, 74
Proline, 80
Propylene glycol, 268
Propylene phenoxytol, 148, 179

189, 267, 268
Prosobranchia, 245
Protein skimmers, 10
Proteins, 10, 26, 74
Proteolysis, 80
Protelytic, 11
Protobranch, 300, 308
Protozoa, 30, 110, 112, 116, 117, 217, 260, 289, 314, 315
Pseudofaeces, 263
Pseudomonadales, 66
Psychrophily, 73
Pumps, 187, 298
Pyncogonids, 159
Pyramidellacea, 247
Pyrrophyta, 125
Quartz gravel, 9
Radioisotopes, 24, 26, 94, 98, 129, 201
Recirculating systems, 3-14, 209, 215, 231
Red blood cells, 11
Red tide, 11, 200
Regeneration, 129, 147
Relaxants, 148, 268
Rhizostomeae (rhizostomes), 145, 152
Rhodophyta, 125
Rhynchocoela, 113
Rila Marine Mix, 57, 152, 218
Rotifera (rotifers), 113, 127, 146, 216, 217
Rubidium, 6
Rubidium chloride (RbCL), 54
Sabellids, 192
Salinity
 culture, bivalve larvae, 277-278, 280-281, 309
 culture, coelenterates, 140, 146, 149, 151
 culture, crustacea, 209, 214, 216, 221, 226
 culture, echinoderms, 231
 culture, polychaetes, 181, 189-190
 effects, bivalve larvae, 276
 environmental factor, 110, 116

GENERAL INDEX

optima, salt marsh organisms, 99
tolerance, bacteria, 73-74
tolerance, bivalve larvae, 265
tolerance, bryozoa, 169-170
Salmonids, 4
Sand dollars, 229, 230, 231
Sandworms, 180, 186
Scallop, bay, 62
Schreiber's solution, 140, 141
SCOR/UNESCO preservative, 268
Scyphistomae, 140
Scyphozoa, 145-154
Sea anemones, 23, 267
Sea cucumbers, 229, 230, 234
Sea hares, 249
Sea lilies, 229, 230
Sea nettles, 140, 146
Sea scallop, 277
Sea stars, 229, 230, 231, 235
Sea urchins, 11, 216, 229, 230, 231, 235
Seawater
  agar, 119
  aged, 76, 77, 102
  artificial, 1, 46, 56, 57, 141, 149, 150, 152, 200, 201, 218, 226
  buffered, 269
  Clark's, 54
  enriched, 46, 53
  filtered, 91, 119, 141, 186, 187, 217, 218, 252, 257, 279, 298
  Kester's, 201
  natural, 20, 120, 141, 179, 187, 188, 199, 200, 216, 218, 265, 276, 277, 278, 298
  seawater-formalin, 269
  sterile, 91, 127, 275
Seitz-Filter, 140
Semaeostomeae, 145
Septibranchia, 297
Serine, 97
Setting, 290, 291
Shell disease, 227

Shipek Sampler, 75
Shipworms, 62
Shrimp, caridean, 218
Sieves, 88, 187, 248, 261, 275, 289, 299, 311, 312
Silicate, 6, 29, 46, 47, 48, 49, 58
Silicon, 9, 55
Silt, 259
Silver nitrate, 126
Snails, 245-256
Sodium (Na), 6, 55, 58, 138
Sodium acetate, 97
Sodium bicarbonate ($NaHCO_3$), 53, 55, 59, 149
Sodium chloride (NaCl), 53, 54, 102, 149
Sodium colistimethate (Coly-mycin), 92
Sodium lactate, 97
Sodium metasilicate (Na_2SiO_3), 51, 54, 55
Sodium molybdate dihydrate ($Na_2MoO_4 2H_2O$), 48, 51
Sodium nitrate ($NaNO_3$), 48, 51, 54, 55
Sodium sulfate (Na_2SO_4), 149
Sodium tetraborate (borax), 268, 269
Sodium thiosulfate, 212, 213
Soil extract, 47, 90, 104
Solemyoidea, 297
Spinach, 181
Spiny lobsters, 218
Spionids, 181, 184
Sponges, 19, 26
Stauromedusae, 145
Stenolaemata, 156, 160
Sterilization, 10, 11, 22, 44, 47, 74, 126, 127, 128, 298, 299
Stone crabs, 217
Streptomycin (Streptomycin sulfate), 20, 69, 111, 112, 113, 114, 115, 116, 117, 118, 120, 121, 122, 123, 125, 126, 127, 129, 314
Stripping, 284, 304
Strobilation, 147, 149, 151
Strontium, 6

Strontium chloride, 54
Sucrose, 268
Sulfa-triple, 99
Sulfamerazine (Sulmet), 111, 113, 314
Sulfur (S), 55, 226
Sulfuric acid, 55, 59
Surf clam, 273, 279, 281
Tackle box culture, 209
Temperature
  culture, algae, 31, 35-37
  culture, bryozoa, 164
  culture, coelenterates, 138-140, 146-149, 151-153
  culture, crustacea, 209, 211-213, 226
  culture, echinoderms, 231, 236
  culture effect, general, 110, 116, 124
  culture mollusks, 250, 258, 264-265, 274-281, 284-285, 288, 290, 304-309
  culture, polychaetes, 179, 181-182, 190
  incubation, bacteria, 76-77, 80
  optima, estuarine bivalve mollusks, 312
  optima, salt marsh organisms, 99-100
  resin preparation, removing trace elements, 200
  tolerance, bacteria, 73-74
  tolerance, crustacea, 209, 221-223, 226
  tolerance, echinoderms, 238
  tolerance, polychaetes, 178
TEPP, 315
Teredinids, 264, 268
Terramycin (oxytetracycline), 125
Tetracycline, 69, 112, 114, 117, 118, 121-123, 125, 127, 129
Thecosomata, 247
Thermophily, 73
Thiamine, 20, 47-48, 53-54, 97, 104

Thiosulfate, 6
Thyroxin, 149
Toluidine blue, 25
Toxic, 7, 10, 39, 43, 119, 124, 126, 310
Toxin, 150, 152, 312
Trace metals, 10, 50, 51, 52, 55, 200, 218, 259
Trichomycin, 117, 118, 119, 123
Triple-sulfa, 99
TRIS (hydroximethyl) amino-methane, 53, 54, 55, 56, 102
Trochophore, 275
Tryosine, 97
Trypaflavin, 112
Tunicates, 19, 22, 26
Turbellarians, 263
Ultraviolet light, 10, 11, 22, 69, 215, 259, 275, 298-299
Urea, 103
Utility-Seven-Seas Mix #156, 57
Valine, 97
Vanadium, 6
Vanomycin, 112
Veligers, 251, 252, 253, 254, 275, 283, 285, 300, 302, 308
Vet-Strep, 69, 314
Viruses, 10
Vitamins, 15, 19, 22, 26, 46-48, 50, 52-53, 55-56, 97, 104
  biotin, 20, 47, 48, 52, 53, 54
  thiamine, 20, 47, 48, 97, 104
  vitamin B_{12}, 20, 47, 48, 53, 54, 58-59, 97
Weather, 35
Worms, 22, 26
Yeast, 119, 264
Zinc, 6, 218, 259
Zinc chloride, 51
Zinc sulfate heptahydrate, 48, 51, 54
Zobell's Marine Broth, 76, 77
Zoeae, 215
Zooplankton (zooplankters), 15, 140, 141, 184, 215, 288, 290
Zoozanthellae, 152